Kunst und Gesellschaft

Series Editor
Christian Steuerwald, Institut für Soziologie, Universität Bielfeld, Bielefeld, Germany

Die Reihe Kunst und Gesellschaft führt verschiedene Ansätze der Soziologie der Kunst zusammen und macht sie einem interessierten Publikum zugänglich. In theoretischen als auch empirischen Arbeiten werden dabei verschiedene Kunstformen wie etwa die Bildenden und Darstellenden Künste, die Musik und die Literatur hinsichtlich ihrer gesellschaftlichen und künstlerischen Bedeutung und Struktur untersucht. Dies beinhaltet nicht nur Analysen zu Kunstwerken und -formen, sondern auch Studien zur Produktion, Vermittlung und Rezeption von Kunst. Neben aktuellen Arbeiten stellt die Reihe auch klassische Texte der Kunstsoziologie vor. Damit sollen zum einen die Zusammenhänge zwischen Kunst und Gesellschaft herausgearbeitet werden. Zum anderen zielt die Reihe darauf, die Relevanz einer Soziologie der Kunst hervorzuheben.

Lisa Gaupp · Alenka Barber-Kersovan ·
Volker Kirchberg
Editors

Arts and Power

Policies in and by the Arts

Springer VS

Editors
Lisa Gaupp
University of Music and Performing Art
Wien – mdw
Wien, Austria

Alenka Barber-Kersovan
Leuphana University
Lüneburg, Germany

Volker Kirchberg
Leuphana University
Lüneburg, Germany

ISSN 2625-1531 ISSN 2625-154X (electronic)
Kunst und Gesellschaft
ISBN 978-3-658-37428-0 ISBN 978-3-658-37429-7 (eBook)
https://doi.org/10.1007/978-3-658-37429-7

Responsible editor: Cori Antonia Mackrodt
This Springer VS imprint is published by the registered company Springer Fachmedien Wiesbaden GmbH, part of Springer Nature.
The registered company address is: Abraham-Lincoln-Str. 46, 65189 Wiesbaden, Germany

Contents

List of Figures

Power and Policies in and by the Arts—Introduction

1

Lisa Gaupp, Alenka Barber-Kersovan and Volker Kirchberg

Abstract

The introduction to this book sketches the sociological preoccupation with concepts of domination and power in society. The following chapters discuss (a) the power of social structures on the arts and (b) the power of the arts on social structures. The newer terms for understanding these interactions come from postcolonial and intersectional sociology, as inequalities in the arts arise from multiple and interwoven processes of discrimination. Sociological analyses of the arts seek to uncover these power-laden processes, deconstruct established structures, and recognize artistic practices as practices of power. In doing so, it is important to distinguish between power and domination. Either the arts are dominated by rulers external to the arts and their institutionalized structures, or the arts are able to turn the tables and exercise domination via their legitimacy advantage, for example by shaping taste. The politics of cultural production, on the other hand, must be examined as a relationship between power and the arts. The breakdown of the book is presented with chapters in three sections, "Establishing and De-establishing Power in the Arts," "Arts and the Power of Social Structures," and "Arts and the Dominance of Political (Dis)Order."

L. Gaupp (✉)
University of Music and Performing Art Wien – mdw, Wien, Austria
e-mail: gaupp@mdw.ac.at

A. Barber-Kersovan · V. Kirchberg
Leuphana University of Lüneburg, Lüneburg, Germany
e-mail: alenka.barber-kersovan@leuphana.de

V. Kirchberg
e-mail: kirchberg@uni.leuphana.de

L. Gaupp et al. (eds.), *Arts and Power*, Kunst und Gesellschaft,
https://doi.org/10.1007/978-3-658-37429-7_1

The reciprocal perspectives of "power in the arts" and "power through the arts" are presented through a two-dimensional model whose first dimension is characterized by "domination versus power" and whose second dimension is characterized by "structural constraint versus volitional agency." All the articles can be placed in this coordinate system of domination (structure) and power (agency), regardless of whether they stand for negative (hegemonic, totalitarian, unjust) or positive (egalitarian, grassroots, balancing) social purposes.

Keywords

Arts and political strategies/political structures · Power through the arts · Power over the arts

1.1 Domination and Power—Pivotal Terms for Sociological Research

A preoccupation with concepts and ideas of domination and power in societal structures has been at the core of sociology since the beginning of the discipline. The emergence of sociological thinking about cohesion and cooperation, as opposed to coercion and constraint, took place at the same time as the transformative crises of early urbanization and industrialization. In the 19th century and at the turn of the 20th century, Marx, Durkheim, Simmel and especially Weber looked at society as a reciprocal and opposing interplay of coercive structures and enabling agencies. In an emerging industrial and urbanized society, new institutions exercised authority and power over social entities such as individuals, collectives and organizations; and these social entities tried, in turn, to use their powers to shape society. In the later 20th century, social theorists such as Gramsci, Adorno, Foucault, Giddens and Bourdieu developed these concepts further, such that most sociological discourses now draw on them. These theorists, often considered postmodern, also discussed more abstract and theoretical concepts of power and domination, divorced from empirical explorations. *Dispositifs*, constraints and violence are relational concepts defined as the enforcement of volition against resistance. This "enforcement of volition", we believe, is also well suited to explain processes in the arts, whether their production, imagination, communication, distribution, critique or consumption.

Alongside these developments in the sociology of culture and the arts as well as in cultural sociology, two other important streams of sociological thinking on questions of power in the arts need to be mentioned, namely, postcolonial and intersectional sociology. While the former examines prevailing (neo-)colonial structures in societies, as well as in sociology as an academic discipline (Bhambra 2007, 2014; Bhambra et al. 2018; Mbembe 2001, 2016; Boatcă et al. 2018; Reuter and Villa 2012; Boatcă and Costa 2016; Gutiérrez 2016), the latter focuses on unequal power structures resulting from multiple interwoven discriminating processes related to overlapping and intersecting social identities such as "race", gender, class, ethnicity, sexual orientation and so on, both in society in general and in the arts in particular (Anderson 2015; Crenshaw 1989; Amīn and Laamiri 2010; Below and von Bismarck 2005; Gaupp 2020). Thus, both sociological approaches seek to uncover power-laden structures and processes, deconstruct dichotomizations, homogenizations and standardizations as well as decolonize and queer the disciplines and fields of (artistic) practice, also on a global level (Caceres et al. 2017; Gaupp et al. 2020; Gaupp 2021). Intersectional approaches thereby regard inequalities as materially produced and determined by social positions by and in, for example, incorporated practices (Bourdieu 2010; Lena 2019). Cultural production is therefore considered to be a power-laden field in an ongoing struggle for social recognition.

The manipulative power of the arts is closely related to Weber's (1958) notion of legitimate rule, as domination goes hand in hand with a widespread belief in its legitimacy. The arts are either dominated by arts-internal or external rulers and their institutionalized social structures, or they rule these structures through their legitimacy, for instance, by defining and shaping art tastes. The arts have no power to rule by coercion,[1] however, they can rule by authority. The rulers of art worlds seek legitimacy for their rule from the ruled: they thus submit to the belief that legitimacy is conferred by the ruled in order that the rulers are able to rule others. The power of the arts is based on a combination of different types of authority, mostly charismatic, sometimes traditional but rarely legal. A sociology of the arts is a sociology of an objectified culture: one that can be created, distributed and received as signifier and signified by the senses, and is thus medium, communication and contents at the same time. The interplay of structure and agency among the actors of arts and culture organizations are strongly affected by the mechanisms of legitimate rule. Describing and understanding these rules of domination and power in and by the arts is the aim of this collection of articles.

---

[1] This statement might be regarded as contradictory to Bourdieu's field theory. However, this power of coercion (i.e., heteronomy) loses influence when and if the actors of the attacked art field retreat into their artistic autonomy and deny the heteronomous, economically or politically usurped culture content the status of "art".

The complexity of relationships between the arts and power is similar to the complexity of the *policies* of artistic production, distribution and consumption. They include issues that can promote or hinder an aesthetic object from reaching its intended public. They also include different forms of cultural and artistic communication which, formally and conceptually, mirror and shape prevailing social, economic and political conditions.

The following chapters deal with this subject from different perspectives—of artists, of organizers and distributors, and of visitors. They are based on different theoretical discourses and make use of different methodological approaches. They consider a wide range of different genres of artistic production[2] such as visual arts, poster art, performance art, "popular" and "classical" music, and they relate to different institutional practices of art organisations (galleries, museums, concert halls, theatres). Other topics addressed in this collection include issues of conservation and curatorship as well as regulatory and evaluative influences on cultural production on the macro, meso or micro level. Geographically, the chapters cover a broad range from Europe (including Eastern Europe), America to Africa as well as some aspects of global communication.[3] Most contributions deal with contemporary issues, but there are also some historical observations which confirm and relativize current assumptions about the field. They reveal the contradictions between the arts and power, and are situated on a continuum between affirmation and critique, aesthetic subversion and political art intervention. The volume is thematically divided into three parts, each containing five chapters.

---

[2]What is acknowledged as an artistic practice, and what is not, is another common example of how brokers and persisting conventions in different art worlds exert a certain power to define the field. In order to point to these processes, we refer to artistic practices or the arts and not to a more open term such as cultural practices, as all described practices in the different chapters are considered and labelled as artistic/art in their respective fields.

[3]We must admit that even though we intended to cover so far under-represented geographical areas and include them in a more or less mainstream research volume such as this one ("mainstream" meaning being published in English by a global publisher such as Springer, covering mostly research of "Western" art practices), the alignment of the chapters appears to be still quite "Western-centric". Reflecting on what it means to curate and publish this volume also entails acknowledging both the constructiveness of an equal "geographical representation" as well as the power-laden processes in the field of international academic publishing.

1.2 Establishing and De-Establishing Power in the Arts

Part 1 of this volume, Establishing and De-Establishing Power in the Arts, looks into ways of gaining power in and by the arts, and into examples of how the arts prove their power or lack of power to affect social issues. Throughout this volume, the authors discuss questions of power that point to an understanding of an environment of artistic practices as social practices. There are two main sociological streams dedicated to such processes of standardization, conventionalization, canonization, institutionalization, valorisation and evaluation: the sociology of conventions (Diaz-Bone 2018) and the sociology of evaluation (Lizé 2016; Peres da Silva and Hondros 2019). The sociology of conventions can be traced back to the *ecónomie des conventions* (Boltanski and Thévenot 1991) with its references to theories of classification (Durkheim and Mauss 1971 [1903]), controlling and sanctioning deviance (Gaupp 2020; Becker 1963), norms and normalisations (Thévenot 1984; von Osten 2003) as well as institutions and institutionalizations in the sociological sense (Luhmann 1970; North 1990; Eymard-Duvernay 1989). Similar issues of (e)valuation are discussed in the field of so-called valuation studies, which again is also closely connected to questions of assigning differences and curating as a practice of power (Gaupp 2021; Gaupp et al. 2020; von Bismarck et al. 2012; Buden 2012; Davida et al. 2019). The work of Pierre Bourdieu is taken as a guidepost for both of these streams of thought, on how power is established and de-established in and by the arts. Other authors concerned with gatekeeping practices that define and exert conventions related to the sociology of art worlds refer back to the work by Howard S. Becker (1982/2008).

How power is gained or lost in the arts is approached in this volume through different perspectives. Julia Rothenberg (New York) looks into the significance of a single artist in shaping a specific neighbourhood in Chicago. Here, art is reified in a star artist who, with the supportive institutional background of a university and the municipality, has the power to shape social and urban development according to his ideas.

Paul Buckermann (Heidelberg) studies a powerful institution that determines the success or failure of contemporary visual artists, the ArtReview's annual Power 100 ranking. The intricate dependencies of art world networks, discourses in external social problems and the dynamics of artwork evaluation create a power ranking by art influencers. This list not only reflects the power of these people (many of them art theorists and "intellectual popstars") but also the power of surrounding social worlds affecting art worlds.

Annelies Fryberger (Paris) likewise looks at the power of evaluation, in this case funding organizations evaluating contemporary art music composers. The power of these gatekeeping institutions is, however, limited due to a tension between a contemporary art music that regards itself as an autonomous art form and potential private donors who lean more towards reproducing heteronomous art tastes.

In Nina Tessa Zahner's (Düsseldorf) contribution on perception and power, she places the changing social interpretation of art in historical and contemporary power terms. Supplementing Bourdieu's field theory with Simmel's conception of artistic forms and contents, she studies the dynamic performativity of artistic meanings that constantly change through time. Dissolving the hierarchy of author and audience and losing confidence in the legitimacy of art experts, artistic meaning remains a varied product of changing powers in art worlds.

Christian Steuerwald (Bielefeld) builds on, one could say, Zahner's contribution. He explores arts organizations' dwindling power to influence the behaviour of their visitors. Today's diverse visitors bring very different "biographical baggages" (Griswold 2013) into museums, theatres and opera halls, and this varied set of expectations and interpretations makes it almost impossible for arts organizations to steer visitors' behaviour patterns. Steuerwald illustrates his thesis by studying "wrong" interactions in theatres, operas and museums, such as applause (at presumably the wrong passages), sing-alongs (during an opera) or triggering acoustic warning barriers (at expensive paintings in art museums). Forceful reactions of arts organizations to imposing their "rules of conduct" and sanctioning unconventional activities as "breaches of accepted behaviour" indicate their fears of losing control over audiences.

With the exception of Rothenberg's chapter, all chapters in Part 1 illustrate the potency of forces outside the artistic core—whether media influencers, public intellectuals or private donors—to shape the arts. This potency of external forces is mirrored by the impotence of art institutions' internal forces, and shrinking power of sense-making institutions (such as experts) that advise how to "see" and "understand" artworks or how to behave in "fine arts" organizations. This lack of authority can be seen as a sign of more democratic and participatory times where art recipients are no longer passive consumers. Or it can be interpreted as a lack of respect and acknowledgement of the efforts that arts organizations make for the benefit of the "higher" objective of *l'art pour l'art*. We prefer the former explanation as it would be a wrongful return to the cliché-like romantic myth of the autonomous artistic genius if we ignored the powerful external institutions of social context that define and exploit the arts for their own purposes.

1.3 Arts and the Power of Social Structures

The second part of this volume focuses on the impact specific social structures exercise on the shaping of artists and art. Already Michel Foucault pointed out the strong means of social control that dominate the body (Foucault 2019), which he understood as a fluid agent shaped by different forces (nutrition, beauty ideals, medical surgery) that allow it to be instrumentalized for commercial or political goals. His ideas on bio-power, the power of discourse and on sexuality have influenced many feminists and queer-theorists, among them Judith Butler, who enlarged his approach with a theory of performativity. She established a theoretical framework which enabled the feminist critique of heteronormativity, in the sense of the hegemonial patriarchal power over the female subordinate position, by deconstructing the allegedly natural identity of sex and gender (Butler 1999). Heteronormativity is based on the dichotomy of the (physically) "strong" man versus the (physically and morally) "weak" woman, which establishes almost automatically a relationship of dominance and subordination on different levels.

Music provides a good example. The grammatical gender of "music", as derived from the Greek *muses*, is feminine in several European languages. Further, following the dictum that "music belongs to the female soul" especially during the 19th century a strong link was assumed between "the other language" that is music and "the other gender" that is women. That, however, did not mean that "the other gender" was allowed to speak "the other language". Just the opposite was the case, because the musical substance (women) was regarded as dead matter which had to be awakened to life through the creative act of the (male) composer/genius (Rieger 1981). Hence the situation can be characterized as ambivalent: on one hand the (female) muse was worshiped as a kind of substitute religion, on the other the access of women to music was, and to a certain extent still is, hindered or even denied. It follows the general pattern of rather rigid gender stereotypes, dividing musical life into male and female spheres.

According to Freia Hoffmann and Eva Rieger these gender-specific attributions are time resistant and not only characteristic for European classical music. They are found also in traditional cultures and include dichotomies such as singing and playing an instrument, performing and composing, quiet and loud instruments, supportive and leading activity, amateurs and professionals, limited and professional training, among others (Hoffmann and Rieger 1992). Though gender polarities, as described by Rebecca Grothjahn and Freia Hoffmann in their critique of music historiography (Grotjahn and Hoffmann 2002), are slowly giving way to feminist and queer-theory approaches, women were for a long time also excluded from music history. The gender-biased scientific canon of the discipline was written as history of (male) composers and their works, and neglected the contributions of women to musical life. Women were not only a faithful public, they also organized musical soirees, acted as patrons and muses and, of course, in many cases were themselves also musicians, singers and dancers. Only recently have their artistic achievements in this field as composers received proper attention. According to Susan McClary, androcentric stereotypes are also found in musicological terminology and allegedly "objective" music analyses, with talk about "feminine endings" and "strong" or "weak" cadences, the first being associated with masculinity and the second with femininity (McClary 2002).

Empirical studies confirm that involvement in musical activities as well as the division of labour in this field is still gender specific. It affects the structure of cultural institutions and is reflected in the underrepresentation of women in crucial positions in musical life. Most professional cultural orchestras are predominantly male, and they are conducted by men; most famous virtuosos—with rare exceptions such as Anne Sophie Mutter and Marta Argerich—are men. But there are many female music teachers and music therapists, which means that women mainly practice those music-related professions that are semi-private and generally underpaid. The dichotomic pattern of gender relationships is also present at all levels of music education. Some instruments, such as piano and flute, are chosen predominantly by women, whereas brass instruments or drums are preferred by men. Though at the primary school level more girls learn to play an instrument, most music competitions are won by boys. Singing seems to be a female domain, while in subjects such as composing and conducting there is a strong social imbalance in the favour of men. Despite efforts to counteract these tendencies, there are still invisible barriers, prejudices and sometimes even inhibitions on the part of women themselves that cause them to be underrepresented in the more prestigious areas of musical life.

How deeply rooted these barriers can still be is exemplified in this volume in the contribution of Sigrún Lilja Einarsdottir about gender-political issues in the Oxford collegiate choral scene. Einarsdottir's study confirms the ongoing existence of segregation and systematic gender discrimination in the choral culture at Oxford colleges, conserving gender-based power structures in order to protect and reinforce a tradition derived from a centuries' old male-oriented power structure.

Einarsdottir depicts a general problem (and a constant theme running through feminist thinking): women's lack of equal participation in all sectors of cultural, social and political life. This disbalance also plays an important role in popular culture, as shown in Magdalena Protte's contribution *Pop Feminism Between*

Hegemony and Subalternity. Protte's point of departure to investigate the relationship between "serious feminism" and "pop feminism" is the success of the popstar Beyoncé Knowles. Beyoncé embodies, through her position as a successful pop star, a neoliberal pro-capitalist style of feminism based on marketing of her body. Protte agrees that pop feminism has been commercialized and depoliticized but still she ascribes it a potential to change the heteronormative matrix.

In their chapter about female bodybuilding, Isabel Fontbona Mola and Ulf Wuggenig transfer feminist theory into the broader field of sports, the arts and body politics. Similar to Protte, they show that women's success—in their case, the exhibition of physical power as muscularity, does not necessarily support feminist ideas and demands. Though some might perceive this act of transgressing conventional heteronormative gender stereotypes as a demonstration of radical feminist or queer empowerment, it can also be interpreted as a reinforcement of heteronormativity, as female bodybuilders are heavily controlled by regulations enforced by a male-dominated culture.

Not gender but racism as an expression of power is reflected in the last two chapters of Part 2. The contribution by Christoph Behnke (Lüneburg) on the development of Blues music in the USA at the turn of the 20th century connects artistic development to the powerful societal development of racist segregation. Behnke's analysis of this music field of cultural production shows the dominant structural effects of the "colour line", a racist demarcation with not only excluding effects but also appropriating effects: as it led to a generation of "racial pride" and the invention of Blues music.

Similarly, Anna-Lisa Klages's (Munich) study of the interconnections between "traditional" visual art production and international development efforts in Uganda shows there are powerful structural interrelations between gender (marginalised women) and class (economic development funded by international agencies). In this case they solidify a global hierarchy of "African" and "European" art forms. But even though "traditional" art forms from the Global South are assigned a lower aesthetic value on a global hegemonic scale, these art forms prove to be powerful actants in Uganda's civil society as they allow societal power structures to be redefined.

All five chapters in Part 2 show the multilevel complexity of social relations as well as the power of social structures, as they manifest themselves as a gender struggle (Einarsdottir, Fontbona Mola and Wuggenig), but also along an intersectional axis such as class (Protte, Klages) and "race" (Behnke, Klages). The complex power-laden relations evolving from unequal intersectional discriminations often have a global dimension in so far as they can also be seen as globally constituted practices of the arts and power (Buchholz and Wuggenig 2005; Behnke et al. 2015; Below and von Bismarck 2005; Beyer and Burkhalter 2012; Buurman et al. 2018), positioning the arts in processes of globalization. The global dimension can also encompass an artistic critique of neo-colonial structures from a Global South (Klages) and an anti-racist perspective (Behnke; Amīn 2014; Caceres et al. 2017).

1.4 Arts and the Dominance of Political (Dis)Order

However, a complete dominance of social structures over the arts is questionable, and there is disagreement about a strong structuralist view on society shaping the arts (Archer 2005). Indeed, there are opportunities for agency in arts production, distribution and consumption. In an analysis of the arts as tools for shaping society, such a (re)appropriation of art as agent also entails a (re)definition of what is considered aesthetic or artistic. However, the prevailing perspective in the following chapters is not of the arts as an enlightening and disempowering tool against the (politically) powerful but more as an instrument in strategies to maintain hegemonic power (of the state). This side of the coin is reflected in the contents of the third and last part of this book.

The relationships between the arts and political power are multiple and ambiguous. As history shows, a substantial part of artistic expression affirms the social order. Cathedrals, palaces, monuments and other objects of colossal architecture are static markers of power. These material signs of the hegemonic order are mostly commissioned and financed by the "powers that be", converting their policies into aesthetic production. A further broad impact on the production and reception of artistic practices is the result of (state-funded) artist training and the aesthetic education in schools. A consequence of this is that the creation of unwanted artistic artefacts can be prevented or even subjected to censorship (Roßbach 2018) and other means of repression with notable consequences for the individuals involved.

A subtler instrument of power is music. Mechanisms of symbolic control and domination are found in the interaction between conductor and orchestra, between stars and fans as well as in musical rituals themselves. But as Fred Prieberg (1991) pointed out, music has always been, and still is, used and misused for political purposes (framing political events, in election campaigns, as propaganda). National anthems are auditory insignia of state hegemony, and musical genres such as marches are closely associated with military power (Prieberg 1991). The strong emotional and physical feedback music triggers can also be misused as an agent of psychological warfare and a means of "no-touch tortur".

This was the case during the Iraq war in Guantanamo (Cusick 2006) as prisoners were exposed for days to loud music they disliked in order to break them.

Music, arts, ideology, political values and economic supremacy can be deployed as a "soft power", a non-aggressive kind of control which, according to Joseph Nye, is deployed to influence "the others" by attraction and co-option rather than by threats and physical force. As such, soft power is an established instrument of foreign policy, making media and the popular arts an important means of domination on a global scale (Nye 2004). Similar mechanisms of power relationships apply to cultural institutions. On one hand, their work can be influenced by prevailing (state) politics and policies, and on the other they implement power themselves. This is especially the case of publicly established museums and galleries, in which curatorial practice is based on value judgements of what to include in an exhibition, and what not, and how to present the chosen artefacts to the public.

Under these premises, in the first chapter in Part 3, *Art Politics' Tendencies of the Documenta Exhibition Series'*, Lutz Hieber (Hannover) analyses transformations of the curational policies of the internationally acclaimed German art fair Documenta in the years 1955, 1968 and 1992. He compares these quinquennial exhibitions with the curatorial practices of the Museum of Modern Art and the Whitney Museum in New York and comes to the conclusion that despite some fundamental differences between the Documenta exhibitions they can be considered as typical for a specific German visual art world. This is due to the lasting effects of the Nazi regime (the murder and expulsion of Jewish and leftist artists, and the repression of so-called "degenerate arts") as well as the perpetuation of the concept of "fine art" (solidifying a 19th century aesthetics).

Contrary to the "autonomous art" paradigm in West Germany's Documentas until the 1980s, we find a highly political poster art in the then still communist People's Republic of Poland. As Karolina Kempa (Hannover) explains in her contribution *The Aesthetic Subversion of Cultural Posters*, the state's totalitarian power over art generated a counter-culture of subversion and latent opposition which aimed at destabilizing the power of the state over the arts. It is surprising that though in totalitarian Poland cultural policy was based on the Socialist Realism art doctrine and the posters were neither aesthetically nor ideologically in line with the official art paradigm they were not censored by the authorities. According to Kempa, this had to do with the fact that the existing art fields were submitted to a hierarchy of cultural relevance, and poster design was deemed relatively unimportant.

That the arts as a tool of agency are not limited to state-hegemonic uses is discussed in the third chapter in this part. Aude Bertrand and Matthias Kettner

(Witten-Herdecke) analyse a public art event called *The Dead Are Coming* that was staged by the Centre for Political Beauty in 2015 as a civil society intervention. The core of the event was the burial ceremony for a Syrian woman who drowned while crossing the Mediterranean in order to migrate to the West. With this radical action, the Centre for Political Beauty appealed to the moral and ethical principles of society and encouraged the audience to take political action. By pointing to the gravity of the "refugee crisis" of 2015, the intervention of the Centre for Political Beauty provoked fierce reactions in a broader public, including mainstream media. Although their spectacle tactics can be criticized, the authors of this chapter found that the means have been justified by the end.

The contribution *Musical Interface Agendas—Music Appropriation via Technological Pre-configuration* by Carsten Wernicke (Lüneburg) examines the power which digital interface objects, such as Roli Seaboard, a digital-material interface, exerts on making music. Wernicke posits that the transition to a "post-digital state" in music includes a strong hidden agenda of a commodification and capitalization. Such digital technology is not a value-free tool for music production but a technological device which, despite its manifest potential to overcome a "Western" cultural hegemony in music production, hides its objective to inscribe the norms and values of a latent "cultural capitalist artefact".

The chapter *Who Plays?* by Glaucia Peres da Silva (Duisburg) deals with the international mobility of musicians in the framework of the World Music market. The study analyses the migration history of musicians who performed at WOMEX (World Music Expo) between 2008 and 2017. Decisions about who can participate in this circle lie solely in the hands of "Western" gatekeepers (labels, booking agencies, journalists), which raises questions of "cultural imperialism". Further inequalities are perpetuated by the fact that not all musicians work under the same conditions concerning such aspects as copyright protection, access to studio technology and travel opportunities. Hence, the study reveals imbalances of power within the World Music market.

The five chapters in Part 3 address different aspects of the dominance of hegemonic political and economic orders on the production and distribution of art. Hieber's text illuminates the on-going political and social power of a subterranean Nazi ideology and the lack of progressive politics in post-war Germany, which affected the conservative orientation and formats of the Documenta exhibitions for decades after the war. In contrast to Hieber's critique, Kempa recognizes in Poland's political dynamics since the 1980s the possibility of reinventing art as subversive counter-culture. The weakening of the totalitarian state allowed a strengthening of a new avant-garde in visual poster art: a new materiality of experiencing art in the artification of posters. Likewise, Bertrand and Kettner's analysis of the 2015 intervention by the Centre for Political Beauty stresses the possibility of the arts to act as a powerful medium to criticize values, presumably set in stone, that is, the conservative and even nationalist response to the refugee trying to enter Germany in 2015.

While Hieber, Kempa, and Bertrand and Kettner focus on political power as a factor and effect of the arts, Wernicke and Peres da Silva look at the technological and infrastructural factors of art production. At second glance, however, the digital materiality of the artefact in Wernicke's case and Peres da Silva's migration and living patterns of musicians in the World Music genre are also highly political. Wernicke shows that a seemingly unpolitical technology in music production undermines the idea of creating "non-Western" music and instead strengthens "Western" cultural hegemony, a highly political fact. Peres da Silva likewise exposes the seemingly global awareness of "World Music" as politically problematic because producers and consumers both live in the "Western" world—and only the creative ideas are mined from the Global South.

1.5 Power in the Arts—Power by the Arts

To summarize, we would sketch out a two-dimensional picture from the bipolarities presented in the contributions collected in this volume. The first dimension is the polarity between domination and power. Domination (or *Herrschaft* to use Weber's original term) is the social mechanism of suppressing and manipulating individuals, collectives and societal systems by coercion, persuasion and infatuation. Power, on the other hand, is not normative; it determines the essence of human socialization from the ground up (Popitz 2017). Agency is a neutral term that understands power as a positive concept (as in the civil rights movement's "power to the people") and as a negative one (as in Foucault's "disciplinary power and subjugation"). The second dimension is the localization of the arts as a societal product and as a cause of societal structures and processes. As Alexander (2020) explains, we can study the arts as a reflection of society or as a device that shapes society. The former sees the arts as a product of societal influences (a superstructure to a materialist base), and the latter understands the arts as a tool to form parts of society (attitudes, values, experiences of its members). These two dimensions, which we label "domination versus power" and "structure versus agency", are at the core of all the chapters compiled here. Kirchberg (2007) argues that "versus" should be replaced by "and" since agency can be transferred into structure, and vice versa. The same can be said for assessing the influences of the arts and society as good or bad; there is a contingency in assessing the mutual effects between the arts and society that makes a simple black and white evaluation difficult.

Chapters in Part 1 emphasize the reflective or structuralist perspective, understanding the characteristics of art as products of society, with the success or failure of artists determined by their rankings in prestigious art journals and by the funding decisions of powerful grant-making organizations. However, this strong structuralist view has been softened by other studies of the arts in Part 1, which attribute to the arts the agency power to define and redefine meanings, and to arts audiences the power to overrule traditional "rules of conduct" when listening to and seeing the "high" arts.

Nevertheless, we cannot overlook the powerful effects of society on the arts, and these have been picked up by the authors in Part 2. Gender and racial discrimination leave their imprint on the arts, whether in the choral culture at Oxford colleges or in the pop music industry as exemplified by the popstar Beyoncé. Both examples also show signs of agency. The centuries-old male hegemony of the Oxford choral societies is finally facing resistance, and Beyoncé—the neoliberal marketing of her body such a sign of the times—is personally changing the heteronormative marketing of her *Gesamtkunstwerk*. The ambivalence of selling "the arts" in a patriarchal, late-capitalist, neo-colonial and racist society is also the topic of the remaining chapters in this part that study female (and feminist) body-building, the African-American struggle against appropriation of Blues music and the efforts of Ugandan artists to create their own arts that do not follow "Western" evaluation criteria.

Agency as a powerful force shaping society is the theme of Part 3. The chapters alternate between the arts pursuing negative, manipulative and hegemonic goals, and positive, justice promoting and enlightening goals. The power of agency is thus not *per se* a positive power as it can be used for negative purposes of domination. The analysis of the first Documenta exhibitions shows how the Nazi regime's aesthetics and artistic values outlasted the end of the Second World War and influenced the signification of (cultural) politics in West Germany in the first decades after World War II. Further, digital technology in music production is not value-free but instead it reaffirms the values of a technology-based capitalist hegemony. And "World Music" is not a revolt of musicians in the Global South against the hegemony of the "Western" music industry but a strengthening of this industry's domination. Other chapters in Part 3 emphasize the positive connotation of agency power. The counter-cultural infiltration of Socialist Realism art in the People's Republic of Poland by an oppositional poster culture weakened the hegemonic stance of the totalitarian rulers towards culture in these years. The more recent political arts interventions by the Centre for Political

Beauty, despite its sometimes questionable use of spectacles, caused at least part of the public to rethink the refugee crisis and extreme right-wing politics in Germany.

Arts and Power—Policies in and by the Arts brings together texts that position the societal functions of art in a force field of domination (structure) and power (agency), whether used for or actively pursuing negative (hegemonic, totalitarian, unjust) or positive (equal rights, grassroots-democratic, just) social purposes. This contributions in this volume are a first exploratory step towards a more systematic empirical and theoretically founded sociological debate on the arts and power, and an invitation to take further steps.

References

Alexander, Victoria D. 2020. *Sociology of the arts: Exploring fine and popular forms*. John Wiley & Sons.

Amīn, Khālid and Laamiri Mohammed (eds.). 2010. *Performing cultural diversity; Critiquing postcolonialism* (1st ed). Tetouan: Université Abdelmalek Essaâdi, Faculty of Letters and Humanities, Research Group of Performance Studies (Series: Conferences and colloquia).

Amīn, Khālid. 2014. Postcolonial Modernity: Theatre in Morocco and the Interweaving Loop. In *The politics of interweaving performance cultures. Beyond postcolonialism*, eds. Erika Fischer-Lichte, Torsten Jost and Saskya Iris Jain, 25–41. Oxford: Routledge Advances in Theatre and Performance Studies Vol. 33.

Anderson, Elijah. 2015. "The White Space". *Sociology of Race and Ethnicity* 1(1): 10–21. https://doi.org/10.1177/2332649214561306.

Archer, Margaret S. 2005. Structure, Culture and Agency. In *The Blackwell Companion to the Sociology of Culture*, eds. Mark D. Jacobs and Nancy Weiss Hanrahan, 17–34. Blackwell.

Becker, Howard S. 1963. *Outsiders: Studies in the Sociology of Deviance*. London: Free Press of Glencoe.

Becker, Howard S. 2008 [1982]. *Art worlds*. 25th Anniversary edition, updated and expanded. Berkeley/Los Angeles/London: University of California Press.

Behnke, Christoph, Cornelia Kastelan, Valérie Knoll and Ulf Wuggenig (eds.). 2015. *Art in the Periphery of the Center*. Sternberg Press.

Below, Irene and Beatrice von Bismarck (eds.). 2005. Schriftenreihe des Ulmer Vereins, Verband für Kunst- und Kulturwissenschaften e. V: N.F., Bd. 1. *Globalisierung, Hierarchisierung: Kulturelle Dominanzen in Kunst und Kunstgeschichte*. Jonas-Verlag.

Beyer, Theresa and Thomas Burkhalter (eds.). 2012. *Out of the absurdity of life: Globale Musik* (1. Aufl.). Traversion

Bhambra, Gurminder K. 2007. *Rethinking modernity: Postcolonialism and the sociological imagination*. Palgrave. https://doi.org/10.1057/9780230206410.

Bhambra, Gurminder K. 2014. *Connected Sociologies. Theory for a Global Age Series*. Bloomsbury Academic. https://doi.org/10.5040/9781472544377.

Bhambra, Gurminder K., Dalia Gebrial and Kerem Nişancıoğlu (eds.). 2018. *Decolonising the university*. Pluto Press.

Boatcă, Manuela and Sérgio Costa. 2016. Postcolonial sociology: A research agenda. In *Global connections. Decolonizing European sociology: Transdisciplinary approaches*, eds. Encarnacion Gutiérrez Rodriguez, Manuela Boatcă and Sérgio Costa, 13–31. Routledge.

Boatcă, Manuela, Sina Farzin and Julian Go. 2018. Postcolonialism and Sociology: E-Mail-Debate. *SOZIOLOGIE* 47(4): 423–438.

Boltanski, Luc and Laurent Thévenot. 1991. *De la justification: les économies de la grandeur*. Paris: Gallimard.

Bourdieu, Pierre. 2010. *Distinction: A social critique of the judgement of taste*. Routledge classics. Routledge

Buchholz, Lara and Ulf Wuggenig. 2005. Cultural Globalization Between Myth and Reality: The Case of the Contemporary Visual Arts. *ART-E-FACT Strategies of Resistance* (4). http://artefact.mi2.hr/_a04/lang_en/theory_buchholz_en.htm.

Buden, Boris. 2012. Towards the Heterosphere: Curator as Translator. In *Performing the curatorial: Within and beyond art,* ed. Maria Lind, 23–46. Sternberg Press.

Butler, Judith. 1999. *Gender Trouble: Feminism and the Subversion of Identity*. Routledge.

Buurman, Nanne, Sarah Dornhof, Birgit Hopfener and Barbara Lutz. 2018. Situating Global Art: An Introduction. Image: Vol. 89. In *Situating Global Art: Topologies – Temporalities – Trajectories*, eds. Sarah Dornhof, Nanne Buurman, Birgit Hopfener and Barbara Lutz, 11–32. transcript.

Caceres, Imayna, Sunanda Mesquita and Sophie Utikal. 2017. Anti*Colonial Fantasies/Decolonial Strategies: A Conversation. In *Kuratieren als antirassistische Praxis*, eds. Natalie Bayer, Belinda Kazeem-Kamiński and Nora Sternfeld, 201–211. Edition Angewandte. De Gruyter.

Crenshaw, Kimberle. 1989. Demarginalizing the intersection of race and sex: A Black Feminist Critique of Antidiscrimination Doctrine, Feminist Theory and Antiracist Politics. *University of Chicago Legal Forum* 1(8): 139–167. http://chicagounbound.uchicago.edu/uclf/vol1989/iss1/8.

Cusick, Suzanne G. 2006. *Music as torture / Music as weapon*. Trans 10, Revista transcultural de Musica, SIBE. http://www.sibetrans.com/trans/articulo/152/music-as-torture-music-as-weapon.

Davida, Dena, Marc Pronovost, Véronique Hudon and Jane Gabriels (eds.). 2019. *Curating Live Arts: Global Perspectives on Theory and Practice*. Berghahn Books Incorporated.

Diaz-Bone, Rainer. 2018. *Die "Economie des conventions"—Grundlagen und Entwicklungen der neuen französischen Wirtschaftssoziologie*. 2. Auflage. Wiesbaden: Springer VS. https://link.springer.com/content/pdf/10.1007%2F978-3-658-21062-5.pdf.

Durkheim, Émile and Marcel Mauss. (1903). *De quelques formes primitives de classification. Contribution à l'étude des représentations collectives*. Année sociologique, 6, wiederabgedruckt in: Mauss, M. (1971). Essais de sociologie, Paris, Ed. de Minuit (Coll. Points). doi: https://doi.org/10.2307/3538416.

Eymard-Duvernay, François. 1989. Conventions de qualité et formes de coordination. *Revue économique* 40(2): 329–359. https://doi.org/10.2307/3502117.

Foucault, Michel. 2019. *Power: the essential works of Michel Foucault 1954–1984*. Penguin UK.

Gaupp, Lisa. 2020. 'The West' vs. 'The Rest'? Festival Curators as Gatekeepers for Sociocultural Diversity, in *Managing Culture: Exchange in a Global World,* eds. Victoria Durrer and Raphaela Henze, 127–153, Palgrave MacMillan, Sociology of the Arts. https://doi.org/10.1007/978-3-030-24646-4.

Gaupp, Lisa. 2021. How to Curate Diversity and Otherness in Global Performance Art, in *Diversity and Otherness. Transcultural Insights into Norms, Practices, Negotiations,* eds. Lisa Gaupp and Giulia Pelillo-Hestermeyer 289–320, De Gruyter. https://www.degruyter.com/document/doi/10.1515/9788366675308/html?lang=en.

Gaupp, Lisa, Anna Abramjan, Frida Merve Akinay, Katharina Hilgert, Anna Catharina Mulder, Rebecca Schmidt, Viviane Schnitzler, Ole Thurich, Lucas Tiemon and Swantje Wurl, 2020. Curatorial practices of the 'Global': Toward a Decolonial Turn in the Museum in Berlin and Hamburg? *Journal of Cultural Management and Cultural Policy, Special Issue: Museum – Politics – Management.* 2020(2), 107–138, doi:https://doi.org/10.14361/zkmm-2020-0205.

Griswold, Wendy. 2013. *Cultures and Societies in a Changing World.* Pine Forge Press.

Grotjahn, Rebecca and Freia Hoffmann (eds.). 2002. *Geschlechterpolaritäten in der Musikgeschichte des 18. bis 20. Jahrhunderts,* Centaurus.

Gutierrez Rodriguez, Encarnacion. 2016. Decolonizing Postcolonial Rhetoric, in *Global connections. Decolonizing European sociology: Transdisciplinary approaches.* eds. Gutierrez Rodriguez, Encarnacion, Manuela Boatcă and Sérgio Costa, 49–67, Routledge.

Hoffmann, Freia and Eva Rieger. 1992. *Von der Spielfrau zur Performance-Künstlerin,* Furore-Verlag.

Kirchberg, Volker. 2007. Cultural consumption analysis: Beyond structure and agency. *Cultural Sociology*, 1(1), 115–135.

Lena, Jennifer C. 2019. *Entitled: Discriminating tastes and the expansion of the arts.* Princeton University Press.

Lizé, Wenceslas. 2016. Artistic work intermediaries as value producers: Agents, managers, tourneurs and the acquisition of symbolic capital in popular music. *Poetics*, 59, 35–49. https://doi.org/10.1016/j.poetic.2016.07.002.

Luhmann Niklas. 1970. Soziologie als Theorie sozialer Systeme, in *Soziologische Aufklärung*, Niklas Luhmann, Vol. 1, 113–136.

Mbembe, Achille J. 2001. *On the Postcolony*, University of California Press. http://hdl.handle.net/2027/heb.02640.0001.001.

Mbembe, Achille J. 2016. Decolonizing the university: New directions. *Arts and Humanities in Higher Education*, 15(1), 29–45. https://doi.org/https://doi.org/10.1177/1474022215618513.

McClary, Susan 2002. *Feminine Endings: Music, Gender, and Sexuality*, University of Minnesota Press.

North, Douglass. 1990. *Institutions, institutional change and economic performance*, Cambridge University Press. doi: https://doi.org/10.1017/CBO9780511808678.

Nye, Joseph. 2004. *Soft power: the means to success in world politics*, Public Affairs.

Peres da Silva, Glaucia and Konstantin Hondros (eds.). 2019. *Music practices across borders: (E)Valuating space, diversity and exchange* (Music and Sound Culture). Transcript.

Popitz, Heinrich. 2017. *Phenomena of power*. Columbia University Press.

Prieberg, Fred. 1991. *Musik und Macht*. Fischer.

Reuter Julia and Paula-Irene Villa (eds.). 2012. *Postkoloniale Soziologie: Empirische Befunde, theoretische Anschlüsse, politische Intervention.* Postcolonial studies, Vol 2, transcript.

Rieger, Eva 1981. *Frau, Musik und Männerherrschaft. Zum Ausschluss der Frau aus der deutschen Musikpädagogik, Musikwissenschaft und Musikausübung*, Ullstein.

Roßbach, Nikola. 2018. *Achtung Zensur! Über Meinungsfreiheit und ihre Grenzen,* Ullstein

Thévenot, Laurent. 1984. Rules and implements: Investment in forms, *Social Science Information,* 23(1), 1–45.

von Bismarck, Beatrice, Jörn Schafaff and Thomas Weski (eds.). 2012. *Cultures of the curatorial.* Sternberg Press.

von Osten, Marion. 2003. Einleitung, in *Norm der Abweichung*, eds. Marion von Osten, (Theorie: Gestaltung), Springer, 7–17.

Weber, Max 1958. The three types of legitimate rule, *Berkeley Publications in Society and Institutions*, 4(1), 1–11.

Part I

Establishing and De-Establishing Power in the Arts

Repurposing the Black Arts Movement on the South Side of Chicago

2

Julia Rothenberg

Abstract

This chapter explores an instance of how the arts and culture led urban rebranding projects so central to the new urban economy play out in African American communities. Under the direction of the University of Chicago in collaboration with Theaster Gates, an African American artist, educator, entrepreneur and Chicago South Side native who has carved out a powerful position as a creative-pacemaker and arts impresario at the university a number of such projects are unfolding around the university and in the neighbouring impoverished Black communities. I place these developments in the context the history of racial segregation in Chicago, the ongoing role of the university as a major growth coalition partner and debates regarding creative development strategies in the neoliberal city. Perhaps most importantly, this chapter attempts to understand how and to what ends the highly political art of the Black Arts Movement is being resignified in the new era of market fundamentalism which, interestingly, owes its ideological underpinnings to the Chicago School of economics. By contrasting the urban context in which the Chicago Black Arts Movement emerged and its deployment today I hope to shed light on, in particular, the racialised dimensions of creative city processes.

J. Rothenberg (✉)
Social Sciences Department, Queensborough Community College,
City University of New York, Bayside, New York, USA
e-mail: jrothenberg@qcc.cuny.edu

L. Gaupp et al. (eds.), *Arts and Power*, Kunst und Gesellschaft,
https://doi.org/10.1007/978-3-658-37429-7_2

Keywords

Creative placemaking · Black Arts Movement · Chicago · University industrial complex · Neoliberal urbanism · Growth coalition · Segregation · Gentrification · Theaster Gates

2.1 Introduction

As reluctance to invest in undervalued Black and Brown neighborhoods fades and Black professionals return to neighborhoods from which their parents fled decades ago (Patillo 2007; Prince 2017), stakeholders deploy culturally specific forms of arts-investment and promotion in upscaling and revitalization projects. Neighborhoods like Harlem in New York and Bronzeville in Chicago derive symbolic capital from their musical, literary and visual urban histories (Chronopoulos 2016; Goldstein 2017). Efforts to highlight the history of Black art and culture help educate the public about these important American cultural movements and provide space for intellectuals, artists and other community members to discuss current issues around race and culture. They also help lure Black professionals to purchase homes and invest in undervalued Black communities. Inevitably, however, as the examples of Harlem, Bedford Stuyvesant, Fort Greene, Bronzeville, and elsewhere demonstrate (Chronopoulos 2016) this investment also increases land value and the accompanying displacement of poor and working-class Black residents. Thus, the ambivalent tendencies inherent in arts-driven revitalization projects[1] take on specific characteristics in the context of African American communities, whose members have a long history of exclusion from urban resources and have been the victims of state and civil violence. Along the way, these communities nurtured art forms that represent a creative response to adverse conditions. In the wake of the Civil Rights era, race has become less relevant than class, at least in terms of access to housing and other urban amenities. In this context, the specific dilemmas raised by arts-led development are further complicated

[1] In the context of neoliberal urban development, the impact of public and private arts-driven investment is always ambivalent. Investment in the arts provides opportunities for artists and their publics to create and disseminate their work. At the same time, enhanced opportunities to consume art and culture attracts tourists, developers, professionals and members of the creative class who are able to pay more for housing and other services, thus driving up rents and displacing artists, small businesses and other lower-income residents from their communities (Lloyd 2010; Mele 2000).

by tensions between intra-race class conflict and ideals of racial cohesion and community, which were, to a degree, reinforced by segregation (Boyd 2008).

Questions about how arts and culture led urban rebranding projects play out in African American communities remain underexplored in the literature on creative placemaking, gentrification and neoliberal urbanism. I investigate one such instance of this phenomenon on the South Side of Chicago, under the direction of the University of Chicago, a key stakeholder on the South Side, in collaboration with Theaster Gates, an African American artist, educator, entrepreneur and South Side native who has carved out a powerful position as a creative place-maker at the university and its surrounding communities. While Gates has an impressive career as an artist in his own right and while his and the university's activities harness a variety of types of art practice, I focus primarily on the influence of the art of the Chicago arm of the Black Arts Movement (BAM)—which emerged in the late 1960s in the United States as a cultural expression of the aims of the Black Power Movement (BPM)—in these placemaking projects.

The emergence of BAM coincided with the period in which the US Welfare State, under the direction of the Johnson administration, crafted policy to address the problems of urban rebellion through federally funded community engagement initiatives. I address the question of how and to what purpose this art has been resuscitated after the decline of the Welfare State and the rise of what scholars beginning in the late 1970s call the "entrepreneurial city" (Harvey 1989; Logan and Molotch 1987). By contrasting the urban context in which the Chicago BAM took hold with its current role in urban restructuring I hope to shed light on the racialized dimensions of creative city processes.

2.2 Culture and Community Development in Hyde Park/Kenwood

Gun violence, poverty and other symptoms of urban blight dominate popular representations of Chicago's South Side. Less is known about its significance as a hotbed of Black Power activism and the Black Arts Movement in the 1960s and 1970s. Manifestations of BAM in Chicago include the Association of for the Advancement of Creative Musicians (AACM), the visual arts group AfriCOBRA (African Commune of Bad Relevant Artists), OBAC (Organization of Black American Culture) and the Chicago mural movement. Earlier, Chicago's Black Renaissance, closely identified with figures such as Gwendolyn Brooks, Richard Wright, Margaret Burroughs and Archibald Motley (Bone et al. 2011) paralleled the rise of Black owned Johnson Publishing company, which, through its history from the 1940s through the first decades of the 21st century published the world-renowned Black lifestyle magazines *Ebony* and *Jet*, was one of the nation's largest and most prestigious Black owned businesses (Green 2009).

I grew up in the 1960s and 1970s on the South Side of Chicago in a neighborhood called Hyde Park/Kenwood (hereafter referred to as *Hyde Park*). Before it developed a national reputation as the home turf of President Barak Obama who lived, taught and organized there, Hyde Park was most often associated with the prestigious University of Chicago, an institution, which dominates the community economically and culturally.

Staid, well-tended and elegant close to the university, Hyde Park became a little seedy around the edges, where the homes of prosperous university professors and administrators gave way to student apartments, SORs, delis, barber shops, liquor stores and beauty salons. The streets closest to the campus were well lit and patrolled by the university's extensive private security force. As the neighborhood fanned out, closer to the mostly Black and low-income communities to the North, West and South, pedestrians were more careful. These streets were disproportionally represented in the crime reports listed in the neighborhood newspapers. Further afield, outside of the boundaries of the neighborhood proper, lay some of the most impoverished and segregated neighborhoods in Chicago. These communities were like another world to middle-class Hyde Park residents, Black and White alike, and forays there were endeavored with extreme caution if not downright fear.

Hyde Park also stands out as one of the most racially integrated communities (Williams 2013) in a city known for extreme levels of racial segregation (Williams and Emamdjomeh 2018). Indeed, despite the university's influence, there remained a strong presence of African American culture in the neighborhood. This was felt during my childhood, when the cultural nationalism of post-civil rights Black politics continued to have an impact on the African American community in Hyde Park, and the Black Panthers were active in the neighborhoods of Hyde Park's surrounding "Black belt."

These communities, referred to as "hyper-ghettos" by the sociologist Loïc Wacquant (2007), were hit hard by several decades of depopulation, retrenchment, unemployment, disinvestment, gang and police violence and the 2008 economic crisis (Semuels 2018; Weber 2015). The conditions with which their residents struggle were exacerbated by one of the most ruthless neoliberal urban regimes in the US, led by Mayor Rahm Emanuel (from 2011–2019), Obama's former White House chief of staff, who enjoyed close ties with both the university and with Gates. Under Emanuel, Chicago residents saw mayoral cover-ups of police violence, the largest number of public-school closings in history, the shuttering of half of the mental health clinics, most of them in poor, African American neighborhoods and the gutting of public housing (Joravsky 2013). Like mayors throughout the US, he has given huge tax breaks to mega corporations, including an offer of a $2 billion break to Amazon to attract the company's new headquarters (Bennett et al. 2017).

Hyde Park avoided change. The bookstores, student restaurants, one or two low-key student bars, a handful of bars and restaurants catering to a primarily Black clientele remained as I remembered them year after year. And then, rapidly, things began to change. While the hotels, expensive restaurants and hip coffee shops that began to populate the neighborhood's commercial districts were jarring to me, what interested me most was the cultural activity and arts investment that had taken hold, much of which was devoted to highlighting African American artists and movements past and present. Almost overnight, it seemed, Hyde Park residents could choose from an array of cultural offerings, many of which were free of charge, almost every day of the week. These offerings were spilling out past the boundaries between Hyde Park and the neighboring Black communities, with a dotting of galleries and film and performance venues amidst the derelict spaces of the depopulating neighborhoods to the South and West.

These offerings were part of a program on the part of the University of Chicago, in collaboration with Gates, to bolster the university and the surrounding neighborhoods' appeal to students, faculty and potential South Side visitors. These spaces have contributed to a resurgence of cultural opportunities for Black artists and Black and White professionals. More complicated is Gates' (and to a degree the university's) claim that this cultural development will ultimately benefit the poor and working-class black residents who have been relegated to under-resourced neighborhoods through a history of racial covenants, redlining, slum clearance and other structural and individual acts of exclusion and violence (Massey and Denton 1993). The development strategies deployed by the University of Chicago in concert with Gates exemplify the recent phenomenon in which Black arts and culture are enlisted in creative city style development and therefore deserve close attention as a case study.

2.3 Who is Theaster Gates?

> Artists don't gentrify places; we bring life through our energy. The arts bring light. The work that has to happen now requires a commitment to equity, opportunity and balance. These values have to be loaded into our fierce commitment to wealth creation.
>
> (https://www.nationalcivicleague.org/national-civic-review/issue/winter-2019-volume-107-number-4/?popup=true).

As a global art star, public speaker, entrepreneur, urban visionary, non-profit director and head of several prestigious institutes and initiatives at the University of Chicago, Gates wears a number of interrelated and powerful hats. Gates' art works, about which I will not go into detail, employ Social Practice principles (https://time.com/5744412/theaster-gates-interview/) of which he considers his real estate activities to be a part[2]), repurposed and found materials and more traditional media and engage themes of racism and the Black experience in America.

In a short period of time, Gates has garnered power, accolades and recognition (Leduc 2019) due in large part to his activities on the South Side of Chicago. He has won numerous prizes and prestigious grants for his art and his urban development projects (e.g. Sweeney 2018) are the subject of feature length articles in art-world gatekeeping journals. He is currently exhibiting his work several high-profile venues.[3] Recently, Gates was commissioned by the City of Chicago to provide a permanent art installation for a Chicago Transit Authority's newly rehabbed South Side station. There, Gates stood next to Mayor Emmanuel in mid-April of 2019 as they introduced the refurbished transit hub, which cost the city $280 million, including the price tag for Gates' instillation "america america," made from two massive fabrications of decommissioned firehose and red paint and "An Extended Song of Our People (AESOP)," a deejay booth providing commuters with soundtracks of house music, jazz and other samples of popular music primarily associated with African American cultural interventions (https://www.transitchicago.com/garfieldgateway/).

Gates has a long personal and professional relationship with Chicago's South Side, where he grew up in a working-class family in the 1970s and 1980s. In 2006, after receiving advanced degrees in ceramics and urban planning, Gates was hired by the University of Chicago to help coordinate its arts programing. In 2008 Gates, using funds from burgeoning art market sales, purchased the first of several properties for $18,000 in the blighted African American community of Woodlawn/Grand Crossing, close to where he was raised, just South of

[2] He has been called "the poster child for socially engaged art" (Moore 2015).

[3] After accusations of racism against the fashion house, Prada, Gates, along with the filmmaker Ava DuVernay, was named co-chair of their rapidly formed Diversity and Inclusion Advisory Council. He is also the curator of "The Black Image Corporation, a project by Theaster Gates" at the Fondazione Prada Osservatorio in Milan where he brings together works by Black fashion photographers, culled from the archives the Johnson Publishing Company, which launched Ebony magazine in 1945 and the now defunct Jet magazine in 1951 (Bertagnoli 2019).

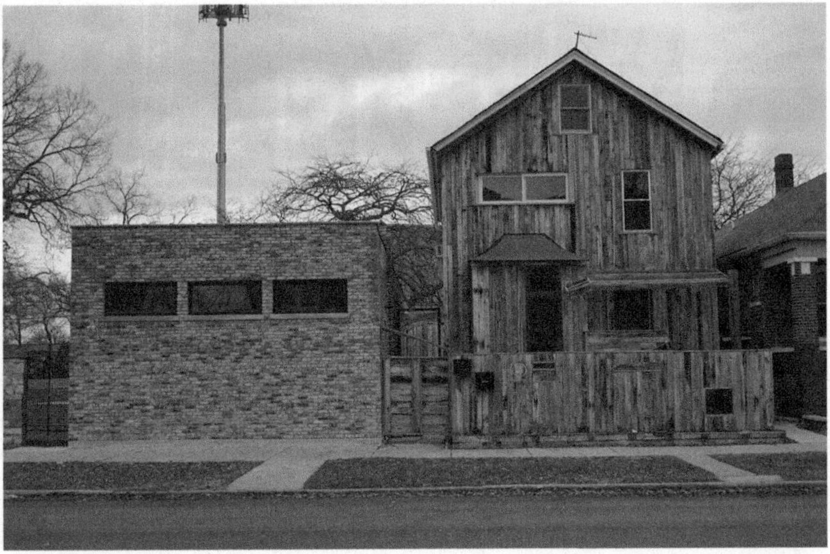

Fig. 2.1 Outside Gates' renovated studio and home of rebuild foundation on Dorchester street, 2016. *Photo* by Geoffrey Berliner

the University of Chicago's pristine, neo-Gothic campus and future neighbor to the Obama Presidential Library. Using repurposed materials, he converted these buildings into studio space, a Black cinema house and an archive and listening space for vinyl albums featuring African American music and quickly won acclaim and arts foundation funding for his creative place-making and community revitalization projects (Figs. 2.1 and 2.2).

Gates received funding for these projects with the familiar argument that arts anchor development, which he articulated in series of speaking engagements, including a TED talk and a speech at the World Economic Forum. Here, he explained how his development ventures grew out of his practice as an artist, familiarizing his audience with his notion of activating blighted or abandoned urban spaces with culture, beauty, meaning and history (Gates 2015). Neighborhood residents, he argued would benefit from his newly created "nodes of cultural activity," public spheres where African American art and culture could be produced, circulated and evaluated and community bonds could be forged and strengthened. Within these reactivated spaces, the community could explore "challenging content" relating to their identities and the "complexities of race

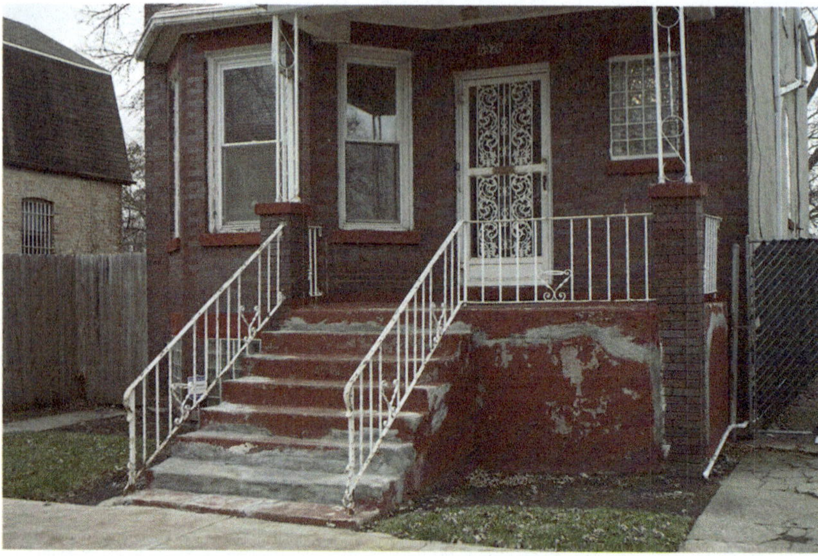

Fig. 2.2 Next door to Gates' house on Dorchester, 2016. *Photo* by Geoffrey Berliner

and class." Through cultural involvement, residents of blighted communities would become "more invested in their place." In a utopian vein, Gates also maintains that "beauty is a basic service" and once people have a taste of it in their neglected communities they will make "a poetic demand, and the political demands that are necessary to wake up our cities …" (Gates 2015).

To administer these projects and apply for grants, Gates founded the non-profit Rebuild Foundation. Rebuild's projects include the studios, events and archives housed on Gates' South Side properties, a mixed-income art and housing collaboration close to these properties and the preservation and rehabilitation of Stony Island Trust & Savings Bank (which he purchased from the city for $1), turning this historic South Side building slated for demolition into a gallery and performance space. Events at Stony Island Bank include regular performances by the current musician in residence, a series of films and music events and an exhibition and discussion series in honor of the Black Panther movement. The bank also houses the Johnson Archives, which contain thousands of important documents from the history of Ebony magazine (Figs. 2.3 and 2.4).

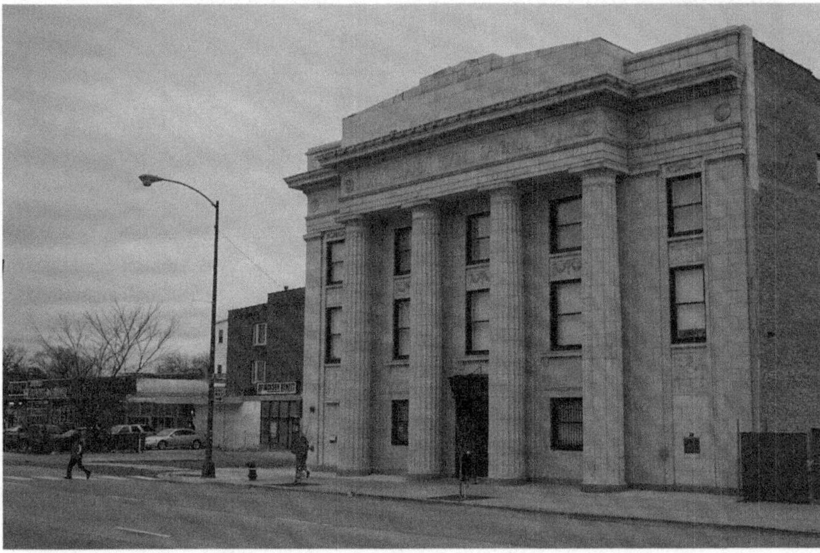

Fig. 2.3 Stony Island Iank, 2016. *Photo* by Geoffrey Berliner

2.4 Gates and the University

Gates' art world power parallels and fuels his influence at the University of Chicago. Long central to local economies, universities have emerged as key players in urban growth as anchors for job creation, urban branding and property development (Baldwin 2015; Fernández-Esquinas and Pinto 2014). In 2011, Gates was appointed director of the University's Arts + Public Life programing to "build creative connections on Chicago's South Side and to administer the University of Chicago's resource-rich art and public programing initiative" (UChicago Arts n. d.). One year later, the university cut the ribbons on its impressive new Logan facility, which received seed funding of $35 million in response to a 2001 report that strongly suggested elevating the arts at the University and creating a multidisciplinary center for the arts on campus (Reva and David Logan Center for the Arts n. d.). The Logan Center, with its myriad facilities, serves as the main site of the university's Arts and Public Life programing. Aside from weekly jazz and blues performances, the center has hosted exhibitions on AfriCOBRA and other exhibitions and performances highlighting Black Arts in Chicago.

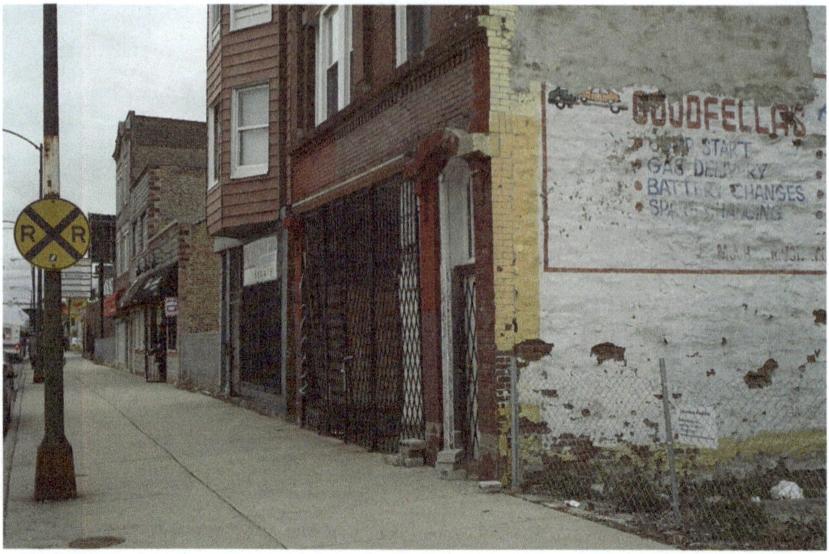

Fig. 2.4 Down the block from Stony Island Bank, 2016. *Photo* by Geoffrey Berliner

In 2013, Gates helped to open the University of Chicago Arts Incubator, which renovated University owned property in the Garfield Park neighborhood just West of Hyde Park, as a gallery and "creativity hub," hosting artist residencies and community-based activities like knitting and yoga, as well as exhibitions, performances and talks. The programing at the incubator and at the Stony Island Bank also includes events aimed at promoting and supporting Black arts entrepreneurship (Rebuild Foundation n. d. b). The ArtPlace Foundation—a joint project funded by a charitable foundations and corporate banks—supported the Arts Incubator with $400,000, administered by the University of Chicago (Fig. 2.5).

More recently, on property leased from the university's holdings, Gates turned several storefronts next to the Arts Incubator into a café, an art bookstore and literary venue, transforming a portion of this shabby block in a blighted African American community into a slim "cultural corridor" or "arts block". The university then announced plans to expand the cultural corridor with a new arts center, the ArtsSpace, a 2,500-square-foot space adjacent to the University's Arts Incubator. In spring 2014 the Knight Foundation awarded the University of Chicago a $3.5 million grant for "The Place Project," to "build on pioneering work by

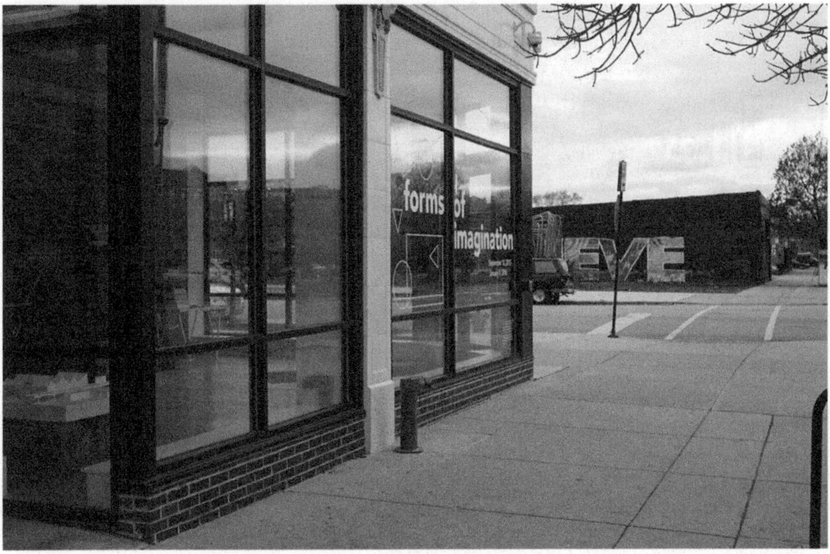

Fig. 2.5 Arts Incubator, 2017. *Photo* by Geoffrey Berliner

Theaster Gates" to "expand and test a community development model that supports arts and culture to help transform communities and promote local growth and vibrancy" (Knight Foundation 2014).

The Arts Space opened in 2018 along with a new 6,600 square foot theatre and performance venue, designed in collaboration with Gates, on the Garfield arts block (U. of C. Holds Grand Opening for Keller Center 2019) (Figs. 2.6, 2.7, 2.8 and 2.9).

Some members of the university community now drive through the stately Washington Park that has long marked a firm border between Hyde Park and the almost exclusively African American community on the other side for arts events. Outside of this small "arts block," the neighborhood remains scary and foreign to most Hyde Park residents accustomed to elegant well-lit streets diligently patrolled by the university security force. Meanwhile, back in Woodlawn's Jackson Park, blocks away from the Stony Island Arts Bank, Dorchester Projects and other Rebuild Foundation holdings, plans are unfolding for the Obama Presidential Library. There is no explicit connection between Gates' projects in Woodlawn and the library. However, when the University of Chicago made a video as part of the pitch to the Obama Foundation for locating the library on the South Side, the

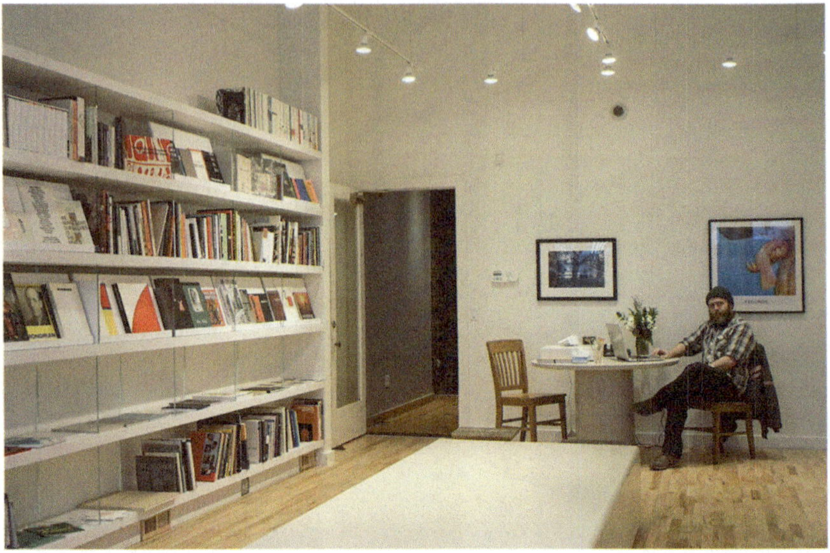

Fig. 2.6 Inside Bing Bookstore and Literary Café, 2017. *Photo* by Geoffrey Berliner

Black cultural life of Chicago's South Side led by Gates played a key role in their marketing campaign.

In addition, there is a new $10.25 million funding initiative, a collaboration among Gates' Rebuild Foundation, the University of Chicago's Place Lab and the City of Chicago to invest in the neighborhood around the Stony Island Arts Bank and Obama Library site. Neighborhood "assets," such as a shuttered elementary school and vacant lots owned by the city, "will be transformed into vibrant civic spaces for public use" (Blackamore 2019). Mere blocks away from the future home of the Obama Presidential Center, then, the University of Chicago will be involved in this large-scale remaking of an entire urban neighborhood, and Gates' Stony Island Arts Bank is considered the "flagship" of this project.

2.5 Race, Class and the University of Chicago

Without knowing the history of the university's role in South Side development and racial segregation, the university, with Gates as its new face of cultural development, appears as a munificent benefactor and font of support for Black cultural

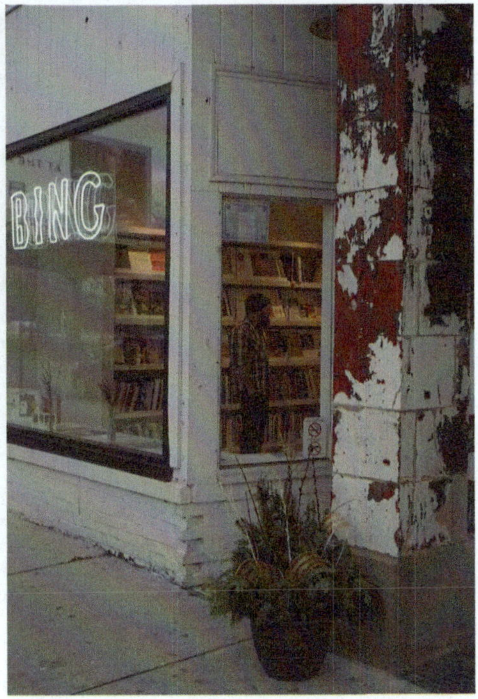

Fig. 2.7 Outside-Bng Bookstore and Literary Café, 2017. *Photo* by Geoffrey Berliner

life in the community. But, a longer view reveals that while Gates needs the university, the university depends on Gates' vision and cultural legitimacy to carry out its mission to close the "rent gap" on Chicago's South Side and to attract students, faculty and increasingly, tourist dollars.

Since its founding in 1892 (Boyer 1999) the university has intervened in the real estate market and public policy to maintain property values in the Hyde Park-Kenwood community and to reinforce the boundaries separating this community from the Black enclaves to the West and North first established during the great migration in the 1920s when African Americans started leaving the South in droves to find work in Northern cities. Early on, the university achieved this goal through financial and legal support of restrictive covenants and through expansion of its own land holdings in the neighborhood (Hirsch 1998).

After WWII the swelling Black population on Chicago's South Side could no longer be contained within the boundaries of Bronzeville, the area just north and West of Hyde Park and a center of Black cultural, economic and social life

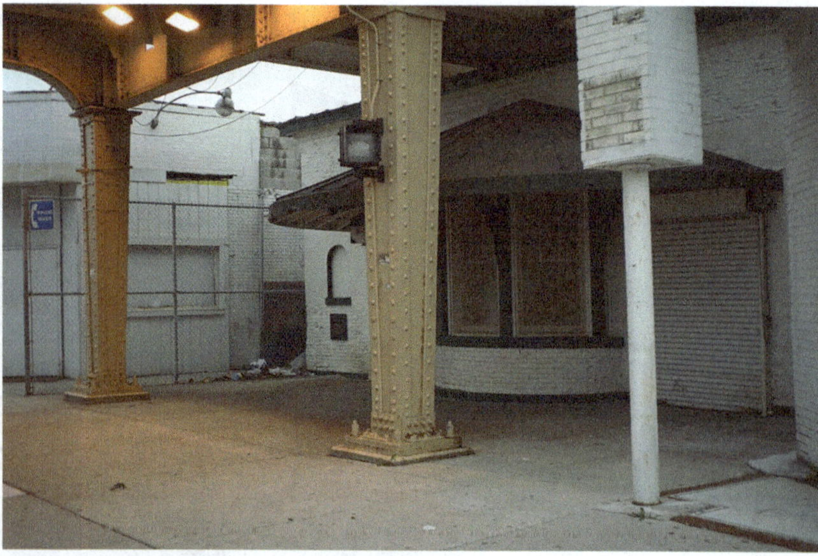

Fig. 2.8 Old Green Line El stop. *Photo* by Geoffrey Berliner

in the 1920s. Throughout the 1940s and 1950s communities South and West of Hyde Park succumbed to succession as White ethnics fled to the suburbs. Fearful of further "darkening" of Hyde Park itself, the university became a key player in forging national urban policy by helping to craft the Urban Community Conservation Act of 1953 in which "slum prevention" became a valid reason for eminent domain and urban renewal. Previous urban renewal policy allowed for the use of eminent domain for slum *clearance* (Hirsch 1998), but this act went further in that properties in areas deemed in danger of *becoming* slums could be confiscated by the city and sold to other agencies or developers as part of an urban renewal plan. This allowed the university to seize control of property in the zone of the encroaching Black belt, along with businesses and housing in Hyde Park itself, and transform it into higher priced faculty housing and university laboratories, classrooms and administration buildings (Baldwin 2015). These strategies contained the encroachment of nearby ghettos into Hyde Park and served to displace lower income Black (and White) residents of Hyde Park.

Also, and this is crucial for understanding the evolution of the community's cultural life, what locals call the university's "Negro Removal Projects" routed out not only residents deemed "undesirable," but the businesses and cultural

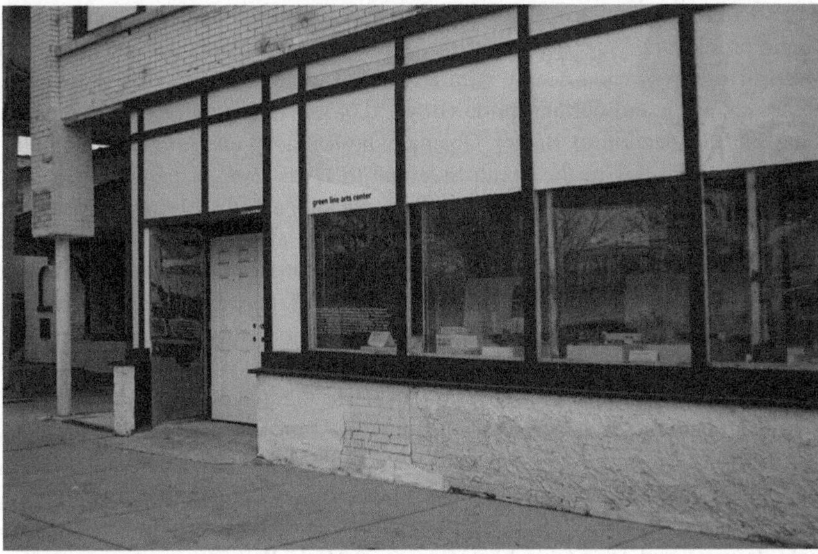

Fig. 2.9 New Green Line Arts Center. *Photo* by Geoffrey Berliner

venues central to Black cultural life on the South Side, including the jazz and blues clubs for which the South Side was famous (Baldwin 2004). In the ensuing decades, Hyde Park developed a reputation as a staid and unglamorous academic outpost, surrounded by impoverished and isolated Black ghettos.

Fast forward to Chicago's push to enter the 21st Century as a globally competitive and relevant city (Betancur 2011; Wilson 2015) and model of postindustrial urban development (Doussard et al. 2009; Bennett et al. 2017). As with postindustrial cities everywhere, forced to find new sources of income in the wake of deindustrialization and federal retrenchment Chicago's growth coalitions embraced strategies of rapid and glamorous renewal of the city's downtown, pro-development policies for gentrifying historically working-class, industrial or artists neighborhoods with easy access to the city's center and generous city spending on creative city formula cultural resources (Lloyde 2010; Wilson 2015). In less than a decade, derelict areas just south of downtown were developed. Finally, as part of a plan to spread to the segregated and stigmatized Black ghettos of the South Side, investment in the historic "Black Metropolis," Bronzeville, was underway. (Anderson and Sternberg 2013; Betancur 2011). This plan—like revitalization in Harlem (Hyra 2006)—would draw on the rich Black cultural legacy of the South Side, enticing Black professionals to invest in a community, which, despite problems of segregation, crime, poverty, divested schools and crumbling infrastructure, resonated deeply with a real and imagined legacy of Black community cohesion and cultural pride (Boyd 2008). With revitalization so close to home and the election of Barack Obama, who has deep and longstanding political and personal ties to the South Side and to Hyde Park in particular (Becker and Drew 2008) the time was ripe for the university, along with the political elite and developers, to capitalize on its enhanced potential. Hyde Park could become not just a racially integrated university neighborhood unfortunately surrounded by untamed urban ghetto, but a desirable tourist destination and residential community. To this end, the university beefed up its cultural offerings, orchestrated by the charismatic and increasingly famous Gates. Gradually, tourists and upper middle-class home-buyers from outside of the university community have become attracted to Hyde Park, overcoming the "fear factor" that has in the past kept them at bay. As just one indicator of neighborhood trends, condos in Hyde Park have seen a 25% increase in value since 2016, with a similar jump in rental prices (Karydes 2019). Thus, while the cultural attractions benefit the new residents, tourists and long-term homeowners in Hyde Park, lower-income renters and prospective residents have been priced out of the offerings (Hyde Park Rental Prices Reach New High n. d.).

Outposts like the arts block on Garfield have yet to fully overcome the stigma associated with these neighborhoods and really see the larger-scale impact of gentrification, but it seems likely that as the university pushes through its cultural regeneration projects this will change (Institute for Housing Studies-DePaul University n. d.). Given what we already know about gentrification patterns (Chronopoulos 2016), if community revitalization occurs, it is likely that the poorest, most disenfranchised members of these communities will be forced to relocate.

Activists in Woodlawn, the community likely to be most affected by development because by 2021 it will house the Obama Presidential Library, are already concerned about the impact of the library on property values (Robinson 2018). While Obama has reassured the community that gentrification there is unlikely, the real-estate website Redfin picked Woodlawn as the city's second-hottest neighborhood for 2017, reporting a 23.3% increase in home values in the first six months of that year (Woodlawn Projected To Be Redfin's Hottest Market of 2017 n. d.)This trend continues despite the fact that 30% of households there live below the poverty rate, and the rate of violent crime is double the city of Chicago as a whole (Chicago Poverty and Crime n. d.).

2.6 Black Arts, Black Power

Themes from the Black Arts movements of the 1960s and 1970s are central to new cultural development projects in and around Hyde Park. These movements were outgrowths of radical, urban-based Black Power politics, which grew in the wake of failures of the Civil Rights movement to dismantle persistent institutional racism, the impact of toxic urban renewal policies and racialized violence in the United States (Taylor 2016a, b). The Black Power movement also emerged from the anti-war movement and increasingly salient connections between United States' imperialist violence against people of color abroad and the ghettoization, neglect, police brutality and economic impoverishment of African Americans at home (Marable 2000). While there were cleavages within the Black Power movement regarding the degree to which Black liberation could be achieved within the framework of capitalism, the most radical wing, which included most members of the Black Panther Party, called for upending imperialism, White supremacy and capitalism itself (*see* Charmichael and Hamilton, 1968).

While the Panther agenda superficially integrated elements of Black capitalism by promoting an ethos of economic self-sufficiency within the Black community, its ends were far more radical than the Black Capitalism advanced by President Richard Nixon's programs or the economic self -sufficiency arguments spanning from Booker T Washington to Malcom X. And, although the Panthers saw community activism as a key element of their struggle, providing free breakfast, health clinics and other services within the Black community (Hilliard 2008), as Panther leader Bobby Seale explained

> [these programs] were part of a revolutionary policy … This is the people's revolutionary program which educates the people that they have to be in opposition and oppose the power structure. That is the great revolutionary character of our program. (Interview with Bobby Seale 1988).

While Black Power ideology drew from Marxism, it was also marked by a Black Nationalism (Van Deburg 1992). This nationalism has a strong cultural and psycho-social dimension, stressing a collective pan-African sensibility and history growing from traditions and philosophy of Africa. It also rejects the internalized racism and oppression through which colonial subjects internalized the colonizer's gaze. From this perspective, liberation is not only economic and political, but must also be psychological and aesthetic (Hooks 2015). This current of Black Power manifested in the work of the BAM but also in the attention paid to self-presentation, spectacle and sartorial practices of Black Power activists, their followers, and eventually as popular fashion within the Black community as Afro hairstyles, African dashikis, and even paramilitary ensembles worn by Black Panthers were embraced. In this regard, the Black Panthers had a keen awareness of the power of aesthetics, visual symbolism and media savvy in coalescing support for social movements in the televisual age (Ongiri 2010).

BAM was officially established when the poet and playwright Amiri Baraka founded the Black Arts Repertory Theater in Harlem in 1965, but it also had an imposing presence in cities throughout the Unites States including Chicago, Detroit, Newark and San Francisco (Neal 1968). To varying degrees, these artists allied themselves with the revolutionary programs of the Black Panthers, the most visible arm of the Black Power movement. Indeed, Larry Neal, a key theorist of the Black Arts Movement argued that "[t]his movement is the aesthetic and spiritual sister of the Black Power concept" (Neal, 1968, p. 1) And, he continues, quoting the his contemporary, the African-American poet Etheridge Knight, it was to the question of a *collective* cultural nationalism that the movement spoke most directly:

> [T]he Black artist, in creating his own aesthetic, must be accountable … to the Black people. Further, he must hasten his own dissolution as an individual (in the Western sense)—painful though the process may be, having been breast-fed the poison of "individual experience". (Night in Neal 1968, p. 2.)

Movement spokespeople outlined a vision of a specific (and political) Black aesthetic which rejected the mainstream art-world in favor of forms and vocabulary that resonated with the experience of the Black community for the purpose of envisioning and promoting radical social change, while insisting on the relevance of art to revolution. In his 1969 manifesto, "Any Day Now: Black Art and Black Liberation," Neal explains: "… a revolution without a culture would destroy the very thing that now unites us; the very thing we are trying to save along with our lives" (Neal 1969, p. 3).

BAM artists in Chicago and elsewhere drew on developments in the literary, musical and visual avant-gardes of the period but they were explicit in their focus on Black experience in the United States. They also drew on the expressivity of traditional or popular African and Afro-American cultural forms and in so doing laid claim to an essential ownership of a particular "Black" aesthetic. Amiri Baraka, writing about the development of jazz and the African American experience, argues that, "[N]egroes played jazz as they had sung blues or, even earlier, as they had shouted and hollered in those anonymous fields, because it was one of the few areas of human expression available to them" (Baraka 1963).[4]

Chicago, as the first large urban metropolis on the path of the great migration, emerged as an epicenter of Black activism and cultural innovation first during the period of the Black Renaissance of the 1930s and 1940s and then later during the 1960s and 1970s. Jeff Donaldson, one of the founding members of AfricCOBRA in the 1970 manifesto, explains the relevance of migration:

> AfriCOBRA, … cannot be understood without understanding the long arc of migration among Black Chicagoans before and during the 1960's. Chicago was a special place in the sense that it was populated by a unique group of Black individuals whose life experiences, up to the time of their migration up South to Chicago, were formed in a crucible of abject repression and subjugation. (Donaldson in Jones-Henderson, 2012 p. 4).

Indeed, the founding members of the AACM, a non-profit organization begun on the south side of Chicago in 1965, all came from families that had moved to Chicago directly from post slavery south (Lewis 2014). While open to a variety of musical influences, AACM excluded non-Black members, and viewed the core of their project as afro-centric and interconnected with the Black community, where they met, staged performances and provided free music lessons to neighborhood youth. AACM musicians pushed the boundaries of free jazz and innovations of the avant-garde art music movement, while incorporating musical forms and instruments from Africa and the East. The Art Ensemble, a group of AACM musicians, was famous for incorporating costumes derived from African robes, jewelry, and face paint into their stage performances. Despite the bold and audacious performances of AACM musicians, they de-emphasized the "great man" or original genius notion so central to Western art, and had a collective understanding that "the power of one pales in the presence of many." As Donaldson explains:

> [We] are a family of image-makers and each member of the family is free to relate to and to express our laws in her/his individual way. Dig the diversity in unity. We can be ourselves and be together, too. (Donaldson quoted in Zorach 2011, p. 106)

[4] Somewhat later, the musicologist Norman Lewis explained how the formation of African American improvisational traditions developed as an expression of racialized experience: "Improvisative musical utterance, like any music, may be interpreted with reference to historical and cultural contexts. The history of sanctions, segregation, and slavery, imposed upon African Americans by the dominant white American culture, has undoubtedly influenced the evolution of a sociomusical belief system that differs in critical respects from that which has emerged from the dominant culture itself" (Lewis, 1996).

The literary, musical and visual artists of Chicago's BAM were deeply collaborative in their aesthetic goals, their community activities and their organizational structures. AfriCOBRA artists each had their own distinct visual style, but they shared an interest in the work of contemporary African artists and favored highly saturated colors, patterned surfaces and rhythmic linear effects. One of the founding members, Barbara Jones-Hogu outlines their aesthetic:

> Black, positive, direct statements created in bright, vivid, singing coolade colours of orange, strawberry, cherry, lemon, lime and grape ... Black positive statements stressing a direction in the image with lettering, lost and found line and shape. (Jones-Hogu quoted in Zorach 2011, p. 110)

Like the Black Panthers, BAM artists felt a profound commitment to the urban communities whose members they hoped to empower and organize. The Wall of Respect, an outdoor mural created on the side of a tavern on Chicago's South Side, provides one important example of their use of art in the service of community empowerment. This 30' x 60' mural was created during the summer of 1967 by Jeff Donaldson and other members of OBAC visual arts workshop. The artists decided not to include signatures in order to highlight the collective, community-based nature of the project. These artists first obtained permission from local gang leaders to utilize the wall and collaborated with these leaders and community members to choose the subjects represented in their homage to the cultural and political heroes of the Black community. Those represented included Billie Holiday, Muddy Waters and James Brown. The statesmen category included Malcolm X, Stokely Carmichael, and Marcus Garvey (Collins and Crawford 2006).

During and after its creation the mural became a gathering place for poets, musicians, artists, dancers and other community members and the site of planned and impromptu performances. At one of the many events, Gwendolyn Brooks, Chicago poet and an inspiration for many of the poets of the Black Arts movement, read a poem inspired by the wall, which included these lines:

> in African images of brass and flowerswirl,
> fist out "Black Power!"--tightens pretty eyes,
> leans back on mothercountry and is tract,
> is treatise through her perfect and tight teeth. (Reardon 2017, para 21).

The Wall was an assertion of Black identity at a key moment in the nation's history when the activism of the Civil Rights Movement was expanding into the expression of Black Power. Indeed, the title of the mural was a direct reference to Aretha Franklin's song, then on the radio, "R-E-S-P-E-C-T" (in Alkalimat et al. 2017).

By the early 1970s the Panthers were the subject of ongoing harassment through the federal government's counterinsurgency program, COINTELPRO. This, combined with the murder of 50 Panthers and the arrest or injury of another 500, convinced some members to abandon their radical strategies and become involved in local electoral politics, forming alliances with the Black middle class and forcing urban political machines to attend to the needs of the Black community and help channel resources into underfunded neighborhoods (Mason 1973). BAM artists suffered similar forms of harassment, surveillance and pressure from local police and federal agents. This pressure, combined with the "institutional incorporation" of a number of BAM artists into academia altered the focus of the movement artists. Installed in arts departments or Black Studies programs formed in the wake of student protests, they achieved stability and influence in the academy, but sacrificed proximity to activism in the trenches and connection to the larger Black community (Fenderson 2019).

Recently there has been a spate of museum and gallery exhibitions both highlighting and historicizing the work of this movement and other Black artists. As with other art movements, especially those with explicit social and political content, the works take on different meanings when they are institutionalized. The curation, display and textual accompaniments to these exhibits invite the viewer to reflect on the political and social aspirations of the period and the relevance to our own. However, the context in which they are displayed and the social, economic and political landscape in which they are being revived should also shape our reflection.

The artists of the BAM self-identified as the cultural arm of a distinctly urban radical social movement which developed in response to, among other factors, an urban political economy and public policy regime which has experienced rapid restructuring since the period when BAM was active in Chicago and elsewhere. This restructuring emerges from larger global processes of neoliberalization, which impact urban fortunes in distinct ways. Below, I will briefly contrast the urban context in which this work first emerged and those that surround its "second life" on Chicago's South Side today.

2.7 First Life: Black Arts Movement, Black Power and Welfare State Urban Policy

The BPM and the BAM emerged from the failure of both the Welfare State and the Civil Rights movement to adequately address the structural dimensions of persistent Black urban poverty. In the decades following the Great Migration, which saw swelling African American populations in Northern cities, federal and local urban renewal initiatives decimated Black communities and entrenched residential segregation which was the legacy of earlier political regimes (Marcuse 1997). In 1964, the Johnson administration, acknowledging the persistent problems of poverty (and related social foment; Levy 2018) even during the period of postwar economic growth, rolled out a series of "War on Poverty" initiatives. These initiatives, which included jobs training programs, pre-school education, food stamps, and legal services were intended, in the words of Johnson, "not only to relieve the symptom of poverty, but to cure it and, above all, to prevent it" (Roy et al. 2015, p. 142).

Despite new federal initiatives, discontent with police brutality, lack of political representation, unemployment and inadequate housing, acerbated by the assassination of Martin L King, exploded into urban riots in New York, Los Angeles, Newark, Baltimore, Detroit, Chicago and elsewhere (Levy 2018) rocking the US political establishment. Responding to these events in 1966 President Johnson established a task force to investigate the problem of urban unrest. These investigations concluded that American cities were plagued by "apartheid," a dire "segregation by race and income" that was generating "civil discontent and potential guerilla warfare" (President's Task Force on the Cities 1966, p. 4, i, vii). Community development was to tackle these questions of race and revolution in American cities (Marcuse 1997).

Johnson's initial initiatives included the "Economic Opportunities Act" (EOA) which, authorized the formation of local Community Action Committees, regulated and funded by the federal government, to address problems of poverty from within communities themselves. The core idea of this 1966 report was that "maximum involvement" of poor communities in articulating their needs and administering services would quell unrest by integrating African Americans into the political system. This new policy approach allowed grassroots community groups to apply for federal funding for community improvement, educational, health and even arts-based projects and helped to politicize and empower members (notably, women) and to train a generation of community activists (Orleck and Hazirjian 2011). The EOA funded more than a thousand Community Action Agencies (CAAs) across the United States with the goal of fostering bottom-up revitalization of the country's poorest areas. In order to receive federal funding, CAAs had to demonstrate "the maximum feasible participation" of residents of the areas and groups served.

Despite the African American community's distrust of the federal government's motives, activists mobilized to receive block grants from the state government. This mobilization served to educate and empower community members and helped to create challenges to the rule of the state (Orleck and Hazirjian 2011).

Some community activists, caught up in the social and cultural politics of the Black Power movement, became radicalized (Roy et al. 2015). Thus, as Benson (1971) argues, there was a direct relationship between the popularization of the Black Power Movement and War on Poverty initiatives prioritizing community involvement in poor, Black communities. BPM, and, by extension the BAM, with its commitment to mobilize the arts in the service of Black community pride and empowerment, were connected at least temporally to federally funded Community Action Programs.

By the late 1960s President Johnson, bending under mounting pressure from mayors who feared the wrath of White constituency, backed away from his commitment to community control of federal funding (Orleck and Hazirjian 2011). This retreat continued through the 1970s, and the seeds for a full-on attack on the Welfare State were in place by the time Ronald Reagan was elected president in 1980. During Reagan's first term he repealed the EOA, in one of the opening moves which dismantled the liberal institutions and ideology of the Great Society. Thus began a downward spiral of conditions affecting the most disadvantaged Americans, including urban African-American poor, who were increasingly demonized and excoriated by the media, the police and large sections of the citizenry (Schram et al. 2003).

2.8 Second Life: Black Arts Movement and Neoliberal Considerations

The BPM and BAM on the South Side of Chicago emerged from and were embedded in a social and political context which included the Great Migration and its impact the social and demographic structure of the city; demands and failures of the Civil Rights Movement; politics of the Black Power Movement and Welfare State policies. Through the creative place-making projects of the University of Chicago and Gates this work and its legacy is once again playing a symbolic role in framing urban developments on Chicago's South Side. The context in which this work is deployed and consumed, however, stands in sharp contrast to that of its "first" life.

As mentioned earlier, recent urban policy in Chicago has followed the blueprint of neoliberal urbanism in postindustrial cities throughout the United-States and elsewhere (Bennett et al. 2017; Doussard et al. 2009). This policy stands in sharp contrast to that of the liberal Welfare State during the height of the War on Poverty initiatives of the 1960s. Neoliberalism as an ideology and approach to governance has come to dominate the world stage since it first emerged during the late 1970s and early 1980s as a response to declining rates of profit, rising inflation, rising unemployment and increasing instability in the international monetary system (Harvey 2007). Rooted in the work of Chicago School scholars like Milton Freidman and Friedrich Hayek, architects of neoliberalism blamed economic stagnation on the Keynesian redistributive and regulatory policies of the Welfare State, which choked the natural growth of the free market (Peck 2010). Repurposing classical liberalism's *laissez-faire* principles, neoliberalism's advocates argued that unfettered markets benefit not only elites but also the great masses of citizens feeling the impact of recession and unemployment. The role of the state was not to provide a safety net or ameliorate social problems, but rather to insure the free flow of markets and provide profitable conditions for capital accumulation. These were the philosophical underpinnings for the deregulation of the financial sector; free trade agreements which buttressed the off-shoring of manufacturing to low wage labor markets in industrializing countries; undermining workers protections and environmental regulations; increasing privatization and defunding of public institutions and services; the implementation of regressive taxation and a host of other policy shifts leading to widening inequalities between households and between capital and labor in the United-States and elsewhere (Kelleher 2019).

The triumph of neoliberalism has had an especially pronounced impact on urban Black, working class and poor populations (Marcuse 1997; Wilson 1987) who, because of racist lending and urban renewal policies of the past, were left stranded in communities reeling from cutbacks in public services, income assistance and loss of manufacturing employment from which their White counterparts had benefited in the post-war period (Goldfield 1997). After his election 1980, Reagan ushered in draconian cuts to welfare and social spending. As Taylor (2016a, b) points out, these cuts had an immediate and profound impact on the well-being of Black families, with rising rates of poverty and unemployment. Cut-backs to social services were legitimized by an ideology of personal responsibility which overturned the post-war social contract. Instead, blame for social problems associated with poverty, unemployment and lack of access to opportunity was placed, through political rhetoric reinforced by media, on the victims themselves. Federal attacks on benefits and services culminated in 1996 when Clinton signed Personal Responsibility Work Reconciliation Act (Schram et al. 2003).

The devastating impact of public service and welfare rollbacks on impoverished Black communities was buttressed by new crime policies and policing practices which ensured that many unemployed youths would be imprisoned, thus exacerbating problems of Black communities (Alexander 2016). More recently, in Chicago, African Americans have been impacted by the revanchist policies of Rahm Emanuel, who oversaw significant public-private development projects downtown and in nearby areas and the closure of more than fifty public schools, almost exclusively in Black and Latino neighborhoods (Taylor 2016a, b). Under Emanuel's reign, the city retained the highest rate of Black unemployment among the nation's most populous cities, and has seen an outmigration of African Americans which has left entire communities abandoned (Bennett et al. 2017). In response to both deindustrialization and shrinking state and federal subsidies, Chicago and other cities, have been forced to become, "growth machines" (Logan and Molotch 2007), competing with one another to attract businesses, elite residents, large-scale real estate development, global investment and tourist dollars.

To this theme of the commodification of urban space under neoliberalism (Smith 2002), three other shifts in urban dynamics are important. First is the new role for and meaning of community development in the neoliberal city. In an urban regime in which place promotion is presented as necessary for survival, a false unity between citizens regardless of class, gender or race is posited by growth coalitions in the common cause of city or neighborhood boosterism. This place promotion is often initiated by non-profits, local Economic Development Corporations and other public-private partnership initiatives (many of which were put in place in order to fill the gaps in social services left by retrenchment) which rely on civic engagement. These initiatives, as Germangeographer Margit Mayer argues, repurpose the institutional critiques and progressive goals of community organizers from the 1960s and 1970s such as "self-management, self-realization and all kinds of unconventional or insurgent creativity …" (Mayer 2013), turning them into a neoliberal governing techniques aimed toward the commodification of urban space.

Here, the discourse of self-help and community activism that drove such initiatives during the 1960s and 1970s is disaggregated from larger demands for structural change on which such activism was based. Community building initiatives like those that emerged as bi-products of both Johnson era policy and the Black Power movement are replaced by new initiatives that blend nostalgia and market imperative. As Defillippis et al. (2009) explain:

> An emphasis on "the bottom line", building "partnerships" with local businesses and corporations, developing "relationships" and focusing on "community assets" has narrowed conceptions of community activism … most contemporary models of community building and development focus exclusively on the local internal community, not the economic, political and social decisions, which rest outside the community and create community needs and concerns. (p. 16)

Another key characteristic of neoliberal urban development is the prominent role played by the arts and culture in urban growth strategies (Kirchberg and Kagan 2013). Art and culture have been central to urban branding and development since at least the 1970s when, as Zukin (1989) described, symbolic or "aesthetic" capital that art and artists brought to SoHo's light manufacturing district was harnessed by developers and city government to transform the neighborhood from a manufacturing to a high end residential and retail district. The symbolic capital that artists, galleries, music and performance venues and the cafes and bars frequented by artists bring to neighborhoods has more recently been identified by Richard Florida (2002). Here, he explains that for postindustrial cities to flourish, they need to appeal to the elite workers of the new economy, the creative class. These creative workers, inspired by a commoditized version of the counter-cultural ethos inherited from the baby boomers (Boltanski and Chiapello, 2005), value creativity, personal expression, diversity and novel experiences and gravitate to cities and neighborhoods that provide opportunities to consume these amenities. Cultural venues and the presence of artists' subcultures are thus integral to the economic health of the city. Florida's mantra has been embraced by city officials everywhere, (although Florida himself [2017] has more recently acknowledged that his "creative city" formula engenders both winners and losers) as growth coalitions plan large and small-scale investment in new cultural institutions and promoting the growth of "cultural corridors" and arts districts.

Finally, scholars have begun to assess the role of universities, in partnership with municipal government and private investors, grassroots community groups, and nonprofit organizations as key players driving creative city type development by casting themselves as institutions that can parcel and repackage "blighted" areas into "destinations" for a safe and profitable urban experience. Here, institutions like the University of Chicago have expanded their profitable business, tech development, health care and medical research facilities, while at the same time rolling out new arts and cultural initiatives. These serve to attract students and residents to college communities, and, eventually, to gentrify surrounding impoverished neighborhoods (Kelly and Silverman 2014; Weiwel and Perry, 2008) and, as in the case of the University of Chicago, to attract further high-profile institutions like the Obama Public Library. However, while universities and other non-profits may fill some of the functions of defunded local governments, unlike government they are not formally accountable to the public nor must they employ democratic decision-making processes. In addition, their own institutional interests may be in conflict with those in the communities over which they have power (Patterson and Silverman 2013).

2.9 Some Concluding Thoughts (for now)

Creative development projects on Chicago's South Side, and their deployment of BAM must be assessed within this more general dynamic of neoliberal urbanism. On the one hand, the climate for arts related development on the South Side was already in place with Chicago's goglobal ambitions. And, the University of Chicago, with multiple interests on Chicago's South Side is uniquely situated to act as an "anchor institution" there. A Report on the "Future of UChicago arts – 2020 and 2025" issued by the Arts Steering Committee who is charged with developing an ambitious vision for UChicago Arts was recently released. The report begins with a proud claim from the provost that: "Over the past decade the University of Chicago has gained increasing recognition as a destination for the arts" (https://provost.uchicago.edu/reports/report-future-uchicago-arts-%E2%80%93-2020-and-2025). The report goes on to detail various successes of recent university arts initiatives, which include expansion of arts programing, infrastructure development, faculty acquisitions, professional training opportunities and, significantly, community engagement. The report is transparent about the motivation behind these arts initiatives:

> … in conjunction with the new Obama Presidential Center, a vibrant South Side arts scene will enhance the University of Chicago's ability to attract local and international visitors, including internationally recognized artists and art professionals. (ReportFutureUChicagoArts n. d., p. 12)

The report then unfolds plans for a "new arts geography" constituted by the campus surrounded by four "art zones" which include the Garfield Arts Block West of Washington Park and the Midway Arts Complex, which borders on the area to the South. "This South Side arts constellation" the report concludes:

> with a transportation system ensuring safe and easy circulation, will emerge as the hottest art scene in Chicago … In conjunction with the building of the Obama Presidential Center, these arts zones will transform the South Side of Chicago into a cultural destination that garners ongoing international attention. (ReportFutureUChicagoArts n. d., p. 2)

Despite the report's cheery predictions that the South Side will emerge as a magnet and even epicenter of art and culture, with all the status and revenue this implies, the reality remains that the university has an image problem in the Black communities which are being enlisted in their strategic arts development plans, especially since these plans are likely to raise property values and hence contribute to displacement. Indeed, in 2010 tensions between the university and the surrounding Black communities erupted around the battle to reopen the adult trauma center at the university of Chicago hospital after the death of 18-year-old Damian Turner, who was shot just a few blocks from the University of Chicago hospital and had to be transported nine miles to the nearest trauma center for treatment. After more than a year of heated battles between community activists and residents and the university, the university finally agreed to open a new trauma center (Schlesinger 2016).

In response to renewed politicization in the Black community, the university has tried to foster an image of positive community involvement. University spokesman Derek Douglas, during a recent interview on Chicago's public radio station, WBEZ, remarks that there are bound to be tensions between "a well-resourced" university and surrounding communities that "are facing challenges," but that the university is trying to find things that are "of mutual benefit" to help bridge the chasm (O'Brien 2019).

Here, Douglas suggests the "flanking" role of non-profits within neoliberal urban governance mentioned by Mayer, DeFillippis and others. The Arts Steering Report also includes present and future plans for "community engagement" and "community outreach," and, without explicitly mentioning race, makes repeated reference to the vitality and creativity of the "local" south side arts community and the importance of providing support and forming alliances with this community. In a subsection on engagement and the South Side the report states that:

> …the infrastructural, intellectual, and financial support for art practiced by South Side residents creates a unique bridge between the University and the cultures that surround it. Efforts should be made to augment that support. (ReportFutureUChicagoArts n. d., p. 7)

Considering the university's history and "image problem" on the South Side, the choice of Theaster Gates, who's own origin story and artwork suggests a strong and authentic connection to the Black communities around Hyde Park, and specifically, the emphasis on the work of African American artists and the legacy of the BAM makes tactical sense. However, what is highlighted in the report is not the structural critique of imperialism and capitalism or even community empowerment in which this work was first embedded. Instead, it functions to frame an identity politics devoid of structural demands or sociological insights. The mission statement of Gates Rebuild Foundation, which receives substantial endowment from the university's Arts and Public Life Initiative, is explicit in its appeal to this depoliticized notion of identity:

> Our mission is to demonstrate the impact of innovative, ambitious and entrepreneurial arts and cultural initiatives. Our work is informed by three core values: black people matter, black spaces matter, and black objects matter. (rebuild-foundation.org)

The work of Mayer and others discussed above sheds light on this appeal to Blackness as a unifying essence to which African Americans, and their White liberal allies on the South Side can get behind, regardless of class or professional interests. In highlighting the cultural nationalist elements of Black Arts, the degree to which class fissures within the Black community have deepened over the past decades is obscured.

As E. Franklin Frazier's (1957) rich sociological account of the Black bourgeois documents, class stratification has been an enduring feature of the African American life since at least emancipation. Nonetheless, it was only during and after the Civil Rights era that a substantial Black middle class emerged in this country. While African Americans, even those situated firmly in the upper and middle classes are significantly less wealthy than their White counterparts, growth of the public-sector employment during the Johnson administration as well as anti-discrimination mandates in federal hiring practices gave rise to an expanded Black middle class. This expansion, Taylor explains, has been important in the rise of a Black political class (Taylor 2016a, b). At the same time, the cultural nationalist dimensions of the Black Power Movement created an essentialist version of community and race unity through which Black politicians could gain the support not only of the Black elite but of the less privileged African Americans whose interests may not have coincided with this expanded Black middle-class. As Adolph Reed (https://nonsite.org/black-politics-after-2016) explains:

> With the rise of a stratum of black public officials and functionaries enabled by the 1965 Voting Rights Act, changing urban racial demography, and expanded opportunity structures, black politics largely settled … into a routine racial interest-group affair, an ethnic politics, that accommodated to the national retreat from redistributive social wage policy by defining a specifically racial agenda limited to challenging disparities and demanding ultimately symbolic appointments and honorific recognition.

The emphasis on racial disparities in opportunity and outcome that emerged in the wake of Black Power set the course, according to Reed, for the identity politics which characterizes much of the popular discussions around Black empowerment today. It is this appeal to racial identity rather than to social and economic policies that might challenge the accumulation of private wealth at the expense of collective needs to which Gates' mission statement and his deployment of Black Arts points. Even when the Black Power Movement embraced cultural nationalism as a source of empowerment, its intellectual spokespeople attempted to understand the complex relationship between capitalism, empire and racism in its shifting historical context. Today, the divergent fortunes of an educated Black middle-class, whose members occupy some of the highest rungs of academia, entertainment and politics in Chicago and elsewhere, a Black working class whose struggles in the face of neoliberal economic restructuring are shared by their White, Latinx and Asian counterparts, and the legions of poor and dispossessed Black families who have been most deeply affected by neoliberal policies cannot be easily reconciled under the symbolic banner of shared Blackness.

Gates' and the university's strategic deployment of Black arts in his place-making activities on Chicago's South Side draws together members of the Black professional class, helping to foster spaces of discussion, celebration and cultural display. His projects raise awareness of the legacy of radical Black politics and art during a time, which although some use the term post-racial, problems of racism have not gone away. At the same time, this deployment of Black Arts as part of a regeneration policy for blighted neighborhoods implicitly draws on the notion that art and beauty will provide a kind of moral uplift to a politically and economically disenfranchised community. The University of Chicago and other stakeholders in the urban growth machine have much to gain from creative development on the South Side. And, given the university's checkered history in the Black community, as well as the growth of a Black professional class with capital to spend in the community's gentrifying neighborhoods, a focus on the legacy of the BAM is expedient. On the most general level, this account of arts related development on Chicago's South Side demonstrates the need to embed studies of urban-based arts development within the context of neoliberal urban development practices and policy.

References

Alexander, Michelle. 2016. *New Jim Crow: Mass Incarceration in the Age of Colorblindness*. Place of publication not identified: New Press.

Alkalimat, Abdul, Romi Crawford, and Rebecca Zorach. 2017. *The Wall of Respect: Public Art and Black Liberation in 1960s Chicago*. Evanston, Illinois: Northwestern University Press.

Anderson, Matthew B., and Carolina Sternberg. 2013. *"Non-White" Gentrification in Chicago's Bronzeville and Pilsen: Racial Economy and the Intraurban Contingency of Urban Redevelopment*. Urban Affairs Review 49(3): 435–467.

Baldwin, Davarian L. 2004. *Black Belts and Ivory Towers: The Place of Race in U.S. Social Thought, 1892–1948*. Critical Sociology 30(2): 397–450. https://doi.org/10.1163/156916304323072161.

Baldwin, Davarian L. 2015. *The "800-Pound Gargoyle": The Long History of Higher Education and Urban Development on Chicago's South Side*. American Quarterly 67(1): 81–103. https://doi.org/10.1353/aq.2015.0001.

Baraka, Amiri. 1963. *An Excerpt from Black Music*. http://www.akashicbooks.com/extra/an-excerpt-from-black-music/. Accessed: 4 June 2019.

Becker, Jo, and Christopher Drew. 2008. *Pragmatic Politics, Forged on the South Side*. New York Times. https://www.nytimes.com/2008/05/11/us/politics/11chicago.html. Accessed: 28 May 2020.

Bennett, Larry, Roberta Garner, and Euan Hague. 2017. *Neoliberal Chicago*. University of Illinois Press. https://doi.org/10.5406/j.ctt1s47658.

Kenneth Benson. (1971). Militant Ideologies and Organizational Contexts: The War on Poverty and the Ideology of "Black Power." The Sociological Quarterly, 12(3), 328–339.

Bertagnoli, Lisa. 2019. *Prada Needed Help Dodging Racial Missteps. It Called in Theaster Gates*. Crain's Chicago Business. https://www.chicagobusiness.com/retail/prada-needed-help-dodging-racial-missteps-it-called-theaster-gates. Accessed: 4 June 2019.

Betancur, John. 2011. *Gentrification and Community Fabric in Chicago*. Urban Studies 48(2): 383–406. https://doi.org/10.1177/0042098009360680.

Blackamore, Lenora. 2019. *Foundation Gets $1.6M to Redevelop Vacant School in Greater Grand Crossing*. Chicago Crusader. https://chicagocrusader.com/local-news/foundation-gets-1-6m-to-redevelop-vacant-school-in-greater-grand-crossing/. Accessed: 11 May 2019.

Bone, Robert, Richard A. Courage, and Amritjit. Singh. 2011. *The Muse in Bronzeville: African American Creative Expression in Chicago, 1932–1950*. Piscataway: Rutgers University Press. http://pi.lib.uchicago.edu/1001/cat/bib/11125610.

Boyd, Michelle R. 2008. *Jim Crow Nostalgia: Reconstructing Race in Bronzeville*. Minneapolis: University of Minnesota Press.

Boyer, John W. 1999. *Three Views of Continuity & Change at the University of Chicago*. Chicago: University of Chicago. Accessed: 4 June 2019.

Carmichael, S., & Hamilton C. V. (1968). Black Power: The Politics of Liberation in America. Jonathan Cape.

Boltanski, Luc, and Eve Chiapello. The New Spirit of Capitalism. London ; New York: Verso, 2005.

Chicago Poverty and Crime. n. d. City of Chicago, Data Portal, Chicago. https://data.cityofchicago.org/Health-Human-Services/Chicago-poverty-and-crime/fwns-pcmk. Accessed: 4 June 2019.

Chronopoulos, Themis. 2016. *African Americans, Gentrification, and Neoliberal Urbanization: The Case of Fort Greene, Brooklyn*. Journal of African American Studies 20(3–4): 294–322. https://doi.org/10.1007/s12111-016-9332-6.

Collins, Lisa Gail, and Margo Natalie Crawford. 2006. *New Thoughts on the Black Arts Movement*. New Brunswick, N.J.: Rutgers University Press. http://pi.lib.uchicago.edu/1001/cat/bib/11149980.

DeFilippis, James, Robert Fisher, and Eric Shragge. 2009. *Community Organizing Theory and Practice: Conservative Trends, Oppositional Alternatives* in The People Shall Rule: ACORN, Community Organizing, and the Struggle for Economic Justice, Robert Fisher ed., Nashville: Vanderbilt University Press.

Donaldson, Jeff R. 1970. *AfriCobra Manifesto (1970)*. 391.Org [blog]. https://391.org/manifestos/1970-africobra-manifesto-jeff-donaldson/. Accessed: 4 June 2019.

Doussard, Marc, Jamie Peck, and Nik Theodore. 2009. *After Deindustrialization: Uneven Growth and Economic Inequality in "Postindustrial" Chicago*. Economic Geography 85(2): 183–207.

Fenderson, Jonathan. 2019. *Building the Black Arts Movement Hoyt Fuller and the Cultural Politics of the 1960s*. University of Illinois Press.

Fernández-Esquinas, Manuel, and Hugo Pinto. 2014. *The Role of Universities in Urban Regeneration: Reframing the Analytical Approach*. European Planning Studies 22(7): 1462–1483. https://doi.org/10.1080/09654313.2013.791967.

Florida, R. L. (2017). The new urban crisis: How our cities are increasing inequality, deepening segregation, and failing the middle class-- and what we can do about it. New York: Basic Books.

Florida, Richard. The rise of the creative class. (2002). Basic Books.

Frazier, E. Franklin. 1957. *Black Bourgeoisie: The Rise of a New Middle Class*. New York: Free Press; London: Collier Macmillan.

Gates, Theaster. 2015. *How to Revive a Neighborhood: With Imagination, Beauty and Art*. TED2015 [Video]. https://www.ted.com/talks/theaster_gates_how_to_revive_a_neighborhood_with_imagination_beauty_and_art. Accessed: 2 June 2019.

Goldfield, Michael. 1997. *The Color of Politics: Race and the Mainsprings of American Politics*. New York: New Press.

Goldstein, Brian D. 2017. *The Roots of Urban Renaissance: Gentrification and the Struggle over Harlem*. Cambridge, Massachusetts: Harvard University Press.

Green, Adam. 2009. *Selling the Race: Culture, Community, and Black Chicago, 1940-1955*. Chicago: University of Chicago Press.

Harvey, David. 1989. *From Managerialism to Entrepreneurialism: The Transformation in Urban Governance in Late Capitalism*. Geografiska Annaler. Series B, Human Geography 71(1): 3–17. https://doi.org/10.2307/490503.

Harvey, David. 2007. *A Brief History of Neoliberalism*. Oxford: Oxford University Press.

Hilliard, David. 2008. *The Black Panther Party: Service to the People Programs*. Albuquerque: University of New Mexico Press. http://pi.lib.uchicago.edu/1001/cat/bib/11169734.

Hirsch, Arnold R. 1998. *Making the Second Ghetto: Race and Housing in Chicago, 1940–1960*. Chicago, Ill.: The University of Chicago Press.

Hooks, Bell. 2015. *Back to Black: Ending Internalized Racism*. Outlaw Culture. https://doi.org/10.4324/9780203822883-21.

Hyde Park Rental Prices Reach New High. n. d. Hpherald.Com [blog] https://hpherald.com/2016/06/30/hyde-park-rental-prices-reach-new/. Accessed: 30 June 2016.

Hyra, Derek S. 2006. *Racial Uplift? Intra-Racial Class Conflict and the Economic Revitalization of Harlem and Bronzeville*. City and Community 5(1): 71–92. https://doi.org/10.1111/j.1540-6040.2006.00156.x.

Institute for Housing Studies-DePaul. n. d. *Analyzing Neighborhoods with Intensifying and Emerging Housing Affordability Pressure*. Institute for Housing Studies - DePaul University. https://www.housingstudies.org/releases/Analyzing-neighborhood-displacement-pressure-2018/. Accessed: 4 June 2019.

Interview with Bobby Seale. 1988. http://digital.wustl.edu/e/eii/eiiweb/sea5427.0172.147bobbyseale.html. Accessed: 4 June 2019.

Jones-Henderson, N. (2012). Remembering AfriCOBRA and the Black Arts Movement in 1960s Chicago. Nka Journal of Contemporary African Art ; Volume 2012, Issue 30, Page 98-103 ; ISSN 1075-716 https://doi.org/10.1215/10757163-1496516

Joravsky, Ben. 2013. *Before the Schools, Mayor Emanuel Closed the Clinics*. Chicago Reader. https://www.chicagoreader.com/chicago/mayor-emanuel-closes-city-mental-health-clinics/Content?oid=9145051. Accessed: 24 April 2019.

Karydes, Medy. 2019. *Chicago Neighborhoods Where Rent Is Increasing the Most*. https://www.rent.com/blog/chicago-neighborhoods-rent-prices-increasing/. Accessed: 31 August 2020.

Kelleher, Kevin. 2019. *Gilded Age 2.0: U.S. Income Inequality Increases to Pre-Great Depression Levels*. Fortune. https://fortune.com/2019/02/13/us-income-inequality-bad-great-depression/. Accessed: 5 June 2019.

Kelly and Silverman, ed. 2014. Schools and Urban Revitalization: Rethinking Institutions and Community Development. New York: Routledge

Kirchberg, V., & Kagan, S. (2013). The roles of artists in the emergence of creative sustainable cities: Theoretical clues and empirical illustrations. City, Culture and Society, 4(3), 137–152. https://doi.org/10.1016/j.ccs.2013.04.001

Knight Foundation. 2014. *Knight Foundation Invests $3.5 Million to Help University of Chicago Expand a Model for Community Revitalization Led by Artist and Urban Planner Theaster Gates*. Knight Foundation. https://knightfoundation.org/press/releases/place-project-theaster-gates. Accessed: 4 June 2019.

Leduc, Marie. 2019. *Defining contemporary art: what the Kunstkompass Top 100 lists can tell us about contemporary art*. Journal of Visual Art Practice, 18(3): 257–274. https://doi.org/10.1080/14702029.2019.1654204.

Levy, Peter B. 2018. *The Great Uprising: Race Riots in Urban America during the 1960s*. Cambridge: Cambridge University Press.

Lewis, George E. 1996. *Improvised Music after 1950: Afrological and Eurological Perspectives*. Black Music Research Journal 16(1): 91. https://doi.org/10.2307/779379.

Lewis, George E. 2014. *A Power Stronger Than Itself. The AACM and American Experimental Music*. Chicago: University of Chicago Press.

Lloyd, R. D. (2010). Neo-Bohemia: Art and commerce in the postindustrial city (2nd ed.). New York: Routledge.

Logan, John R., and Harvey Luskin Molotch. 1987. *Urban Fortunes: The Political Economy of Place*. Berkeley; London: University of California Press.

Logan, John R, and Harvey Luskin Molotch. 2007. *Urban Fortunes: The Political Economy of Place*. Berkeley; London: University of California Press.

Marable, Manning. 2000. *How Capitalism Underdeveloped Black America: Problems in Race, Political Economy, and Society*. Updated ed. Cambridge, Mass.: South End Press.

Marcuse, Peter. 1997. *The Enclave, the Citadel, and the Ghetto: What Has Changed in the Post-Fordist U.S. City.* Urban Affairs Review 33(2): 228–264. https://doi.org/10.1177/107808749703300206.

Mason, B. J. 1973. *A Shift to the Middle*. Ebony 28(10): 80.

Massey, Douglas S., and Nancy A. Denton. 1993. *American Apartheid: Segregation and the Making of the Underclass*. ACLS Humanities E-Book. Cambridge, Mass.: Harvard University Press. http://pi.lib.uchicago.edu/1001/cat/bib/10923885.

Mayer, Margit. 2013. *First World Urban Activism: Beyond Austerity Urbanism and Creative City Politics* McCartney, John T. City 17(1): 5–19. https://doi.org/10.1080/13604813.2013.757417.

Mele, C. (2000). Selling the Lower East Side: Culture, real estate, and resistance in New York City. Minneapolis: University of Minnesota Press.

Moore, Natalie. 2015. *How Theaster Gates Is Revitalizing Chicago's South Side, One Vacant Building at a Time*. Smithsonian Magazine. https://www.smithsonianmag.com/innovation/theaster-gates-ingenuity-awards-chicago-180957203/. Accessed: 2 June 2019.

Neal, Larry. 1968. *The Black Arts Movement*. The Drama Review: TDR 12(4): 28–39. https://doi.org/10.2307/1144377.

Neal, Larry. 1969. *Any Day Now: Black Art and Black Liberation*. Ebony 24(10): 54.

O'Brien, Katie. 2019. *It's Complicated: The University Of Chicago's Relationship With Its Neighbors*. WBEZ CHICAGO. https://www.wbez.org/stories/its-complicated-the-university-of-chicagos-relationship-with-its-neighbors/4882b290-40d8-4730-bb68-2eaea92be0bc. Accessed: 5 June 2019.

Ongiri, Amy Abugo. 2010. *Spectacular Blackness: The Cultural Politics of the Black Power Movement and the Search for a Black Aesthetic*. University of Virginia Press.

Orleck, Annelise, and Lisa Gayle Hazirjian. 2011. *The War on Poverty: A New Grassroots History, 1964–1980*. Athens: University of Georgia Press.

Pattillo, M. E. (2007). Black on the block: The politics of race and class in the city. Chicago: University of Chicago Press.

Patterson, K. L., & Silverman, R. M. (2013). Urban Education and Neighborhood Revitalization. Journal of Urban Affairs, 35(1), 1–5. https://doi-org.proxy.uchicago.edu/10.1111/juaf.120

Peck, Jamie. 2010. *Constructions of Neoliberal Reason*. Oxford: Oxford University Press.

Prince, Sabiyha Robin. 2017. *African Americans and Gentrification in Washington, D.C.: Race, Class and Social Justice in the Nation's Capital.* Abingdon; New York: Routledge.

Reardon, 2017. *The Wall of Respect: The Story of a South Side Mural — and of Black Power*. Chicago Tribune. https://www.chicagotribune.com/entertainment/books/ct-books-wall-of-respect-1008-20171003-story.html. Accessed: 15 May 2019.

Rebuild Foundation n. d. b.https://m.facebook.com/CivicCommonsUS/photos/supporting-local-talent-through-the-civiccommons-in-chicago-thirty-local-artisan/394224957643819/?locale2=ko_KR. (Accessed January 5, 2022).

ReportFutureUChicagoArts. n. d. https://provost.uchicago.edu/sites/default/files/documents/reports/ReportFutureUChicagoArts20202025.pdf. Accessed: 5 June 2019.

Reva and David Logan Center for the Arts. n. d. *About the Logan Center*. https://arts.uchicago.edu/logan-center/about-logan-center. Accessed: 4 June 2019.

Robinson, Lyletta. 2018. *Fear of Gentrification in Woodlawn: Why It Matters in the Obama Center Discussion*. Chicago Cityscape. https://blog.chicagocityscape.com/fear-of-gentrification-in-woodlawn-why-it-matters-in-the-obama-center-discussion-595feedd3013. Accessed: 4 June 2019.

Roy, Ananya, Stuart Schrader, and Emma Shaw Crane. 2015. *"The Anti-Poverty Hoax": Development, Pacification, and the Making of Community in the Global 1960s*. Cities 44(April): 139–145. https://doi.org/10.1016/j.cities., p.,2014.07.005.

Schlesinger, Sonia. 2016. *Community Members Discuss Goals for Trauma Center at Town Hall Meeting*. https://www.chicagomaroon.com/2016/05/20/community-members-discuss-goals-for-trauma-center-at-town-hall-meeting/. Accessed: 5 June 2019.

Schram, Sanford, Joe Soss, and Richard C. Fording. 2003. *Race and the Politics of Welfare Reform*. Ann Arbor: University of Michigan Press. http://pi.lib.uchicago.edu/1001/cat/bib/11212689.

Semuels, Alana. 2018. *Chicago's Awful Divide*. The Atlantic. https://www.theatlantic.com/business/archive/2018/03/chicago-segregation-poverty/556649/. Accessed: 5 June 2019.

Smith, N. (2002). New globalisation, new urbanism: gentrification as global urban strategy. Antipode, 34(3), 427–450.

Sweeney, Gretchen. 2018. *2018 J.C. Nichols Prize Winner—Theaster Gates*. ULI Americas. https://americas.uli.org/nichols-prize-winners/2018-j-c-nichols-prize-winner-theaster-gates/. Accessed: 5 June 2019.

Task Force on the Cities. (n. d.). Report to the President by the Task Force on the Cities. December 5, 1966. [With Task Force papers. Located in: Urban Task Force Reports, 1964–1965, vol. 2] OCLC: 6113121.

Taylor, Keeanga-Yamahtta. 2016a. *From #BlackLivesMatter to Black Liberation*. Chicago, Ill: Haymarket Books.

Taylor, Paul C. 2016b. *Black Is Beautiful: A Philosophy of Black Aesthetics*. Vol. 6. Foundations of the Philosophy of the Arts; Chichester, West Sussex.

UChicago Arts. n. d. The University of Chicago. https://arts.uchicago.edu/artsandpubliclife.

U. of C. Holds Grand Opening for Keller Center. 2019. Hpherald.com [blog]. https://hpherald.com/2019/05/08/u-of-c-holds-grand-opening-for-keller-center/. Accessed: 4 June 2019.

Van Deburg, William L. 1992. *New Day in Babylon: The Black Power Movement and American Culture, 1965–1975*. Chicago: University of Chicago Press.

Wacquant, Loïc. 2007. *Urban Outcasts: A Comparative Sociology of Advanced Marginality*. 1 edition. Cambridge: Polity.

Weber, Rachel. 2015. *From Boom to Bubble: How Finance Built the New Chicago*. University of Chicago Press.

Wim Wiewel, & David C. Perry. (2008). Global Universities and Urban Development : Case Studies and Analysis. Routledge.

Williams, Jakobi. 2013. *From the Bullet to the Ballot: The Illinois Chapter of the Black Panther Party and Racial Coalition Politics in Chicago*. The John Hope Franklin Series in African American History and Culture. Chapel Hill: University of North Carolina Press.

Williams, Aaron, and Armand Emamdjomeh. 2018. *America is more diverse than ever — but still segregated*. Sec. Analysis. Washington Post. https://www.washingtonpost.com/graphics/2018/national/segregation-us-cities/. Accessed: 3 June 2019.

Wilson, David. 2015. *Chicago Goes Global: Redevelopment, Culture, and Fear*. Journal of Cultural Geography 32(3): 251–269. https://doi.org/https://doi.org/10.1080/08873631.2015.1067391.

Wilson, William J. 1987. *The Truly Disadvantaged: The Inner City, the Underclass, and Public Policy*. Chicago: University of Chicago Press.

Woodlawn Projected To Be Redfin's Hottest Market of 2017. n. d. Steele Consulting. http://whatasteele.com/news/woodlawn-redfins-hottest-chicago-neighborhood. Accessed: 4 June 2019.

Zorach, Rebecca. 2011. *'Dig the Diversity in Unity': AfriCOBRA's Black Family*. Afterall: A Journal of Art, Context and Enquiry 28(September): 102–111. https://doi.org/10.1086/662975.

Zukin, 1989. Loft living. (1989). Rutgers Univ. Press.

The ArtReview's Power 100: On the Concept and the Constitution of Power in Contemporary Art

3

Paul Buckermann

Abstract

Why is there a ranking of the most influential agents in contemporary art? This article contributes to a deeper understanding of epistemologies in art worlds and does so by focusing on paradigmatic worldviews. I reconstruct the ranking *Power 100*'s understanding of art's role in society, its assumed internal mechanisms and the function of power in it. My results show how the list makes sense within its own terms, because the analysis highlights the consistency of the list's paradigmatic assumptions about a world-to-ranked and its concrete production. Furthermore, I discuss sociological concepts of power (Weber, Bourdieu, Luhmann) to argue that the list can be described as an internal perspective on the art world rather than an external understanding of it. The *Power 100*'s epistemology combines a deep belief in art's positive role in society with a proto-sociological conception of symbolic (power) structures and the autonomous reproduction of the arts. Referring to research on evaluation and comparison, my empirical and theoretical results have implications for both a sociology of the arts and research on rankings.

Keywords

Ranking · Evaluation · Contemporary art · Power · Sociology of the arts

P. Buckermann (✉)
Heidelberg University, Heidelberg, Germany
e-mail: paul.buckermann@uni-heidelberg.de

L. Gaupp et al. (eds.), *Arts and Power*, Kunst und Gesellschaft,
https://doi.org/10.1007/978-3-658-37429-7_3

3.1 Why a Ranking? Why Power?

The British art magazine *ArtReview* has been publishing an annual ranking of the one hundred "most powerful people of the art world" (ArtReview 2019a)[1] since 2002. Every year the *Power 100* draws criticism from those art world professionals who see the list as just another symptom of the total marketization and instrumentalization of the arts. In 2017, however, reactions to the *Power 100* differed significantly. While high-ranking gallerists and private collectors had been dominating the list (and were still strong in 2017), that year it was topped by Hito Steyerl. Art magazines and intellectuals praised *ArtReview*'s surprising result because the German artist and theorist is a distinguished voice in contemporary art who is constantly exposing the dubious connections between the art world, authoritarian politics, tax fraud and arms manufacturers (Steyerl 2017). As one commentator put it, having "a woman so critical of exploitation and the brute force of money within the art world top the *Power 100* list feels like a stand for an alternative understanding of power rooted in intellect and integrity" (Judah 2017). *ArtReview*'s argument for Steyerl's top position already laid the ground for such responses: "Art is powerful. Or at least it's the construct of powerful forces, not always of the positive kind. … Steyerl makes the top slot on this list because she actively attempts to disrupt this nexus of power" (ArtReview 2017, p. 102).

As soon as Steyerl and other intellectual popstars like Donna Haraway (No. 3), Bruno Latour (No. 9) and Judith Butler (No. 48) were ranked among the power elite of the arts, the ranking's meaning shifted for a brief moment. The review device no longer offered a perspective from the worlds of cultural and economic commerce, both of them external and hostile to the art world (Velthuis 2005, pp. 25 ff.). By acknowledging Steyerl's critical stance and extraordinary reputation among decidedly critical curators, scholars and artists, the list seemed to be adopting evaluative criteria taken from autonomous art, such as artistic novelty, critical intellect and a coherent relation between form and content, rather than following market logic, popular taste or political instrumentalization. However, in 2017 *ArtReview* altered neither its mission nor its methods. From a Bourdieusian field theoretical perspective (Bourdieu 1992/1996; Bourdieu 1983/1993), only certain results seemed to match the established symbolic logic of autonomous

[1] All Power 100 lists since 2002 and the current profiles of ranked units are online. The *ArtReview* was founded in 1949 and describes itself as one "of the world's leading international contemporary art magazines" (ArtReview 2019b).

art instead of an "allodoxia" (Bourdieu 1984/1988) distorted by market and mass culture as a phenomenon of heteronomic forces in the arts.

I understand such discursive shifts as symptoms of profound epistemic conflicts over symbolic values, hierarchies, and privileges in the arts. Whether they are deeply rooted in a common belief in the arts (Bourdieu 1977/1980) or lingering on its symbolic margins, such debates are a proper starting point for taking a closer look at what art actually means for those involved in the arts. Although it is a variable and contested concept, power seems to be a useful epistemological tool for certain art world agents to understand the world they inhabit. But who cares about power in the arts, and for what reason? Why map the art world using power structures instead of aesthetic or art historical criteria that are independent of power relations? What does power in contemporary art actually mean? I assume that paradigmatic conceptions of the art world, its internal structures and its social embeddedness make power both an appropriate navigating category and a perspective worth challenging.

To argue for this sociological perspective, I present a reconstruction of the *Power 100*'s underlying understanding of art's role in society, its assumed mechanisms of reproduction and the function of power in it. First, my results show how the list makes sense within its own terms, because the analysis highlights the consistency of the list's paradigmatic knowledge and concrete production. Second, I use sociological concepts of power (Weber, Bourdieu, Luhmann) to argue that the list can be described as an internal perspective on the art world rather than an external understanding of it. Power fits into this perspective, because the *Power 100*'s fundamental worldview combines a deep belief in art's positive role in society with a proto-sociological conception of symbolic (power) structures and the autonomous reproduction of the arts.

Rankings concerned with contemporary art have neither been a prominent object for ranking studies nor for the sociology of the arts. Researching the fundamental knowledge on which a ranking is based contributes towards a broader understanding of rankings of the arts and of rankings in a "society of rankings" in general (Esposito and Stark 2019), and does so by focusing on how they are produced. In contrast to the *effects* of rankings (Martins 2005; Espeland and Sauder 2007; Brandtner 2017), little is known about their production. I complement such research as there is on producers (Rindova et al. 2018) by reconstructing the epistemological foundation of rankings.

Moreover, an inquiry into their epistemological foundation constitutes a new departure in the sociology of the arts. Instead of approaching rankings as driving forces of order and change that are both powerful and selective (Buckermann

2018, 2020, 2021), scholars have mainly treated them as sources of data. Using ranking results for quantitative studies, they have developed centre-periphery models that show how certain regions are unequally represented in the global art world (Quemin 2006, 2013, 2018; Buchholz and Wuggenig 2005; Buchholz 2008, 2018). By contrast, the constructivist perspective on the production of rankings proposed here addresses the plurality of epistemologies inside this symbolically autonomous sphere (Buckermann 2020).

Generally, I propose asking how agents in the art world perceive and observe their own complex ecology, which is the social phenomenon called "the arts" with all its material, symbolic, semantic, professional and institutional infrastructures. In particular, sociologists are familiar with such models of a distributed and multi-layered form of art in modernity from their own work. As the arts have become progressively autonomous in the long transition from feudal to modern society (Bourdieu 1992/1996; Luhmann 1995/2000; White and White 1965/1993), no royal academy, no prince, no priest, no university and no guild has been able to legitimately stop anybody becoming an artist, anything becoming an artwork, or anything being called a museum—at least in the official script of liberal Western modernity.

The sociology of the arts is in turn interested in how structures and semantics have emerged in modernity that both stabilize art and keep it in flux. The large-scale decentralization and deregulation of the collective production of art has in fact led to the emergence of critical tensions between abundance and scarcity. These conditions of "*too little, enough, and too much*" (Abbott 2014, p. 2) mean a vast number of artworks, artists, cultural institutions and discursive possibilities constitute an abundant supply relative to limited resources, exhibition spaces, institutional positions, magazine pages, awards, professorships, scholarships, public attention, art historical publications, art fair booths and collection shelves. The sociology of the arts empirically describes and conceptually explains how certain patterns of selection and evaluation have necessarily emerged in reaction to multiple pressures to select. By investigating the multiple perspectives within the arts, I aim to contribute towards a general programme of understanding how art in modern society is subject to both stabilization and change. Although partially congruent with various ways of sociologically modeling the arts (see Zembylas 2006), this approach also goes beyond them, focusing instead on epistemological machineries situated within these structures: how do actors within a world perceive and model a world shaped by abundance and scarcity? How do they navigate through a complex and deregulated world and establish models for decision-making based on these worldviews?

In general, I assume that participants embedded in and emerging from autonomous social structures need a holistic cognitive map to make sense of what is going on around them. Here, theoretical research needs to discuss the implications for autonomy of such multiple worldviews in different sociological approaches.[2] In this paper, however, I only show that analysing the concept of power in the *Power 100*'s ranking opens up an analytical pathway to such a comprehensive worldview. I propose taking a publication as a starting point, and in a process of "reverse engineering" working backwards from it to arrive a reconstruction of how the list was produced, with all its contingent steps. These decisions should then consistently relate to a set of basic underlying assumptions about the world-to-be-ordered. The specific production and representation of order is assumed to be tightly intertwined with such deep paradigmatic assumptions.

Rankings like the *Power 100* almost cry out for such a fundamental reconstruction. Systematic review tools for reputation, resonance, quality or power still amount to a rather odd way of understanding contemporary art. By claiming the position of the "outsider", the producers of rankings are obliged to promote their tools as what might be termed "disclosure devices" (Hansen and Flyverbom 2015), capable of revealing facts where only loose assumptions or total uncertainty had existed before. To gain legitimacy, novel devices must expose their methodological architecture in a manner more transparent than that of established authorities. A qualitative reconstruction does not aim to expose methodical flaws or an absurd conception of "art-as-it-really-is". In approaching all forms of communicated order with an analytical "agnosticism" (Callon 1984), I follow a common hypothesis in contemporary sociological inquiries into commensuration (Espeland and Stevens 1998), comparisons (Heintz 2016), quantification (Espeland and Stevens 2008), evaluations (Lamont 2012) and other means of producing order: rankings do not and cannot represent a given world but necessarily produce their own reality through an internally consistent architecture of observations. In the case of the *Power 100*, I assume that this production of a tight order is shaped by paradigmatic assumptions about art in society, its reproductive mechanisms and internal logics.

This paper will proceed in three steps. In 3.2, I reconstruct the *Power 100*'s worldview and how it has been methodologically constructed. In 3.3, I present

[2] See Buckermann 2020 for such a discussion in the sociology of art's seminal approaches ranging from interactionist (Becker 1982/2008) to field theoretical (Bourdieu 1992/1996), and systems theoretical (Luhmann 1995/2000) traditions.

different forms of power analytically extracted from the *Power 100*. In the final part 3.4, I use sociological concepts of power and influence to conceptualize *ArtReview*'s understanding of art on a general level.

3.2 Making the List: The World of the Power 100

The worldview informing the *Power 100* can be analytically unpacked by reconstructing how the list was produced. While compiling the *Power 100, ArtReview* does not actually count, measure or quantify anything. In fact, the list relies exclusively on expert knowledge (see Quemin 2013, Chap. 3) because the magazine assembles a panel of up to thirty experts from the art world. All of them remain anonymous "so that power-crazed art people don't hassle them or enact various types of cruel revenge" (ArtReview 2017, p. 100). The panel should represent a high degree of diversity in regional and professional backgrounds. Each person points out who is influential in their art world. There are basic criteria for this decision, and the ensuing debate is described as "a battle of opinion as much as fact" (ArtReview 2016, p. 101). To make it onto the list, a person must have been active over the last twelve months, must have influence over what kind of art is produced and made visible, must have an effect on the general status of art in society, and this influence must be recognized at an international level. After the first panel has decided on a first selection, a second panel of anonymous experts checks the list for "obvious" regional biases (ArtReview 2015, p. 121). Even though this process is supposed to be based on a certain procedure with double checks and fixed criteria, the selection is in effect an aggregation of qualitative judgments by individual persons who are perceived as experts by the magazine. Alongside the Top 100, *ArtReview* publishes an editorial comment, an introduction, secondary quantitative evaluations of the power list (gender, nationality, profession) and a personal profile for each person among "the most influential people in the art world [*sic*]" (ArtReview 2019a).

Any reconstruction of the paradigmatic assumptions behind the list must start by identifying the main set of problems the review aims to address. The producers justify the ranking by claiming it solves these. Because such a "call for a list" (Stäheli 2012) legitimizes the need for the ranking, as it also implicitly opens up a certain perspective on methodology and epistemology. Even though the problem needs to be addressed as a collective issue in a specific community, I argue that it primarily originates in a selective understanding of world. This paradigmatic worldview must contain comprehensive assumptions about art, important symbolic orders, institutional and professional infrastructures, and art's general social embeddedness. I argue that these understandings influence the methods used to derive order from chaos. This chaos, however, only takes shape in light of a proto-assumption of order. Only an initial process of categorization, evaluation and comparison can make manifest the chaos and abundance of artistic production and its related art world organizations.

The problem addressed by the *Power 100* emerges from a particular understanding of art's place in society, for *ArtReview* promotes a particularly positive image of art's social potential. An analysis of the results of the list reveals that it is based on a quasi-cybernetic model, in which two separate spheres communicate with each other in feedback loops: artistic production on the one hand and the rest of society on the other. The information exchange between art and society follows a circular process, driven by catalysing mechanisms within both artistic production and society (Fig. 3.1): artistic production "reflects and interrogates the world around it" (ArtReview 2015, p. 121), and cultural products make "things visible" (ArtReview 2016, p. 101). Works of art emerge as materialized "ideas" (ArtReview 2015, p. 121) initiated by broader social themes, issues, problems, discourses, agendas and structures, which inform both the form and content of art. However, artworks are not determined by society because they always emerge through reformulation and translation within a creative process. In turn, this exclusive "habit of proposing or analysing alternatives to reality or social norms"

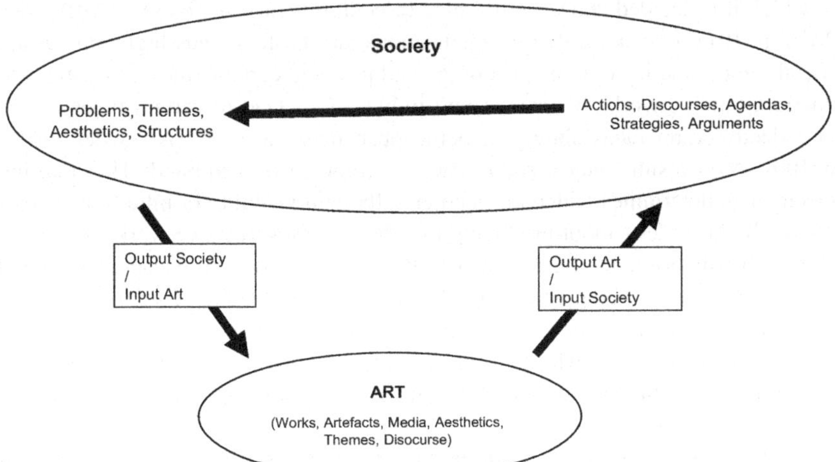

Fig. 3.1 Analytical reconstruction of the *Power 100*'s general model of art in society

(ArtReview 2015, p. 121) enables artistic information to feed back into society. Because art "tells us something about the world around us that we've overlooked", it is potentially able to spark speculation "on what directions that world can or will follow" (ArtReview 2016, p. 101). Art can play "a positive role … in the building of a society" (ArtReview 2016, p. 101) because it is able to "provoke discussion, conversation and debate" (ArtReview 2016, p. 38) on a general level, to make invisible things visible, generate new ideas, and produce speculative knowledge for the past, the present and the future of a society.

Thus the list addresses a set of problems that arise from a particular social form of this general social ecology of art. According to *ArtReview*, artistic production and general discourses are not directly related, but "artists, and the art they make, participate in an ever-shifting, often tense game played by wealthy collectors, corporations, public functionaries, biennial curators, art-fair directors, … academics and governments" and many more (ArtReview 2015, p. 121). Certain filters, translators and catalysts have emerged at the intersections of the cyclical process between artistic production and society: these constitute the institutional art world, ranging from tiny off-spaces to global festivals. The art world consists of those networks and structures in which exchanges between art and society are filtered and translated by all kinds of "operators and fixers" (ArtReview 2015, p. 121).

In this art world, different perspectives on "how the world values art (monetarily, aesthetically, instrumentally)" encounter each other and build a "framework" in which it is decided "how … the value of contemporary art [is] set" (ArtReview 2016, p. 101). Because cultural organisations and professionals highlight certain social topics, and in so doing encourage and promote certain kinds of artistic production, "the networks of interest and influence … lead to certain types of art, or indeed certain ideas about art, being pushed to the fore" (ArtReview 2017, p. 100). As a result, only certain artworks come to be exhibited. This filtering mechanism determines which ideas potentially gain visibility or influence in society at all. Art is less about producing the "best or most incisive work", but more about "distribution and circulation" (ArtReview 2014, p. 126) because in the end it is "a question of being able to publicize that idea" (ArtReview 2015, p. 121). *ArtReview* claims to offer explanations for "why this person gets an exhibition rather than that person; why this type of art is valued more highly in monetary or intellectual terms than that; and in some cases, just plain 'Why?'" (ArtReview 2015, p. 121).

From this perspective, the institutional art world has an essential role for artistic production *and* society (Fig. 3.2). *ArtReview* urges us to take a closer look at the inner structures of the art world because inside these mediating and selective

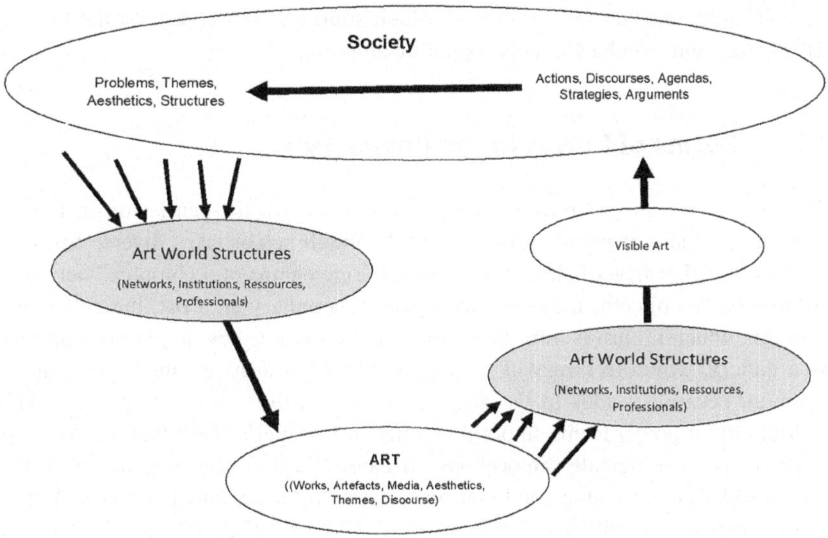

Fig. 3.2 Analytical reconstruction of the *Power 100*'s model of the contemporary art world

structures lies the very problem to which the list is reacting. The list reacts to the problem by offering an "analysis of how the structures of art work to make some and not other artworks more public" (ArtReview 2016, p. 38). In doing so, "the list allows for readings that suggest some deeper truths" (ArtReview 2017, p. 101) about the inner workings of the art world and potential threats of contamination or instrumentalization.

Understanding how the art world works means understanding not only how art's potential to create a better society is enabled and fostered, but also how it is obstructed or abused. Even if "ArtReview's *Power 100* aims to provide you with a 'picture of the system'", and "it's up to you what you do with it" (ArtReview 2017, p. 12), the list's scope goes far beyond the art world. If we accept *ArtReview*'s diagnosis and its understanding of art's role in a society, then the list becomes about the fate of the whole world. To understand what is actually going on in the art world, *ArtReview* uses power as its central epistemological concept. The premise of this approach is that the art world has a certain structure and that it can be scrutinized and represented. Reducing the complex structure of the art world to a single ordering factor makes it possible to see it in an extensive comparative perspective (Heintz 2016), which potentially relates everyone on a global scale (Heintz and Werron 2011), and which supposedly influences what kind of art is made, and which of it receives public attention.

3.3 Forms of Power in the Power 100

The particular forms of power implicit in *ArtReview's* list can be reconstructed from the portraits of people ranked in it. Although power takes diverse forms in the *Power 100,* it nevertheless flows through long chains of a complex "network" of the arts: "a collector may support a gallery, a gallery an artist, the collector a museum, which in turn is introduced to an artist, whose show might be supported by a gallery, which in turn will be supported by the artist having a show at the museum, because it adds to the artist's cachet" (ArtReview 2015, p. 121). This multiplicity of power forms smoothly aligns with Weber's claim that the "concept of power is sociologically amorphous" because "[a]ll conceivable qualities of a person and all conceivable combinations of circumstances may put him in a position to impose his will in a given situation" (Weber 1922/1978, p. 53). Such an open understanding of power covers the many different forms of power displayed in the list. However, developments in the theory of social differentiation can produce a more sophisticated understanding of *ArtReview*'s worldview and broader symptoms of power. Using Bourdieu's and Luhmann's writings on power and influence, we will see that the *Power 100* refers to a certain form of power that both emerges from, and is deeply embedded in, the arts in society. Underlying the list can be found a consistent understanding of *art as an autonomous symbolic sphere.*

Two major indices of a person's influence in the arts appear repeatedly in the *Power 100.* First, the list highlights different professions, especially different forms of network connections to artists and curators on the one hand and non-commercial cultural organizations on the other. Second, it focuses on temporary and permanent collaborations among those artists, curators, and institutions that are part of the list themselves. Their importance is assessed in a number of different ways, depending on their different professions. Thus for example: (1) to show the influence and impact of individual artists, their connections to specific institutions are listed; (2) museum directors are seen as representatives of their institutions, whose influence in turn is demonstrated by their collaborations with certain artists; (3) the same criteria are used for curators, whose power is concretely demonstrated by means of temporary time-based events (biennales etc.), the artists they have exhibited, and their permanent institutional positions on boards and selection committees. (4) To show the power of selected gallerists, *ArtReview* regularly avoids referring to their commercial success because the power of gallerists derives from having multiple locations where they can exhibit artists and collaborate with curators. Such links to artists are also displayed through a gallery's roster of artists. (5) Collectors are powerful if they run private exhibition spaces where certain artists and curators appear publicly. These connections between gallerists, collectors and artists are used to extend collaboration to public institutions, and show whether an artist has been considered by a reputable museum before being exhibited by a collector. (6) Scholars and theorists have a major influence on the production and discursive shaping of art. A theorist's concrete impact is demonstrated by exhibitions in public art institutions that openly refer to his or her theoretical work. Terms like *gender, gaia, cyborg* or *object-oriented ontology* serve as examples of an individual scholar's broader intellectual influence.

Connections among artists, curators and institutions demonstrate the power of core professionals. Such connections also demonstrate the power of scholars, collectors and art market agents. Genuine art world professionals not only confer power directly onto other agents, but also act as mediating figures. When collaborating with gallerists or collectors, curators and artists add the cachet of their own reputation, along with specific connections to public cultural institutions. That certain institutions should play this fundamental role in the symbolic structures of the arts comes as no surprise to sociologists. The symbolic value of art is not derived from essential qualities of artefacts, aesthetics or technical skills, but is produced by the recognition of recognized institutions. The symbolic power of institutions and professionals themselves is in turn displayed by an institution's connections to artists. Because the *Power 100* makes implicit use of these arguments, a circular model of symbolic power production in the arts emerges: the reputation of artists derives from recognition by prestigious institutions, which in turn have to enjoy the respect of certain artists and other institutions. This model of self-referential and autocatalytic symbolic power shapes the ranking's basic logic. In the next and final section, I offer a sociological discussion of power in the *Power 100* that identifies its worldview as being symbolically internal to the art world.

3.4 Power in the Power 100 from a Sociological Perspective

The specific logic of power reviewed in the *Power 100* is deeply embedded in its understanding of art as an autonomous symbolic machine. Power is not understood as an objective property of a person, power relations are not categorically understood as conflictual, power does not enable direct control, and power in the art world is not a zero-sum game, meaning that more power for one person does not imply a reduction in the power of another. On the contrary, power in the arts emerges relationally through cooperation, and it increases on a general scale because receiving power always implies ascribing power. This concept of a sympoietically emerging, relational and catalytic power is an integral part of *ArtReview*'s foundational worldview. It is a worldview that is internal to an autonomous sphere of the arts, because it both refers to and observes its very own symbolic order.

The underlying mechanisms of the emergence of power and its logics cannot be reduced to external or material factors. Financial means and exhibition space, political power or academic publications can indeed enable dealers, collectors or scholars to gain power in the arts. However, these resources do not translate directly or immediately into influence in the art world. To be able to validate, consecrate and confirm power in the arts depends on being recognized by symbolically recognized persons and institutions within the art world. According to Bourdieu (1992/1996, p. 170), this *symbolic power* is both the product and engine of a "symbolic alchemy", in which not only works and their authors are collectively produced but also a comprehensive symbolic order. This symbolic order both structures and is structured by *symbolic capital*, which "is common to all members of a group" (Bourdieu 1994/1998, p. 103) and thereby productively interconnected with a fundamental belief in the arts. Neither a collector's net value, nor a gallery's sales numbers, nor a museum's square metres of exhibition space, nor a scholar's academic reputation is a decisive factor for influence in the art world. The indispensable means for accruing and legitimizing power in the arts is recognition by recognized art world agents "endowed with the categories of perception and appreciation permitting them to perceive, know and recognize it" (ibid., p. 102). This symbolic capital is "*a percipi*, a being-perceived functioning as positive or negative symbolic capital" (ibid., p. 104); it is therefore displayed by network contacts operating as quasi-institutionalised proxies for symbolically reshaped *social capital* (Bourdieu 1986) within an autonomous sphere.

This understanding of power in the *Power 100*'s general worldview seems almost functional for an art that has historically been deregulated and emancipated from religious or political authorities (see 3.1). Power emerges where conditions of abundance and scarcity produce pressure for selections and for the legitimation of these selections. This constellation is crucial for Luhmann's theory of *symbolically generalized communication media* (Luhmann 1975; 1997/2012, pp. 190–238), such as truth, money, love, and *power*. More alternatives (e.g. art works) than possible realizations (e.g. exhibitions) lead to incongruences between interests and selections. This fosters the evolution of social techniques regulating the stable acceptance of selections (Luhmann 1969, p. 168). The relative abundance of academic works, political interests or works of art necessitates selections that always imply contingency and possible opposition to them.

In *ArtReview*'s worldview, not all themes and topics of a society are reflected in works of art and not all works of art can influence broader discourses. The necessary selections between artistic production and society are made in the intermediary art world. At the same time, relative scarcity is not only about limited exhibition space, financial resources, art market commodities, collections shelves, roster positions or art fair booths. All these resources are also deeply intertwined with symbolic value. The scarcity of these symbolic resources and their reproductive mechanisms prevent money or political authority flooding the art world with (relevant) private museums, art fairs and public institutions. Therefore, *ArtReview*'s understanding of the art world is ambivalent with respect to its filtering and translating function in a world with multiple interests. On the one hand, the art world's intermediary position exposes the arts to the risk of being contaminated or colonized, because its organizations and personnel depend on resources from external agents who have their own interests (economic profit, art-washing, religious or political propaganda, etc.).

On the other hand, the particular symbolic order of the art world shelters the autonomous self-regulation of art because the specific reputational order of the arts emerges within the collective production of the arts. In fact, this conception of order is reminiscent of sociological theories of social differentiation: different social realms have different evaluative logics, influence structures and resource regimes. Bourdieu's forms of capital, their different logics of accumulation, and their convertibility (Bourdieu 1986) are fundamentally interwoven and produced within the relational structure of fields (Bourdieu 1981/1993). Luhmann's theory of different forms of symbolically generalized communication media is an integral part of his theory of functional differentiation (1997/2012, pp. 190–238). In a strict sociological reformulation, *ArtReview's* institutionalized art world functions as a machine mediating contact between autonomous art and society. The mechanisms within this machine are shaped by symbolic power, which in turn translates different financial means, political power or academic publications. Placed in a complex landscape shaped by different abundances and scarcities, the art world generates not only recognition and reduces complexity but also enables artistic production to find a variety of resources. In turn, social discourses and innovations benefit from these reductions because new perspectives, speculative visions and creative solutions are prepared and displayed.

Although this image seems suspiciously familiar to sociologists, *ArtReview*'s worldview is fundamentally different to a sociological perspective on the autonomy of the arts. The list is based upon and reproduces a fundamental belief in the arts that ascribes a unique potential to artistic creation and its innovative effects on society. However, sociologists analyse such beliefs, structures, and semantics as socially produced, historically situated and contingent. Any concept of the nature of art and how it is created is therefore a historical phenomenon emerging from collective and contested social processes. Complex institutional and symbolic infrastructures may indeed stabilize those processes that make art possible at all today, but at the same time they are inescapably contingent. Although "[s] ociology and art do not make good bedfellows" (Bourdieu 1981/1993, p. 139), *ArtReview*'s epistemological foundation nevertheless assumes basic factors that recognize how artistic productions is deeply socially embedded. I assume it is the paradigmatic worldviews of other art world professionals that explain why they accuse the *Power 100* of embodying marketization or instrumentalization instead of committing the sacrilege of *sociologization*.

References

Abbott, Andrew. 2014. The problem of excess. *Sociological Theory* 32(1): 1–26.

ArtReview. 2014. *ArtReview* November 2014. Vol. 66 no. 8.

ArtReview. 2015. *ArtReview* November 2015. Vol. 67 no. 8.

ArtReview. 2016. *ArtReview* November 2016. Vo. 68 no 8.

ArtReview. 2017. ArtReview November 2017. Vol. 69 no. 8.

ArtReview. 2019a. Power 100. The annual ranking of the most influential people in art. https://artreview.com/power-100/. Accessed: 17 October 2019.

ArtReview. 2019b. About. https://artreview.com/about/. Accessed: 17 October 2019.

Becker, Howard. 2008 [1982]. *Art worlds*. 25th Anniversary edition, updated and expanded. Berkeley/Los Angeles/London: University of California Press.

Bourdieu Pierre. 1980 [1977]. The production of belief: Contribution to an economy of symbolic goods. *Media Culture Society* 2: 261–293.

Bourdieu, Pierre. 1986. The forms of capital. In *Handbook of Theory and Research for the Sociology of Education*, ed. John Richardson, 241–258. New York: Greenwood.

Bourdieu, Pierre. 1988 [1984]. The hit parade of french intellectuals, or who is to judge the legitimacy of the judges? In *Homo Academicus*, Pierre Bourdieu, 256–270. Stanford: Stanford University Press.

Bourdieu, Pierre. 1993 [1981]. But who created the 'creators'? In *Sociology in Question*, Pierre Bourdieu, 139–148. London: Sage.

Bourdieu, Pierre. 1993 [1983]: The field of cultural production, or: The economic world reversed. In *The field of cultural production: Essays on art and literature*. Pierre Bourdieu, 29–73. New York: Columbia University Press.

Bourdieu, Pierre. 1996 [1992]. *The rules of art: Genesis and structure of the literary field.* Stanford: Stanford University Press.

Bourdieu, Pierre. 1998 [1994]. *Practical reason: On the theory of action.* Stanford: Stanford University Press.

Brandtner, Christof. 2017. Putting the world in orders. Plurality in organizational evaluation. *Sociological Theory* 35(3): 200–227.

Buchholz, Larissa. 2008. Feldtheorie und Globalisierung. In *Nach Bourdieu. Visualität, Kunst, Politik*, Eds. Beatrice von Bismarck, Therese Kaufmann, and Ulf Wuggenig, 211–238. Wien: Turia+Kant.

Buchholz, Larissa. 2018. Rethinking the center-periphery model: Dimensions and temporalities of macro-structure in a global field of cultural production. *Poetics* 71: 18–32.

Buchholz, Larissa and Ulf Wuggenig. 2005. Cultural globalisation between myth and reality: The case of the contemporary visual arts, *Glocalogue* 4.

Buckermann, Paul. 2018. Reactions, reactivity, rampant reductions. On self-fulfilling prophecies, the Matthew effect, and global potentials of a new gallery ranking. *Creative Industries Journal* 11(2): 197–202.

Buckermann, Paul. 2020. *Die Vermessung der Kunstwelt. Quantifizierende Beobachtungen und plurale Ordnungen der Kunst.* Weilerwist: Velbrück.

Buckermann, Paul. 2021. Ranking Art: Paradigmatic Worldviews in the Quantification and Evaluation of Contemporary Art. *Theory, Culture & Society* 38(4): 89–109.

Callon, Michel. 1984. Some elements of a sociology of translation: Domestication of the scallops and the fishermen of St Brieuc Bay. *Power, action and belief: A new sociology of knowledge? The Sociological Review* 32/S1, Ed. John Law, 196–233. London: Routledge.

Espeland, Wendy, and Michael Sauder. 2007. Rankings and reactivity: How public measures recreate social worlds. *American Journal of Sociology* 113(1): 1–40.

Espeland, Wendy and Mitchell Stevens. 1998. Commensuration as a social process. *Annual Review of Sociology* 24: 313–343.

Espeland, Wendy and Mitchell Stevens. 2008. A sociology of quantification. *European Journal of Sociology* 49(3): 401–436.

Esposito, Elena and David Stark. 2019. What's observed in a rating? Rankings as orientation in the face of uncertainty. *Theory, Culture & Society* 36(4): 3–26.

Hansen, Hans Krause and Mikkel Flyverbom. 2015. The politics of transparency and the calibration of knowledge in the digital age. *Organization* 22(6): 872–889.

Heintz, Bettina. 2016. "Wir leben im Zeitalter der Vergleichung." Perspektiven einer Soziologie des Vergleichs. *Zeitschrift für Soziologie* 45(5): 305–323.

Heintz, Bettina and Tobias Werron. 2011. Wie ist Globalisierung möglich? Zur Entstehung globaler Vergleichshorizonte am Beispiel von Wissenschaft und Sport. *Kölner Zeitschrift für Soziologie und Sozialpsychologie* 63: 359–394.

Judah, Hettie. 2017. What Does Hito Steyerl's Power 100 Ranking Say About the Art World? https://news.artnet.com/art-world/hito-steyerl-power-100-1139549. Accessed: 17 October 2019.

Lamont, Michèle. 2012. Toward a comparative sociology of valuation and evaluation. *Annual Review of Sociology* 38: 201–221.

Luhmann. Niklas. 1969. Klassische Theorie der Macht: Kritik ihrer Prämissen. *Zeitschrift für Politik* 16(2): 149–1970.

Luhmann, Niklas. 1975. *Macht*. Stuttgart: Enke. Englisch edition: Luhmann, Niklas. 2017. *Trust and Power*, Part 2. Cambridge: Polity Press.
Luhmann, Niklas. 2000 [1995]. *Art as a social system*. Stanford: Stanford University Press.
Luhmann, Niklas. 2012 [1997]. *Theory of Society, Vol. 1*. Stanford: Stanford University Press.
Martins, Luis. 2005. A model of the effects of reputational rankings on organizational change. *Organization Science* 16(6): 701–720.
Quemin, Alain. 2006. Globalization and mixing in the visual arts. An empirical survey of ‚high culture' and globalization. *International Sociology* 21(4): 522–550.
Quemin, Alain. 2013. *Les stars de l'art contemporain. Notoriété et consécration artistiques dans les arts visuels*. Paris: CNRS Éditions.
Quemin, Alain. 2018. The uneven distribution of international success in the visual artist among nations, according to the rankings of the top 100 artists in the world. In *Art and the Challenge of Markets, Vol. 1*, ed. Victoria Alexander, Samuli Hägg, Simo Häyrynen, and Erkki Sevänen, 249–275. Cham: Palgrave.
Rindova, Violina, Luis Martins, Santosh Srinivas, and David Chandler. 2018. The good, the bad, and the ugly of organizational rankings. A multidisciplinary review of the literature and directions for future research. *Journal of Management* 44(6): 2175–2208.
Stäheli, Urs. 2012. Listing the global: Dis/connectivity beyond representation? *Distinktion: Scandinavian Journal of Social Theory* 13(3): 233–246.
Steyerl, Hito. 2017. *Duty free Art. Art in the Age of Planetary Civil War*. London: Verso.
Velthuis, Olav. 2005. *Talking prices. Symbolic meanings of prices on the market for contemporary art*. Princeton: Princeton University Press.
Weber, Max. 1978 [1922]. *Economy and Society. An Outline of Interpretative Sociology*. Berkeley/Los Angeles/London: University of California Press.
White, Harrison and Cynthia White. 1993 [1965]. *Canvases and careers: Institutional change in the French painting world*. Chicago: University of Chicago Press.
Zembylas, Tasos. 2006. Modelle sozialer (Un) Ordnung. Überlegungen zur Konstitution der Forschungsgegenstände der Kulturbetriebslehre. In *Kulturbetriebsforschung*, eds. Tasos Zembylas and Peter Schmuck, 17–45. Wiesbaden: VS.

How Organisations Exercise Power Through Peer Review: A Case Study from Contemporary Art Music

4

Annelies Fryberger

Abstract

Gatekeepers exercise power by granting access to resources in their respective fields, and they typically distribute these resources based on evaluations of quality. Arts funding organizations act as gatekeeping organizations, and they use evaluation to determine who belongs in a given field and who does not. The case presented here is an American organization whose role is to fund contemporary art music. This type of music, as in academia, is typically evaluated via peer review. This organization must balance the conflicting priorities of its public and private donors: the former require openness and diversity, the latter aim to reproduce cultural hierarchies. To work around this inherent conflict, the organization chooses to not define publicly the type of music it supports. However, by managing the peer review process in a specific way, the organization can make the evaluation produce a result coherent with the definition of contemporary art music espoused by its private donors. With this case, we can see how arts funding organizations may exercise the power of defining the possibilities and limits of art production fields.

Keywords

Boundary work · Contemporary art music · Evaluation · Peer review

A. Fryberger (✉)
IRCAM/CRAL, Paris, Frankreich
e-mail: annelies.fryberger@gmail.com

L. Gaupp et al. (eds.), *Arts and Power*, Kunst und Gesellschaft,
https://doi.org/10.1007/978-3-658-37429-7_4

4.1 Introduction

Gatekeepers exercise power by granting access to resources in their respective fields, and they typically distribute these resources based on evaluations of quality. In some fields, such as academia (Lamont 2009), the evaluative systems in place rely primarily on peer review, meaning that a person being evaluated may be the one doing the evaluating at a different occasion. This system is also used in some parts of the music world, notably when we move into the realm of contemporary art music. In this part of the "pure pole" (Bourdieu 1998) of the field of music, which is not market-oriented, gatekeeping organisations typically call on composers and other actors in the contemporary art music domain to judge the work of their peers. This means that funding agencies, whether they be public (such as the Ministry for Culture in France or the National Endowment for the Arts in the USA) or private (such as a foundation—the Koussevitsky foundation, the Fromm foundation, for example), will typically convene a panel of peers to decide whom they should fund. The organisation thus delegates some of its gatekeeping powers to members of the field itself, which has the advantage of making these funding decisions legitimate to actors within it.

The objective of the present chapter is to look precisely at how this process works, and to see how a gatekeeping organisation can use (and even manipulate) the peer review process in order to obtain results that are coherent with the organisation's mission. I will present a case study of an American organisation which focuses exclusively on funding contemporary art music. This case is pulled from a larger study on the evaluation of contemporary art music in the United States and France (Fryberger 2016). The goal here is to show how an arts funding organisation exercises its power by imposing a definition of contemporary art music: with its funding decisions, it can shape what is considered contemporary art music and what is seen to lie outside this domain. This is a clear example of boundary work (Lamont 2009). What we will see in this case study is how this organisation, by limiting the scope of who could be considered a peer, coopts the peer review process to obtain results in line with the vision of contemporary art music shared by its primary funders.

In fields where singularities are present—meaning goods whose fundamental value is their uniqueness and no "objective hierarchy" can be used to evaluate them (Karpik 2007, p. 39)—formalised evaluation processes strive to make the incomparable comparable, by quantifying certain aspects, for instance. Lucien Karpik states that the public presentation and marketing of singularities necessarily implies an "arbitrary selection of certain dimensions to the detriment of others" (Karpik 2007, p. 41). Thus "all valuation hinges on relations of visibility"

(Espeland and Lom 2015, p. 35). What is hidden or ignored cannot be compared, and evaluation processes seek precisely to organise attention through the use of formal criteria or quantification, for example, in order to obtain results that are coherent with a certain ideology or other standard. Our attention is by no means neutral and fair in its distribution—noticing is indeed "a complex cognitive, political, and professional process that is powerfully mediated by the form we give to valuation" (Espeland and Lom 2015, p. 33).

Given that evaluation is fundamentally a question of comparison, the crux of the issue is what, precisely, can be compared: where is attention directed. Evaluating can be seen as a sub-function of classification, or as form of boundary work (Lamont 1992, 2009). We reify categories that are, at root, social conventions (Zerubavel 1991). Thus, part of the work of evaluation is to discern which of the things being evaluated actually fit the category in question, and there is a penalty for those objects which do not fit established categories (Zuckerman 1999; Bielby and Bielby 1994). Sociologist Stefan Hirschauer calls this the "foundational framing activity" (2010, p. 92) of peer reviewers, who must first ask basic questions such as: "Is it science? What discipline? What genre?" (ibid.). Depending on the evaluator's place in the field in question, her way of drawing disciplinary boundaries will change (Lamont 2009)—and the judgments made on the works to be evaluated are also implicit judgments of one's own work and that of one's colleagues (Hirschauer 2010, p. 92).

4.1.1 What is Contemporary Art Music?[1]

In every art world, determining what artistic production can be considered part of the art world in question and what stands outside occupies considerable attention within the art world in question (Becker 1982, p. 36). That being said, the following general definition gives the context necessary to understand what is at issue in the present case study:

> [Contemporary art music comprises] pieces that [are] composed or preplanned reflectively, fixed in some sort of notation for a performer or creator to interpret or execute, and intended to be listened to by an attentive, informed, and critical audience. We might add that it is a style of music that traces its primary lineage back to the courts and churches of pre-Renaissance Europe, and although those courts

[1] This section is adapted from Fryberger (2019).

> and churches are today mostly long defunct or culturally marginal, contemporary art music maintains an important relationship with their modern-day descendants and the structures of production and listening that they represent. (Rutherford-Johnson 2017, p. 3)

To situate this more precisely, some composers who have become more or less household names working in this genre include, for example, historically, John Cage, Karlheinz Stockhausen, and Pierre Boulez—and for composers working currently, we might cite John Adams, Philip Glass, or Wolfgang Rihm. In terms of funding, government structures and non-profit organizations have primarily taken on the role once played by courts and churches, and the world of contemporary art music, not just in the United States, is indeed highly subsidised. It is made up of a majority of actors who do not make a living from their artistic practice, as most composers and performers working in this domain have a "day job", so to speak, often as teachers (Jeffri 2008; see also Lena and Lindemann 2014). In the US, funding for this type of work is not centralised, and various non-profits and university structures provide support for the creation and performance of this type of music.

However, it must be stated that "contemporary art music" is not the term that the organisation in this case study uses to describe the music it funds. This is a field where idiosyncratic terminology proliferates, with different preferences expressed for terms such as "new music", "contemporary music", "present music"—to cite only examples in English. I have chosen to use this term nonetheless, for two reasons. Firstly, to protect this organisation's identity, and secondly, because this term is relevant in the sociological literature on art, where scholars have studied extensively how distinctions between art and craft (Becker 1982) or highbrow and lowbrow cultural production (Levine 1990) have emerged. The term "contemporary art music" is widely used in the field of musicology and other fields which study music,[2] and the use of this term indicates where this production is situated in terms of its cultural and symbolic legitimacy.

[2] See, for example, Born (1997), El-Ghadban (2009), and Gidal (2010), among many others.

4.1.2 Methods

The choice of this organisation[3] for this case study could be qualified as an example of "emergent sampling" (Patton 2002, p. 244), as this organisation was mentioned by all of the American composers I interviewed when beginning this research (between 2010 and 2012, N=8) and it was evident that it was a key funding organisation in the field. There is also an element of opportunism here: when I asked the organisation if I could observe their peer review panels, the response was positive. This was not the case for the two other organisations to whom I asked the same question. To introduce the organisation briefly: it is one of the largest funding bodies for contemporary art music in the United States, distributing approximately one million dollars annually directly to composers, orchestras, and performers. The organisation traces its roots back to the 1930s, but exists in its current form only since the early 2010s. It is non-profit and semi-public, meaning it has funding from government bodies, such as the National Endowment for the Arts (the only federal arts funding agency in the United States), as well as private donors. Private funds constitute the vast majority of its endowment, which is not untypical for this type of organisation in the American context (Martel 2011). Private and public funders often have conflicting rationales (Kallman and Clark 2016, p. 77), as we will see, and this organisation has to find a way to please both. Public funding bodies, such as the National Endowment for the Arts, have traditionally held a view that culture must be for all, and so arts organisations must find ways to make their production accessible to the masses (Martel 2011). However, the private donors for this American organisation espouse a more elitist view, as we will see, and want their money to be spent supporting "consecrated" (Bourdieu 1968) culture.

I conducted an ethnography of this organisation in 2012–2015, in the context of a larger study on the evaluation of contemporary art music in the United States and France. For this organisation, I specifically collected data on its peer review process. This involved directly observing two peer-review panels (in 2012 and 2013, with 3 external experts per panel, in addition to staff members from the organisation), and conducting semi-structured interviews with the evaluators in the days following the panel discussions, with an aim to understand how they saw

[3] I keep the identity of the organisation and the individuals I quote confidential, in keeping with "the presumption ... that the privacy of research subjects should always be protected" (Patton 2002, p. 411).

their role in the discussion and the roles of the other panelists. My interview protocol for interviews with evaluators was inspired by the protocol used by Michèle Lamont in her research on peer review in academia[4] (Lamont 2009). In addition to these interviews, I conducted interviews with all members of staff of the organisation (N = 8).

Contents analysis was also an important part of this research: I systematically analysed the content of this organisation's blog from its founding until 2015, in an effort to understand how the organisation circumscribed the concept of contemporary art music.

This text will focus on the year 2013, when this organisation changed both its application and review processes. The applications moved from a paper-based (or pdf-based) format to the creation of an online profile, and evaluations are now conducted by individuals remotely. The organisation no longer convenes its reviewers for face-to-face meetings, whether online or off. This change, which could be seen as purely bureaucratic, actually revealed many of the underlying evaluative schemas of the organisation. This is indeed what makes it such a compelling case for studying what interests us here: how power is exercised in artistic fields.

4.2 The Power to Define

The organisation publicly presents its work by explicitly avoiding defining the music it supports. This is one way that they can please both their public and private funders. Here is the way they discuss the definition of contemporary art music on their website:

> In our day-to-day work, we specifically aim to not define too precisely what we mean by contemporary art music. To define is to limit. The music we can hear being made around us in such abundance is definition enough. We simply want it to thrive.[5]

[4] Michèle Lamont graciously shared her interview protocol with me, and I will happily provide the interview protocol I developed on request.

[5] I have kept the essence of this quote, but reworded it so as to keep the organisation anonymous.

This is an idealised point of view, which conveniently elides the necessary selection process inherent to any funding organisation—it must make choices as to who shall receive the money it offers. This organisation indeed receives diverse applications: jazz, musicals (à la Broadway), electronic music similar to what would be heard at an institution like the IRCAM, and contemporary classical music running the gamut from tonal symphonic music to atonal and/or conceptual music for small and large ensembles. In an interview with the president of this organisation, he discusses this avoidance of a clear definition:

> We don't define [this] music here, so that is extremely difficult to get around because it's like, well, what is it? Well, we are not going to tell you what it is, but sort of the folks that we listen to and hang around with, and do stuff for and give money to, it's—variety and diversity are the characterising traits. ... And that's variety and diversity in terms of, you know, some of the obvious stuff, like musical materials, instrumentation, genre, style, all these lovely words, aesthetic perspective, and to a slowly increasing degree extends to background, although we are still very heavily into the kind of people who go to college and then do interesting things afterwards. (interview with the organisation president, 2013)

With this quote, it becomes evident that there is an implicit definition at work and that it involves the social network of contemporary art music, which is indeed restricted ("the folks we listen to and hang around with"). It is also clear that the backgrounds of these composers are not diverse, although are slowly becoming more diverse, and that these are highly educated individuals. This puts some very clear markers on this definition of contemporary art music, which is explicitly not made explicit in the way the organisation presents itself.

The obvious question then becomes: how does this organisation narrow down the field of contestants, and how does it define quality when it does not want to explicitly define the kind of music it supports?

4.2.1 Evaluation and Boundary Work

The stated reason for the change in application and review processes in 2013 was that the organisation wanted more diversity and wanted to fund more innovative projects. According to the members of the organisation that I interviewed, the main thing standing in their way was the dynamics of the review committees themselves, which tended to skew toward safer choices because panelists had to find a consensus (interview with staff member, 2012).

Due to the newly simplified application process, the number of applicants tripled the first year the new system was in place. The kinds of applicants also changed, in that the organisation received a larger share of applicants working in popular music genres. These applicants were also present in the previous system, but in far fewer numbers. This posed a number of problems. The organisation wanted the definition of contemporary art music to come from the field, but this new applicant pool seemed to show that contemporary art music was something other than what the organisation wanted it to be. Since a public, explicit definition of the applications they desired was impossible, what were the options available? The person responsible for the evaluation process said the following shortly after the release of the new system:

> One of the things we saw immediately, right after we launched, almost within a day, we had 20 projects already submitted and these were people who were looking for recording funding and they are more hip-hoppy, sort of pop-oriented than what we … we've always had [people like that interested in applying], especially in our recording programs: "oh I'm doing this album," like, "yeah sure, you can apply", but the application process was so, I don't know, it was like a big wall to get over, so we never ended up getting a lot of applications from people that were broader, but we've made the process so much simpler and what we saw, like I mean immediately, is that this is a platform that people who are out there trying to do this stuff are familiar with. … well, we wanted more diversity, it's something that we've been shooting for anyway. (interview with staff member responsible for applicant evaluation processes, 2013)

We see here that the emergence of a definition of contemporary art music from the field was previously constrained by the application process itself. The previous application required skills that one learns when studying composition in an institution of higher education—it included project descriptions, written scores, and excerpts from recordings—whereas the new application simply asks applicants "to display their work for the public in the best possible way" (interview with staff member, 2013) on an online platform. The prior "wall" was sufficient to keep out the more "hip-hoppy, pop-oriented" projects that now find their way into the new system because of its simplicity. The job of the organisation then became to build a new wall to keep these applications from being funded. With this observation, we can see that this was indeed an exercise in boundary making, and less, or not at all, an exercise in quality assessment.

4.2.2 Rebuilding the "Wall"

Since the application itself was no longer a barrier, the organisation had to find a way to make the evaluation process produce the narrow results it wanted. Previously, the organisation convened meetings of external experts to evaluate their applications and make funding decisions. When I observed two of these review panels, some patterns emerged: the more "traditional" composers tended to dominate the discussions (by "traditional," I mean those who were conservatory trained, working with established new music ensembles, with an emphasis on the written note over improvisation). Their opinions were more often adopted than not, and individuals with a background in improvisation and sound art, for example, had less of an influence on the decisions being made. To sum up these observations briefly, I established three evaluative archetypes in these panels, which I will call traditionalists, improvisers, and mediators. A summary of the salient behaviours of these archetypes for the analysis here is included in Table 4.1.

Staff members of the organisation observed that panels tended to go for safer choices, where consensus was easy to establish—and this consensus was often led or forged by the more traditional composers. In short, this was the conservative dynamic of consensus seeking, which the organisation wanted to avoid in the new system. In principle, they wanted to allow for more "experimental" applications (interview with staff member, 2012) to come through, such that the traditionalist view would not systematically predominate. This intention, however, is hobbled by the way scores are counted under the new system, as we will see now.

Table 4.1 Traits of the three evaluative archetypes, derived from peer review panels observed in 2012 and 2013

	Average score given to applications (out of 6, 6 being the best score)	Opinion adopted by group, % of contested decisions	Average number of times individual was explicitly called on to express opinion	Evaluative style
Traditionalists	3.65	50.5%	18.5	Directive, decisive
Improvisers	5	33%	10.5	Engaged, frank
Mediators	3.95	16.5%	13	Retiring, willing to compromise

In my observations of peer review panels prior to the change in the application system, I noticed patterns in the way these individuals scored applicants—notably, improvisers tended to score everyone higher, while the traditional composers tended to be sparing with their high scores (see Table 4.1). Under the new application and evaluation system, the organisation receives evaluators' scores and compiles them to create a ranked list. However, they do not perform this operation without intervening: the new system actually gives more weight to high scores coming from individuals who give very few, and less weight to high scores coming from individuals who tend to score everyone high. Additional weight is also given to scores coming from individuals with specific expertise in the application being evaluated. Essentially, they have created a system which gives more weight to the scoring behaviour associated with traditional composers than that associated with improvisers or mediators, and given that these individuals also tended to score traditional applications higher, they have thus created a system which inherently favours these kinds of applications. In essence, they have shot their stated intentions in the foot.

This has obvious consequences for the results of the evaluation. As Michèle Lamont has shown, we tend to see work that is similar to our own as being of better quality than work that is very different from ours—this is called homophilia (Lamont 2009). This behaviour is also present here: the composers in the peer review panels I observed tended to like the work of other composers who did work similar to their own. Some even went so far as to take an activist stance, feeling the obligation to defend styles that were similar to their own when they were perceived as being underrepresented. For example:

> My role on the panel was to kind of speak about the applicants who were maybe a little more left of center. I mean, somebody like [fellow panelist X] obviously can speak pretty well towards people who are writing string quartets and orchestral music, and so forth. But [fellow panelist Y] and I, I felt like we had a little more diverse background, and so there was an opportunity for me to suggest that some of the improvisers were doing very good work. … I frankly don't think we need a whole lot more string quartets in the world, but we need other things to happen that can breathe life into new music, and I think that's the helpful role that I was able to play. … The majority of my work is in improvised music and in jazz. So I felt was important for me to speak boldly at those times when maybe those voices weren't getting heard well. (interview with composer from 2012 panel)

This individual fit the profile of a high-scoring improviser, so we can deduce that if the scoring behaviour associated with composers with a traditional background is given more weight, as discussed above, the quoted panelist's desire to "speak boldly" in defense of applicants who are more "left of center" would not be heard under the new evaluative system. This means that applications that have a more "traditional" feel will have an easier path to success than less traditional applicants. In this way, the effect of homophilia for the traditional applicants is reinforced, and the organisation can express its preferences and shape the contemporary art music world to fit them, all the while portraying this boundary-making work as an exercise in quality assessment. The "wall" is reconstructed and remains hidden.

4.2.3 How this Power to Define is Seen as Legitimate

With the money that gets distributed based on these processes, the "space of possibles" (Bourdieu 1998; Lane 2005) of contemporary art music is defined. We can see here how this organisation exercises its power to determine this space, using a tightly constrained peer review process to conduct this boundary work, without ever having to publicly define the music it funds. It recruits "external" experts—people who are recognised within contemporary art music—and it is their task to give the organisation an accurate judgment based on their experience. Generally speaking, peers must be sufficiently similar for a comparison to be possible (White 2002; Zuckerman 1999). In this case, to qualify as an external expert, a "peer", to be asked to participate in "peer review" for this organisation, the minimal requirement is to have received funding in the past from this organisation. In this way, the organisation and its definition reproduce themselves: if the organisation recruited evaluators with expertise in hip-hop, the situation would be very different. In short, only some of the applications are actually "peer reviewed", in that no experts, no peers, evaluate the "hip-hoppy" applications. These applications are essentially invisible (Espeland and Lom 2015) to the recruited experts. The experts recruited can be seen as peers only for a specific subset of the applications, and these are the applications which will be successful.

The mechanisms described here do an excellent job of meeting the conflicting priorities of the organisation's funders: on the one hand, an obligation toward openness and non-elitism, as public funding bodies will have it, and on the other, a desire to reproduce a hierarchy that has benefitted their private funders. In the words of the organisation's president:

> I spend a lot of my time talking to people who give us money, and I spent the past year very gently talking to all the people, especially people who gave us endowment money, ... you know, by changing like this we are risking their goodwill, and so I

> feel that particular risk very directly. … The most nervous among them are watching very carefully, you know, if this round, instead of funding what we normally do, [we] fund, you know, a hundred hip-hop projects, that would cause major problems. … It's not going to happen just because … we are… I mean, we will see if there are actually some hip-hop projects that come through. … And it will be very interesting to see, bringing them through the panel that we have constructed, what that will look like. [But] the system that you use is of course directly related, or the people who come through and get funded, is directly related to the system you use. (interview with organisation president, 2013)

The organisation's president is aware of the issue, that the field defines contemporary art music differently from the organisation's private donors, but he is also not willing to take responsibility for the way the organisation deals with this problem. The "system" he mentions is at once their own and something external to the organisation, because it brings in experts from the "outside". In short, they had to build a new system that would continue to exclude hip-hop from their definition of contemporary art music, but they lay the responsibility for this exclusion in the hands of the system and the external experts that are part of it. The organisation is thereby able to maintain a non-elitist identity all the while funding elitist music. They situate themselves as external to the field, in the sense that they want the field to give them something which they deny their participation in: a definition of contemporary art music.

4.3 Concluding Remarks

What can we learn from this case study? It shows the kind of power that an arts funding organisation has to draw boundaries around the field of which it is a part, and how it can do so without declaring this to be its intention. In short, quality assessment can hide boundary-making work. The power to evaluate is power indeed, and this case clarifies how funding organisations influence the boundaries of artistic genres, and more generally, how peer review can be used to reproduce the status quo. The power that this organisation implicitly holds, therefore, is the power to define. What is left out of its definition of contemporary art music is invisible—quite literally, in this case, as successful applicants have their online profiles made public whereas those who are unsuccessful are only visible to evaluators and staff members of the organisation. Contemporary art music, as defined through the processes described here, is made visible in the form that this organisation wishes it to have.

One interpretation would be to see this organisation as acting as a seawall against innovation, and instead reinforcing existing hierarchies and boundaries. It is not our task as researchers to say whether this is good or bad (Becker 1982), and we also have to recognise that the cultural hierarchies that we have discussed here (the discrepancy between "legitimate" culture and hip-hop) tend to reproduce themselves (Bourdieu 1968). Any institutionalised field experiences forces like those described here. That these forces are restrictive is a matter of course, and establishing new conventions is both risky and expensive (Becker 1982). My point in this chapter was to describe in precise detail how these boundaries can be maintained, despite a push for openness. Peer review is by nature conservative, in that its very essence is for experts to judge production that is close to their own. Thus, the expertise of a field is inclined to reproduce itself, with today's priests resisting the work of tomorrow's prophets (Bourdieu 1968).

References

Becker, Howard S. 1982. *Art Worlds*. Berkeley: University of California Press.

Bielby, William T. and Denise D. Bielby. 1994. "All Hits Are Flukes": Institutionalized Decision Making and the Rhetoric of Network Prime-Time Program Development. *American Journal of Sociology* 99(5): 1287–1313.

Born, Georgina. 1997. Modernist Discourse, Psychic Forms, and Agency: Aesthetic Subjectivities at IRCAM. *Cultural Anthropology* 12(4): 480–501.

Bourdieu, Pierre. 1998. Les règles de l'art: genèse et structure du champ littéraire. Paris: Le Seuil.

Bourdieu, Pierre. 1968. Outline of a sociological theory of art perception. *International Social Science Journal* XX(4): 589–612.

El-Ghadban, Yara. 2009. Facing the music: Rituals of belonging and recognition in contemporary Western art music. *American Ethnologist* 36(1): 140–160.

Espeland, Wendy Nelson, and Stacy E. Lom. 2015. Noticing Numbers: How Quantification Changes What We See and What We Don't. In *Making Things Valuable*, by Martin Kornberger, Lise Justesen, Anders Koed Madsen and Jan Mouritsen, 18–37. Oxford: Oxford University Press.

Fryberger, Annelies. 2016. De l'évaluation en musique contemporaine en France et aux Etats-Unis: Jurys de pairs, commandes d'oeuvres, et médias sociaux. Phd Dissertation. EHESS, Paris, France.

Fryberger, Annelies. 2019. Valuation in a reversed economy: the case of contemporary art music in France and the United States. In *Music practices across borders. (E)valuating Space, Diversity and Exchange*, by Glaucia Peres da Silva and Konstantin Hondros, 41–60. Transcript Verlag.

Gidal, Marc. 2010. Contemporary Latin American Composers of Art Music in the United States: Cosmopolitans Navigating Multiculturalism and Universalism. *Latin American Music Review* 31(1): 40–78.

Hirschauer, Stefan. 2010. Editorial Judgments: A Praxeology of "Voting" in Peer Review. *Social Studies of Science* 40(1).
Jeffri, Joan. 2008. *Taking Note: A Study of Composers and New Music Activity in the U.S.* Ann Arbor: Inter-university Consortium for Political and Social Research.
Kallman, Meghan Elizabeth and Terry Nichols Clark. 2016. *The Third Sector: Community Organizations, NGOs, and Nonprofits.* Champaign: University of Illinois Press.
Karpik, Lucien. 2007. *L'économie des singularités.* Paris: Gallimard.
Lamont, Michèle. 1992. *Money, Morals, and Manners: The Culture of the French and American Upper-Middle Class.* Chicago: University of Chicago Press.
Lamont, Michèle. 2009. *How Professors Think: Inside the Curious World of Academic Judgment.* Cambridge: Harvard University Press.
Lane, Jeremy F. 2005. When Does Art Become Art? Assessing Pierre Bourdieu's Theory of Artistic Fields. In *The Sociology of Art. Ways of Seeing*, by David Inglis and John Hughson, 30–42. New York: Palgrave-McMillan.
Lena, Jennifer C., and Danielle J. Lindemann. 2014. Who is an artist? New data for an old question. *Poetics* 43: 70–85.
Levine, Lawrence W. 1990. *Highbrow/Lowbrow: The Emergence of Cultural Hierarchy in America.* Cambridge: Harvard University Press.
Martel, Frédéric. 2011. *De la culture en Amérique.* Flammarion.
Patton, Michael Quinn. 2002. *Qualitative Research & Evaluation Methods. 3rd Edition.* Thousand Oaks: Sage.
Rutherford-Johnson, Tim. 2017. *Music after the Fall: Modern Composition and Culture since 1989.* Oakland: University of California Press.
White, Harrison C. 2002. *Markets from Networks: Socioeconomic Models of Production.* Princeton: Princeton University Press.
Zerubavel, Eviatar. 1991. *The Fine Line. Making Distinctions in Everyday Life.* Chicago: The University of Chicago Press.
Zuckerman, Ezra W. 1999. The Categorical Imperative: Securities Analysts and the Illegitimacy Discount. *American Journal of Sociology* 104(5): 1398–1438.

Art Perception and Power. A Plea for Relational Sociological Aesthetics

5

Nina Tessa Zahner

Abstract

Questions of art perception touch issues of participation, justice and the legitimacy of worldviews and have thus always been subject to power struggles. Research on art perception is deeply involved in these issues as it promotes specific worldviews over others. The paper discusses resent developments of the research on art perception in terms of power relations. It asks how different research traditions prefer a specific conception of art and art viewing to others and how they integrate these ideas into their concepts. I will argue that it is still helpful to turn to Pierre Bourdieu's field analytic research tools in order to grasp processes of power reproduction in the context of art viewing, but plead to supplement Bourdieu with Simmel's conception of the interplay of form and content as this helps to avoid Bourdieu's anti-sensualistic bias that is not very helpful when it comes to contemporary art.

Keywords

Art · Sociology of the arts · Sociology of perception and the senses · Art perception · Exhibition studies · Audience studies · Relational sociology · Pierre Bourdieu · Georg Simmel · Erving Goffman

N. T. Zahner (✉)
Kunstakadmie Düsseldorf, Düsseldorf, Germany
e-mail: nina.zahner@kunstakademie-duesseldorf.de

L. Gaupp et al. (eds.), *Arts and Power*, Kunst und Gesellschaft,
https://doi.org/10.1007/978-3-658-37429-7_5

5.1 Introduction

The production, distribution and consumption of the arts have always been subject to power struggles for cultural representation (DiMaggio 1982, 1987; Wolff 1993; Blau et al. 1985; Bourdieu 1970, 1987, 1993, 2005). Nowadays these struggles are discussed in terms of economization (Alexander et al. 2018; Boltanski and Esquerre 2018), de-enchantment, massification (Schultheis 2015; Schultheis et al. 2016) and democratization (Lena 2019; Zahner 2006a, 2009) of the arts. At the heart of all these issues lie questions on how artistic works are experienced by artists, cultural intermediaries and visitors (Zahner 2018). Against this background, it is surprising how little we still know about how people with different biographic, professional, social and cultural backgrounds actually engage with the arts and make sense of art works (Zahner 2021a). This is even more startling when we keep in mind that these questions are highly political and touch issues of participation, justice and the legitimacy of worldviews. They address questions of power and social inequality with respect to seeing, viewing, watching, receiving, perceiving and experiencing the arts (Zahner 2017, 2021c, d). Different concepts of viewing art lead to differing opinions on the programming of art institutions such as the inclusion of blockbuster offerings, displays of commercial or minority cultures and the role of curators, programmers and art expertise in the staging of art exhibitions, plays, performances etc. (Lena 2019; Zahner 2006a, 2018). Questions of power are crucial to all issues related to perception and perceiving.

Art exhibitions have evolved to major visitor attractions over the past 30 years. Art museum attendance in Europe and elsewhere has multiplied, with leading museums today being visited by several million people each year (Themed Entertainment Association 2019). To visit art museums and art exhibitions in order to see and view historic and contemporary works of art has become a leisure activity of major social, cultural and economic relevance and is thus of growing sociological interest when it comes to questions of power and the arts. Therefore, it is all the more surprising how little we know about what people actually *do* in art exhibitions, *how* the look at art works, *how* they make sense of art, *how* they experience art.

In this article, I discuss current sociological research on the visiting of art exhibition and how some of these studies proceed certain power implications. In a final chapter, I will turn to the sociology of Georg Simmel and will show how the sociology of art experience could massively benefit from his conception of the interplay of form and content.

5.2 Social Science Research and the Visiting of Art Exhibitions

In sociology there is a huge body of research on aesthetic judgment and taste (Coulangeon 2016; Halley and Sonolet 2018; Quinn et al. 2018b). For decades, these discussions have been shaped by the influential work of Pierre Bourdieu (1987; Bourdieu and Darbel [1966/1969] 2006, 1970) on taste and cultural participation. This research tradition draws heavily on Kantian aesthetics (Hennion 2013; Shusterman 2002; Zahner 2012), focuses mainly on constant and consistent aesthetic orientations and promotes a notion of the viewing and seeing of art as a process of *decoding*. Investigations of museum audiences have thus for the last 30 years mainly applied a semiotic perspective on people's experience and interpretation of works of art. These studies often come to the result that participation in and understanding of the arts is influenced by people's socialisation, upbringing and art specific knowledge, that is, by them belonging to particular social classes (Quinn 2018a; Hanquinet et al. 2013; Hanquinet 2018; Zahner 2012). They often draw on concepts implicitly integrating enlightenment assumptions on the emancipating character of aesthetic experience (Behnke 2012; Rössel 2009) and thus often distinguish an adequate and an inadequate way of viewing art:

> I do mean to point out that relying on a semiotic scheme to explain spectatorship is an incomplete procedure that … imagines the spectator chiefly as a recipient. Too often semiotically based analysis predicts an ideal spectator as reader, and thus can trap itself inside a moral conviction about proper or correct reception. (Kennedy 2009, pp. 12–13)

In recent years, perspectives and approaches informed by Bourdieu's work have been criticised for losing relevance in contemporary society and for ignoring how people encounter and experience works of art in the actual social and spatial situation in museums or galleries (Quinn et al. 2018b). The cognitive sciences and psychology are currently undertaking extensive research into the relationships between aesthetic sensibilities, the structure of the brain and the visual apparatus. By exploiting the facilities of portable eye scanning equipment (Brieber et al. 2014; Tröndle et al. 2014; Santini et al. 2018), this research examines, for example, the organisation of people's looking at abstract art (Brinkmann et al. 2014) or the relationship between aesthetic stimuli like "symmetry" and people's aesthetic judgment (Brinkmann 2021). This research remains experimental in character and primarily focuses on the individual spectator.

Some works of Empirical Visitor Studies aim to actually investigate the experience in the "real" situation of the exhibition by using questionnaires to conduct pre- and post-visit surveys (for an overview see Kirchberg and Tröndle 2012). These studies produce interesting results but are not able to observe *the immediate experience* of art works in exhibition settings as they can only measure verbalized, cognitively processed reactions to art exhibitions (Kirchberg and Tröndle 2012, p. 448). In order to avoid these pitfalls some researchers use methodologies that involve the deployment of elaborated technical setups to merge a diverse body of data, including movement tracking, heart rate, skin conductance in combination with pre- and post-surveys to generate information cartographies of visitors, visualizing their spatial behaviour, physiological responses and their experience of the exhibition (Tröndle et al. 2014). These methodologies focus on the experience in exhibition setting in terms of physiological reactions to works of art and merge these data with pre- and post-interrogations. They have thus little to say on how people actually *engage* with artworks in the "real" exhibition setting (Lamont and Swidler 2014). In addition, they apply an individualist perspective to the study of seeing, viewing, experiencing and making sense of art. To take an individualistic perspective on the visiting of art exhibitions is a rather normative reduction that draws on bourgeois ideas of art viewing as a contemplative practice. In contrary to this conception recent studies show that visiting art museums is an immensely social and not a solipsistic practice: it takes place in a public space and even if you go on your own, you will react to or interact with the actual spatial and social conditions of the exhibition (Heath and vom Lehn 2004; Tzortzi 2014).

Dirk vom Lehn and his colleagues (vom Lehn 2006; vom Lehn et al. 2001; vom Lehn 2018) focus their research on experiencing art in the "real" situation of the museum or the exhibition. They conduct video ethnography in order to understand the ways in which people, in ordinary circumstances, discriminate, evaluate and experience art works in exhibitions spaces.

5.3 The Practical Aesthetics of Experiencing Art and Art Exhibitions—Gar-Finkel, Goffman and Vom Lehn

The studies by Dirk vom Lehn and his colleagues are committed to the tradition of *Ethnomethodology* following Harold Garfinkel and draw on Erwing Goffman's *frame analysis*. They thus investigate the *interaction order* that structures the viewing, seeing and understanding of art works in museum and gallery exhibitions. Vom Lehns video-based research aims to show "how visitors coordinate their viewing and inspecting of artworks, and how they experience art through verbal, physical, and material actions and interactions" (vom Lehn 2006, p. 84). The focus of these studies is on the *social practices* of the viewing of art, thus processual, physical and material aspects of the visiting of art exhibitions come into view. Research in the tradition of Garfinkel propagates a view that remains attached to everyday life and finds order in practical daily action (vom Lehn 2012, pp. 36–37). The *Sociology of the Everyday Experience* focuses on processes of constituting meaning and sets out to search for the meaningful structure in which man and his world are anchored. With this perspective, actors emerge not as human beings with a specific history or disposition but as *meaning constructions*, created only in the context of action. Ethnomethodology is thus not interested in human *actors*, but in *actions* and the structure of action. Similar to Goffman it is primarily concerned with the normative order that exists within and between these units: "the codes of conduct that exist everywhere where people are" (Goffman 2013, p. 7; my translation).

Following these perspectives, the studies of vom Lehn and his colleagues show how exhibition visitors interactively coordinate their perspectives on works of art through expressions and gestures and how they instruct one another to see in a specific way. The way of perceiving develops—according to these studies—in and through the in situ interaction between the visitors:

> The observations [show] in particular that the perception of art is not influenced by the structure of the works or cognitive dispositions, but is carried out in specific, local and social situations. (Heath and vom Lehn 2004, p. 57)

Vom Lehn and his colleagues move the social and spatial *context*, the situation of perception, into the focus of research. In their point of view, it is not socially learned perception schemes or dispositions that structure the perception and the experience of art works: How people see and experience an art work is in their view progressively emerging through a complex configuration of bodily and verbal action (Heath and vom Lehn 2004, p. 52).

In their research they aim at showing how social interaction and the visual, vocal, tactile conduct of others provide resources for experiencing art exhibitions. In this view it is the monitoring of the actions of others that influences what art works visitors look at, how they approach, appreciate and understand art (vom Lehn et al. 2001, p. 206). Based on these ideas, these researchers show that interactions in museum and gallery spaces follow a certain *script*; that they are organized according to specific aesthetic ideals:

> In and through their talk and bodily conduct, the participants transpose action to the object and encourage the co-participant(s) to see and respond to the *enlivened* exhibit. … The spectator or *recipient* is encouraged not so much to respond to the action of the other, but rather the en-livened object, the exhibit in and through which action is embodied. (Heath and vom Lehn 2004, p. 60)

According to these studies, interactions in museum or gallery exhibitions create an interaction situation that is focused on—to use a famous term of Theodor W. Adorno here—*careful sensing* (behutsames Abtasten) of the object: Visitors in art exhibition situations follow in their interactions a script that places the presented object at the centre of interaction and ascribes to it a dominant position in the experiencing process. They thus perform a specific idea of aesthetic perception, one that aims at a confrontation with and a reflection on the object in its own right. It is this perspective that ensures that the "object retains an element of surprise, of curiosity" (Heath and vom Lehn 2004, p. 61), an *object-guided experience*:

> The very ways in which an exhibit is revealed provides the spectator with a unique and seemingly uncontaminated encounter with the object; the co-participant confronts the object and its particular characteristics as if independently of the very ways in which they are being *instructed* to look. (Heath and vom Lehn 2004, p. 62)

What these studies do not see, is the power relations and mechanisms that are virulent in exhibition situations. Because of the overly microscopic methodology they apply, they are drawing an all too harmonious, somewhat naive picture of art perception and tend to reproduce and affirm the power structures operating here. One has to see that the practices observed in exhibition situations are not *made* by those who are in the situation, as Ethnomethodology suggests, but on the contrary, are historically grown *interaction orders* that structure—as Goffmann points out—what the situation should be and how to behave accordingly (Goffman 2008, p. 9). Research in New Museology shows that and how exhibition spaces are places of debate about worldviews (Bennett 1994a; DiMaggio and Useem 1978; Fyfe 2000; Hetherington 2005) and how the staging of exhibitions sets certain types of knowledge dominant and marginalizes others (Macdonald 2007; Fyfe 2011a, b, p. 35). New Museology therefore investigates how the potential heterogeneity of experiences in exhibitions is disciplined by the museum space, the way objects are presented and how the potential experience is tailored according to the approaches of established visitor groups (Bennett 1994b, 2011; Hetherington 2012). By these means museums and their exhibitions become visible as agents that reproduce a dominant ideology and existing power relations (DiMaggio 1982; Fyfe 2000, 2011a; Prior 2005; Wolff and

Seed 1988). A growing body of research thus focuses on the social backgrounds and worldviews of curators and museum staff (Baur 2010; Ziese 2010; Zolberg 1981). Against this background Pierre Bourdieu's late field analytic research is very helpful: It shows how our reality is the result of historical debate about the enforcement of specific worldviews and their continuous activation in our everyday practice.

5.4 The Field Analytic Aesthetics of Experiencing Art Exhibitions—Pierre Bourdieu

In his late field analytic studies, Bourdieu investigates how reality is the product of the interplay of two states of history: *objectified history*, accumulated in buildings, monuments, artworks, books, theories, customs, law, language, etc. and *incorporated history,* inscribed in the body and the *habitus* (Bourdieu 2011, p. 26). This conception of reality shows great similarities with that of Georg Simmel, developed more than fifty years earlier. Simmel promoted against Plato's idea of two separate realities—a sphere of transcendental unchangeable ideas and an empirical sphere of sensual perception—a notion of knowing based on the experience of objects in the world: "It is not the objects in themselves and not a sovereign mind in us that determines the veracity of our imaginations; but life itself" (Simmel 1922, p. 124; my translation). It is again everyday situations, that are the focal point of consideration here, but the everyday for Simmel is *a complexion of hidden relations* and not just the mode of perception of the everyday man. Social science has to identify these relations in order to analyse how different elements determine one another's position, "just as the masses of matter do by virtue of gravity" (Simmel 2016, p. 81; my translation). Simmel's method of *understanding* (*Verstehen*) thus reaches through the acting subject to the objective constellations in order to identify meaning structures—underlying interests, feelings, aspirations—imposed on the individual (Simmel 2016, p. 522). Simmel sees these meaning structures—*conceptual formations* as he calls them—emerging of the *interplay of historical conceptual formations and individual biographic experiences.*

Bourdieu uses an identical concept. He also sees history as a double-sided process that works through the installation into things, the *habitat*, and the bodily incorporation in the *habitus* of the actor. But in Bourdieu's view—and here lies the big difference to Simmel—this double-sided process has a tendency to *reproduce* social power relations: Language, things, buildings, traditions, bodies, etc. are for Bourdieu above all instruments that smoothly perpetuate social domination (Bourdieu 2011, p. 48). This conception also lies at the heart of his theory of perception: Even so Bourdieu updated his earlier theory of art reception in *The Rules of Art* and replaced the semiotic idea of an *adequate* decoding of the art work—that he has been heavily criticised for—with a more open conception, he held on to the idea that a true understanding of an art work is based on the "harmony of habitus and artwork" (Bourdieu 2005, p. 494; my translation), an understanding of the practical schemes on which the artwork is based. This version of his theory of perception is historically more variable than the earlier one and refers to a wide variety of artistic concepts, but it holds on to the idea of the necessity of social prerequisites in order to understand a work of art (Zahner 2021a, c, d). Even so, Bourdieu replaces here an overly static conceptualized methodological toolbox with a less normative ethnological field analytic tool, his research still aims at identifying the means of *understanding* an artwork *adequately*. It seems as if it does not even cross Bourdieu's mind that *understanding* an artwork might *not* be the only way to approach, perceive or experience art. Bourdieu is instead still concerned with identifying cognitive schemes that people apply when looking at art works, when *re*cepting them. He does not arrive at a conception of art *per*ception as a purely sensual activity as Sophia Prinz and Omar Lizardo want to understand (Lizardo 2011, p. 39; Prinz 2013, pp. 302–303).

Simmel's conception is here much more elaborated and offers—in my view—a fruitful integration of practical and field analytic aesthetics.[1]

5.5 The Relational Sociological Aesthetics of Experiencing Art and Art Exhibitions—Georg Simmel

Simmel promoted a notion of knowing based on the *experience* of objects as an interaction of form and content, a synthesis of the immediate sensory impressions to the unity of form (Geßner 2003, p. 34). The unity of form (*Gestalt*) arises through the creative form finding activity of the knowing subject. Simmel thinks of *form* as of a purely heuristic tool. He connects here with Kant, according to whom the handling of sensual reality requires the interplay of sensory experience and pure thinking (Helle 2001, p. 34). For Simmel this processing is carried out

[1] Simmel influenced Goffman a great deal but fell—in my conviction—victim to an all too narrow interpretation of his theoretical insights. Simmel is not so much focused on interaction *or* form as he is often accused of, but is much more interested in their *interplay*, as we will see. He offers thus an interesting perspective on questions of power.

by an active subject standing amidst the stream of life: Perception is then utterly connected to the worldly situation of the perceiver and his motifs (Simmel 1922, 113 and 124). It is thus formed by the worldly-sensual situatedness of the perceiver and the "dull pressure of things" (Hübner-Funk 1982, p. 187; my translation).

In order to find truth, the social scientist has to escape the worldly situatedness of his perception by distancing himself from it by a refined interest in finding *pure forms* (Simmel 2017a, p. 101), in identifying the objective constellations of human action, the forms that both express and shape it (Simmel 2017b, p. 260). To identify these objective constellations of action the social scientist performs an intellectual activity that forms from the scientific material something that the material does not hold in itself. This act of forming identifies meaning structures that are not identical to the ones of the situated worldly actor, but are imposed on the material by the researcher and his research question. Simmel's relational method thus imposes meanings and values to the material that the worldly actors themselves are not aware of: "underlying interests, feelings, aspirations" (Simmel2017c, p. 522). These meanings and values are not imposed on the material by psychological methods but by the reconstruction of *Wechselwirkungen,* the interplay of social forms that stand beyond the individual will and individual action (Simmel 2017c, p. 159). These *social forms* are shaped by the actions of worldly actors, who are the product of their individual histories or trajectories. Society for Simmel is thus made up of the interplay of individual conceptual constructions that are shaped by the biographic experiences of their actors. It is then the interplay of *individual* conceptual constructions—history on the level of the individual—that forms the conceptual constructions on the level of society. Simmel's method does *not* push for some kind of socio-psychological *interactionism,* but focuses on culture (objektive Kultur) as a dynamic, constant interplay of forms. This is not a *formal sociology*—as put forward by Leopold von Wiese—but a *sociology of formation,* a relational method that focuses on the complexion of supraindividual dynamic relations as the forces of social change (Böhringer 2018, p. 849; Helle 2001, pp. 126–130).

This conception is very similar to Pierre Bourdieu's conception of social fields and his method of field analysis. The fundamental difference between these two conceptions is that Bourdieu applies his field analytical tool quite the same way to all kinds of social fields, while Simmel is a truly relational thinker and sees his methods shaped by the material it is used upon. Contrary to Bourdieu, Simmel adapted his analytical tool, his form of reflection, to the content he was dealing with. When Bourdieu's notion of art perception keeps a certain "intellectualism" (Bourdieu 2005, p. 490) to it, that's because he applied his notion of *scientific perception* to the perception of art. This resulted in a notion of perception that aimed at *understanding* the meaning, the content of an artwork. In clear contrast to this, Simmel thought of art as a cultural form, a specific form of knowledge gained from sensuality. Works of art for Simmel represent reality from the perspective of the individual and with focus on the singular. In epistemological terms, Simmel thus anticipated the idea of *Symbolic Conciseness* (*Symbolische Prägnanz*) of Ernst Cassirer: For Simmel aesthetic experience arises from the interplay of viewing subject and viewed object, while there is no right or adequate way to perceive the work. The viewer is just as much the creator of the work as the artist (Orth 1998, p. 748).

The conceptualization of the interplay of form and content by Simmel offers thus an interesting perspective to the study of audiences. It allows to integrate insights in how people actually engage with art works, how they experience them with the perspective of critical sociology and the analyses of power relations in art exhibitions and the art field.

5.6 Conclusion

Vom Lehn and colleagues have developed interesting insights in the *interaction orders* of art exhibitions. Unfortunately, these studies do not reflect on the social situatedness of the frames of perception they discover. This research thus tends to reproduce existing power relations within the art context rather. To grasp these power relations and the mechanisms of their reproduction, Pierre Bourdieu's late field analytic research is very helpful: It shows how our reality is the result of historical debate about the enforcement of specific worldviews and their continuous activation in our everyday practice. Unfortunately, Pierre Bourdieu's conception promotes a notion of art perception that focuses all too much on the understanding of art works and on processes of meaning making. It promotes thus a notion of art perception that sees art experience as a process of understanding, based on specific art related knowledge, distinguishes between adequate and an inadequate ways of viewing art works and prefers expert knowledge to everyday knowledge (Zahner 2021a).

This concept is a historically specific one. It is a perspective that was dominant in the art field in the 1960s, the time when Bourdieu carried out his studies on art audiences (Zahner 2012) and that he, as a man of his time, was obviously not able to escape:

> Pierre Bourdieu ... rigorously exposes the hidden economy and veiled interests of the so-called disinterested aesthetic of high culture, but none the less remains too enchanted by the myth he demystifies to acknowledge the existence of any legitimate popular aesthetic. (Shusterman 1991, p. 204)

With the end of the 1960s, this focusing on "reading" an art work was increasingly replaced by a growing appreciation of the everyday and an idea of unconditional perception of art (Behnke and Wuggenig 1994; Graw 2008; Hanquinet 2018; Zahner 2006a, b). The *performativity of meaning* now increasingly came into focus. Meaning was no longer conceived as emerging in relation to a stable context, but rather was thought of as a process-related phenomenon that arises in a process between the perceiver and the perceived. These new conceptions of art perception relate to ideas introduced by romanticism, materialism, sensualism, American pragmatism and the early avant-gardes of the 20th century (Fähnders 2014, pp. 208–209; Reckwitz 2012). Processes of individualization promote a fundamental reconceptualization or dissolution of the hierarchical author-audience relationship (Gerhards 2001; Zahner 2021a), an increased distrust of the legitimacy of (expert)knowledge and a growing appeal to questions of authenticity (Beck 1986; Giddens 1996; Nassehi 2006, p. 61). The sociological studies of art experience have up until today not lived up to these fundamental transitions. Georg Simmel's sociological aesthetics offer a promising methodological approach here. They present a notion of perceiving that does not privilege one conceptual form of perception over another, and installs—for the social science, and only for the social science—a relational, historical informed methodology of research that can serve as a mighty counterpart to all sorts of metaphysics and substantial thinking that currently gain ground again (Zahner 2021b; Zahner and Schürkmann 2021; Escher and Zahner 2021).

References

Alexander, Victoria D., Samuli Hägg, Simo Häyrynen, and Erkki Sevänen (eds.). 2018. Art and the Challenge of Markets Volume 1: National Cultural Politics and the Challenges of Marketization and Globalization. Cham: Springer International Publishing.

Baur, Joachim. 2010. Museumsanalyse: Zur Einführung. In Museumsanalyse: Methoden und Konturen eines neuen Forschungsfeldes, ed. Joachim Baur, 7–14. Bielefeld: transcript-Verl.

Beck, Ulrich. 1986. Risikogesellschaft: Auf dem Weg in eine andere Moderne. Frankfurt am Main: Suhrkamp.

Behnke, Christoph. 2012. Gründe für den Besuch von Ausstellungen und Fragen der Kunstvermittlung. In Das Kunstfeld. Eine Studie über die Akteure der zeitgenössischen Kunst, ed. Ulf Wuggenig and Heike Munder, 125–141. Ennetbaden: Lars Müller Verlag.

Behnke, Christoph and Ulf Wuggenig. 1994. Heteronomisierung des ästhetischen Feldes: Kunst, Ökonomie und Unterhaltung im Urteil eines Avantgardekunst-Publikums. In Das symbolische Kapital der Lebensstile, 229–252. Frankfurt/Main [u.a.]: Campus-Verl.

Bennett, Tony. 1994a. The Exhibitionary Complex. Culture, power, history 123–154.

Bennett, Tony. 1994b. The Reluctant Museum Visitor: A Study of Non-Goers to History Museums and Art Galleries. Redfern, NSW: Australia Council for the Arts.

Bennett, Tony. 2011. Culture, Choice, Necessity. A Political Critique of Bourdieu's Aesethetic. Poetics 39:530–546.

Blau, Judith R., Peter Michael Blau, and Reid M. Golden. 1985. Social Inequality and the Arts. The American Journal of Sociology: AJS.

Böhringer, Hannes. 2018. In der Unentschiedenheit des Lebens. Simmels Lebensphilosophie. In Simmel-Handbuch: Begriffe, Hauptwerke, Aktualität, ed. Hans-Peter Müller and Tilman Reitz, 844–853. Berlin: Suhrkamp.

Boltanski, Luc and Arnaud Esquerre. 2018. Bereicherung: Eine Kritik der Ware. Berlin: Suhrkamp Verlag.

Bourdieu, Pierre. 1970. Elemente zu einer soziologischen Theorie der Kunstwahrnehmung. In Zur Soziologie der symbolischen Formen, ed. Pierre Bourdieu, 159–201. Frankfurt am Main: Suhrkamp.

Bourdieu, Pierre. 1987. Die feinen Unterschiede: Kritik der gesellschaftlichen Urteilskraft. Frankfurt am Main: Suhrkamp.

Bourdieu, Pierre. 1993. The Field of Cultural Production: Essays on Art and Literature. New York: Columbia University Press.

Bourdieu, Pierre. 2005. Die Regeln der Kunst: Genese und Struktur des literarischen Feldes. Frankfurt am Main: Suhrkamp.

Bourdieu, Pierre. 2011. Der Tote packt den Lebenden. In Der Tote packt den Lebenden, 17–54. Hamburg: VSA-Verl.

Bourdieu, Pierre and Alain Darbel. 2006 [1966/1969]. Die Liebe zur Kunst: Europäische Kunstmuseen und ihre Besucher. Konstanz: UVK.

Brieber, David; Nadal, Marcos; Leder, Helmut; Rosenberg, Raphael. 2014. Art in Time and Space: Context Modulates the Relation between Art Experience and Viewing Time. PLoS ONE (6): e99019.

Brinkmann, Hanna. 2021. Auf den Spuren von Seh-Modi beim Kunstbetrachten. In Wahrnehmen als Soziale Praxis. Künste und Sinne im Zusammenspiel., ed. Christiane Schürkmann and Nina Tessa Zahner. Wiesbaden: Springer.

Brinkmann, Hanna; Commare, Laura; Leder, Helmut; Rosenberg, Raphael. 2014. Abstract Art as a Universal Language? Leonardo 47 (3): 256–257.

Coulangeon, Philippe. 2016. The Sociology of Cultural Participation in France Thirty Years after Distinction. In Routledge International Handbook of the Sociology of Art and Culture, ed. Laurie Hanquinet and Mike Savage, 26–37. London & New York: Routledge.

DiMaggio, Paul. 1982. Cultural entrepreneurship in nineteenth-century Boston. Media Cultura and Society 4:1.

DiMaggio, Paul. 1987. Classification in Art. American Sociological Review 52 (4): 440–455.

DiMaggio, Paul and Michael Useem. 1978. Social Class and Arts Consumption: The Origins and Consequences of Class Differences in Exposure to the Arts in America. Theory and Society 5 (141–161).

Escher, Cornelia and Nina Tessa Zahner. 2021. Einleitung. Begegnung mit dem Materiellen in Architekturgeschichte und Soziologie. In Begegnung mit dem Materiellen., ed. Cornelia Escher and Nina Tessa Zahner. Bielfeld: trancript.

Fähnders, Walter. 2014. Der avantgardistische Künstler. In Avantgarde und Modernismus: Dezentrierung, Subversion und Transformation im literarisch-künstlerischen Feld, ed. Wolfgang Asholt, 201–220. Berlin/Boston: de Gruyter.

Fyfe, Gordon. 2000. Art, Power and Modernity: English Art Institutions, 1750 - 1950. London: Leicester Univ. Press.

Fyfe, Gordon. 2011a. Sociology and the Social Aspects of Museums. In A Companion to Museum Studies., ed. Sharon Macdonald, 33–49. Malden, Mass.: Blackwell.

Fyfe, Gordon. 2011b. Sociology and the Social Aspects of Museums: 33–49. In A Companion to Museum Studies, ed. Sharon Macdonald, 33–49.

Gerhards, Jürgen. 2001. Der Aufstand des Publikums.: Eine systemtheoretische Interpretation des Kulturwandels in Deutschland zwischen 1960 und 1989. Zeitschrift für Soziologie 30 (3163–184).

Geßner, Willfried. 2003. Der Schatz im Acker: Georg Simmels Philosophie der Kultur. Weilerswist: Velbrück Wiss.

Giddens, Anthony. 1996. Leben in einer posttraditionalen Gesellschaft. In Reflexive Modernisierung: Eine Kontroverse, ed. Ulrich Beck, Anthony Giddens and Scott Lash, 113–194. Frankfurt am Main: Suhrkamp.

Goffman, Erving. 2008. Rahmen-Analyse: Ein Versuch über die Organisation von Alltagserfahrungen. Frankfurt am Main: Suhrkamp.

Goffman, Erving. 2013. Interaktionsrituale: Über Verhalten in direkter Kommunikation. Frankfurt am Main: Suhrkamp.

Graw, Isabelle. 2008. Der große Preis. Kunst zwischen Markt und Celebrity Kultur. Köln: DuMont.

Halley, Jeffrey A. and Daglind E. Sonolet (eds.). 2018. Bourdieu in Question. New Directions in French Sociology of Art. Boston: Brill.

Hanquinet, Laurie. 2018. 'J'adore!' Aesthetics in Bourdieu's Account of Taste. In The Persistence of Taste: Art, Museums and Everyday Life after Bourdieu, ed. Malcolm Quinn, Dave Beech, Michael Lehnert, Carol Tulloch and Stephen Wilson, 141–152. Milton: Routledge.

Hanquinet, Laurie; Roose. Henk; Savage, Mike. 2013. The Eyes of the Beholder: Aesthetic Preferences and the Remaking of Cultural Capital. Sociology 48 (1): 111–132.

Heath, Christian and Dirk vom Lehn. 2004. Configuring Reception: (Dis-)Regarding the 'Spectator' on Museums and Galleries. Theory, Culture & Society 21(6):43–65.

Helle, Horst Jürgen. 2001. Georg Simmel. Einführung in seine Theorie und Methode/Introduction to His Theory and Method. Oldenbourg: de Gruyter.

Hennion, Antoine. 2013. "Dinge, die dauern...". Objekte, Vermittlung, Soziologie. In Akteur-Medien-Theorie, ed. Tristan Thielmann, Erhard Schüttpelz and Peter Gendolla, 81–105. Bielefeld: transcript.

Hetherington, Kevin. 2005. From Blindness to Blindness: Museums, Heterogeneity and the Subject. In Actor Network Theory and After, 51–73. Oxford: Blackwell.

Hetherington, Stephen. 2012. Epistemology: The Key Thinkers. London: Continuum International Publishing.
Hübner-Funk, Sibylle (ed.). 1982. Georg Simmels Konzeption von Gesellschaft: Ein Beitrag zum Verhältnis von Soziologie, Ästhetik und Politik. Köln: Pahl-Rugenstein.
Kennedy, Dennis. 2009. The Spectator and the Spectacle: Audiences in Modernity and Postmodernity. Cambridge: Cambridge Univ. Press.
Kirchberg, Volker and Martin Tröndle. 2012. Experiencing Exhibitions: A Review of Studies on Visitor Experiences in Museums. Curator: The Museum Journal 55 (4): 435–452.
Lamont, Michèle and Ann Swidler. 2014. Methodological Pluralism and the Possibilities and Limits of Interviewing. Qualitative Sociology 37 (2): 153–171.
Lena, Jennifer C. 2019. Entitled: Discriminating Tastes and the Expansion of the Arts. Princeton, N.J., Oxford: Princeton University Press.
Lizardo, Omar. 2011. Pierre Bourdieu as a Post-Cultural Theorist. Cultural sociology 5 (1): 25–44.
Macdonald, Sharon. 2007. Interconnecting: Museum Visiting and Exhibition Design. CoDesign 3 (1): 149–162.
Nassehi, Armin. 2006. Der soziologische Diskurs der Moderne. Frankfurt am Main: Suhrkamp.
Orth, Ernst Wolfgang. 1998. Georg Simmel. In Ästhetik und Kunstphilosophie: Von der Antike bis zur Gegenwart in Einzeldarstellungen, ed. Julian Nida-Rümelin and Monika Betzler, 745–752. Stuttgart: Kröner.
Prinz, Sophia. 2013. Die Praxis des Sehens: Über das Zusammenspiel von Körpern, Artefakten und visueller Ordnung. Bielefeld: transcript.
Prior, Nick. 2005. A Question of Perception: Bourdieu, Art and the Postmodern. The British Journal of Sociology. 56 (1): 123–139.
Quinn, Malcolm. 2018a. Introduction: Taste, Hierarchy and Social Value after Bourdieu. In The Persistence of Taste: Art, Museums and Everyday Life after Bourdieu, ed. Malcolm Quinn, Dave Beech, Michael Lehnert, Carol Tulloch and Stephen Wilson, 1–18. Milton: Routledge.
Quinn, Malcolm, Dave Beech, Michael Lehnert, Carol Tulloch and Stephen Wilson (eds.). 2018b. The Persistence of Taste: Art, Museums and Everyday Life after Bourdieu. Milton: Routledge.
Reckwitz, Andreas. 2012. Die Erfindung der Kreativität: Zum Prozess gesellschaftlicher Ästhetisierung. Berlin: Suhrkamp.
Rössel, Jörg. 2009. Kulturelles Kapital und Musikrezeption: Eine empirische Überprüfung von Bourdieus Theorie der Kunstwahrnehmung. Soziale Welt 60 (3).
Santini, Thiago; Brinkmann, Hanna; Reitstätter, Luise; Leder, Helmut; Rosenberg, Raphael; Rosenstiel, Wolfgang; Kasneci, Enkelejda. 2018. The Art of Pervasive Eye Tracking: Unconstrained Eye Tracking in the Austrian Gallery Belvedere. In Proceedings of the 7th Workshop on Pervasive Eye Tracking and Mobile Eye-Based Interaction, 1–8. Warsaw, Poland: ACM Press.
Schultheis, Franz. 2015. Kunst und Kapital: Begegnungen auf der Art Basel. Köln: König.
Schultheis, Franz, Erwin Single, Raphaela Köfeler and Thomas Mazzurana. 2016. Art Unlimited?: Dynamics and Paradoxes of a Globalizing Art World. Bielefeld: transcript.
Shusterman, Richard. 1991. Form and Funk. The Aesthetic Challenge of Popular Art. The British Journal of Aesthetics 31 (3): 203–213.

Shusterman, Richard. 2002. Surface and Depth: Dialectics of Criticism and Culture. Ithaca, NY: Cornell Univ. Press.
Simmel, Georg. 1922. Über eine Beziehung der Selektionslehre zur Erkenntnistheorie. In Zur Philosophie der Kunst: Philosophische und kunstphilosophische Aufsätze, ed. Georg Simmel and Gertrud Simmel, 111–125. Potsdam: Kiepenheuer.
Simmel, Georg. 2016. Soziologie: Untersuchungen über die Formen der Vergesellschaftung, 8th edn. Frankfurt am Main: Suhrkamp.
Simmel, Georg. 2017a. Hauptprobleme der Philosophie, philosophische Kultur. Frankfurt am Main: Suhrkamp.
Simmel, Georg. 2017b. Kant. Die Probleme der Geschichtsphilosphie zweite Fassung (1905/1907). Berlin: Suhrkamp-Verlag.
Simmel, Georg. 2017c. Philosophie des Geldes. Berlin: Suhrkamp.
Themed Entertainment Association. 2019. Leading 20 Museums in Europe in 2017. https://www.statista.com/statistics/747942/attendance-at-leading-museums-in-europe/. Accessed 2019. Chart 23 May 2018.
Tröndle, Martin; Greenwood, Steven; Kirchberg, Tschacher, Wolfgang. 2014. An Integrative and Comprehensive Methodology for Studying Aesthetic Experience in the Field: Merging Movement Tracking, Physiology and Psycho-logical Data. Environment and Behavior 46 (1): 102–135.
Tzortzi, Kali. 2014. Movement in Museums: Mediating between Museum Intent and Visitor Experience. Museum Management and Curatorship 29 (4): 327–348.
vom Lehn, Dirk. 2006. Die Kunst der Kunstbetrachtung.: Aspekte einer pragmatischen Ästhetik in Kunstausstellungen. Soziale Welt 57 (1): 83–99.
vom Lehn, Dirk. 2012. Harold Garfinkel. Konstanz: UVK.
vom Lehn, Dirk. 2018. Tasting Rembrandt. In The Persistence of Taste: Art, Museums and Everyday Life after Bourdieu, ed. Malcolm Quinn, Dave Beech, Michael Lehnert, Carol Tulloch and Stephen Wilson, 124–140. Milton: Routledge.
vom Lehn, Dirk, Christian Heath and Jon Hindmarsh. 2001. Exhibiting Interaction: Conduct and Collaboration in Museums and Galleries. Symbolic Interaction 24 (2): 189–216.
Wolff, Janet. 1993. The Social Production of Art.: Auszüge: 95–136. Basingstoke: Macmillan.
Wolff, Janet and John Seed (eds.). 1988. The Culture of Capital: Art, Power and the Nineteenth-Century Middle class. Manchester: Manchester Univ. Press.
Zahner, Nina Tessa. 2006a. Die neuen Regeln der Kunst: Andy Warhol und der Umbau des Kunstbetriebs im 20. Jahrhundert. Frankfurt/Main: Campus-Verl.
Zahner, Nina Tessa. 2006b. Kunst zwischen Kulturindustrie und Hochkultur: Andy Warhol und die Transformation Kunstfeldes den 1960er Jahren. In Sonderheft "Kunstmarkt", 189–218.
Zahner, Nina Tessa. 2009. Die Kunst der Inszenierung. Die Transformation des Kunstfeldes der 1960er Jahre als Herausforderung für die Kunstfeldkonzeption Pierre Bourdieus. In Mediale Erregungen?: Autonomie und Aufmerksamkeit im Literatur- und Kulturbetrieb der Gegenwart, ed. Christian N. Joch Markus Wolf, 289–308. Tübingen: Niemeyer.
Zahner, Nina Tessa. 2012. Zur Soziologie des Ausstellungsbesuchs. Positionen der soziologischen Forschung zur Inklusion und Exklusion von Publika im Kunstfeld. 209–232.

Sociologia Internationalis. Europäische Zeitschrift für Kulturforschung 50 (1/2): 209–232.

Zahner, Nina Tessa. 2017. Die Heteronomien des Marktes: Eine Rekonstruktion der Diagnosen zur Ökonomisierung der Kunst. In Autonomie der Kunst?: Zur Aktualität eines gesellschaftlichen Leitbildes, ed. Uta Karstein and Nina Tessa Zahner, 139–164. Wiesbaden: Springer VS.

Zahner, Nina Tessa. 2018. The Economization of the Arts and Culture Sector in Germany After 1945. In Art and the Challenge of Markets Volume 1: National Cultural Politics and the Challenges of Marketization and Globalization, ed. Victoria D. Alexander, Samuli Hägg, Simo Häyrynen and Erkki Sevänen, 95–124. Cham: Springer International Publishing.

Zahner, Nina Tessa. 2021a. Das Publikum als Ort der Auseinandersetzung um legitime Formen des Kunst- und Weltwahrnehmens. In Wahrnehmen als Soziale Praxis. Künste und Sinne im Zusammenspiel., ed. Christiane Schürkmann and Nina Tessa Zahner. Wiesbaden: Springer.

Zahner, Nina Tessa. 2021b. Interpassivität, Affekt und Desubjektivierung – Fragmente einer Theorie aus Gegenbegriffen: Rezension zu Seyfert, Robert: Beziehungsweisen. Elemente einer relationalen Soziologie. Weilerswirst: Velbrück 2020, 278 Seiten. In Das postpoietische Paradigma: Sonderband der Zeitschrift für theoretische Soziologie, ed. Christian Dries and Takemitsu Morikawa, (im Erscheinen).

Zahner, Nina Tessa. 2021c. Kunstwahrnehmen im Ausstellungskontext. In Begegnung mit dem Materiellen., ed. Cornelia Escher and Nina Tessa Zahner. Bielfeld: trancript.

Zahner, Nina Tessa. 2021d. Sinnlichkeit und Sinn. Kulturen der Kunstbetrachtung. In Kulturen der Bewertung, ed. Oliver Berli and Nicolae, Stefan, Schäfer, Hilmar. Wiesbaden: VS.

Zahner, Nina Tessa and Christiane Schürkmann. 2021. Einleitung. Wahrnehmen als Soziale Praxis. Soziologische Perspektiven auf Wahrnehmen. In Wahrnehmen als Soziale Praxis. Künste und Sinne im Zusammenspiel., ed. Christiane Schürkmann and Nina Tessa Zahner. Wiesbaden: Springer.

Ziese, Maren. 2010. Kuratoren und Besucher: Modelle kuratorischer Praxis in Kunstausstellungen. Bielefeld: transcript.

Zolberg, Vera L. 1981. Conflicting Visions in American Art Museums. Theory and Society 10 (1): 103–125.

Art Organisations and Their Visitors. About Standardisation Methods of Non-members

6

Christian Steuerwald

Abstract

This study examines how art organisations can succeed in influencing visitor behaviour. As empirical studies show, art organisations have only limited possibilities to do this. Rather, it is above all the side effects of the social areas of family and school which shape the behaviour of visitors in art organisations, with the result that striking similarities can be observed in visitor behaviour in general.

Keywords

Art organisations · Visitor behaviour · Art audience · Crisis experiments

Art sociological visitor research can in a simplified way be reduced to two sets of questions: 1) who goes to exhibitions, museums, operas or theatres and why and 2) what do visitors do when they are in art organisations? Sociological research, for example, shows that, despite differences in detail, the artgoing public is characterised above all by a high level of education, comes predominantly from bourgeois social groups, and is interested in art for reasons of distinction, cultural stimulation, and/or as a specific form of enjoyment and experience (Behnke 2012; Bourdieu 1999; Bourdieu and Darbel 2006; Glogner-Pilz and Föhl 2016; Klein

C. Steuerwald (✉)
Universität Bielefeld, Bielefeld, Germany
e-mail: christian.steuerwald@uni-bielefeld.de

L. Gaupp et al. (eds.), *Arts and Power*, Kunst und Gesellschaft,
https://doi.org/10.1007/978-3-658-37429-7_6

1990; Kirchberg 2005; Reuband 2018; Rössel et al. 2002; Schulze 1992; Steuerwald 2012). In addition, further studies examine specific features and focus on non-visitors, occasional visitors or the general public, or concentrate on historical configurations (cf. Behnke 2012; Daniel 1995; Renz 2016; Zahner 2015).

In contrast to the frequently examined questions about the social-structural characteristics and motives of visitors, there are comparatively fewer studies about the interaction order in art organisations and about visitor behaviour. For example, there are only a few studies which examine visitors' self-presentation, clapping, the amount of time spent in front of exhibits, which way visitors choose to go in exhibitions, the aesthetics of perception and types of perception, or decisions about where to sit (Bourdieu 1974; Gebhardt and Zingerle 1998; Graf and Treinen 1983; Jenniches 1969; Martin 1998; Robinson et al. 1928; Schulze 1992; Steuerwald 2012; Tröndle et al. 2014; vom Lehn 2006).

Following on from the classical complex of questions and based on our own empirical studies in theatres, operas and museums, the present study tries to deal with art sociological visitor research in terms of organisational sociology. It is not about people and their organisations, but rather about organisations and their people. Thus, the questions of who goes to art organisations and why and what they do there are of less interest in the present research. Rather, the research focuses on how art organisations succeed in influencing and standardising the behaviour of their visitors, aligning them with the goals of the organisation, and thereby minimising disturbances of the orders of interaction. It is about art and power and how organisations exercise power on visitors. In order to answer the research questions, the first section sets out the organisational sociological problem regarding non-membership of visitors in art organisations. In the second section, the results of empirical studies which investigate behavioural patterns and show how visitors behave remarkably similarly are presented. Finally, the third section seeks to answer the question of why these behavioural patterns are similar. The paper ends with a summary and a conclusion.

6.1 Art Organisations and the Problem of Non-Membership of Visitors

Sociologically, organisations can be described as institutions which are oriented towards at least one purpose and which are characterised by systematical rules in processes oriented towards this purpose or purposes (Endruweit 2004; Kühl 2011; Luhmann 1976; Mayntz and Ziegler 1977; Preisendörfer 2005). Despite observable irrationalities, organisations are more or less rational (Powell and

DiMaggio 1991). In order to achieve their organisational purpose, organisations not only have a division of labour, which increases with their size, and consequently a hierarchical structure, but also need members who take on specific tasks.[1] As a rule, organisations themselves decide on membership and thus on the entry and exit of persons, i.e. who is admitted and receives such membership and who does not, and who has to leave the organisation (Luhmann 1976).

Since this membership is linked to the organisation's internal position structure, membership in organisations is associated with various role requirements whereby members are instructed in their tasks and behaviour. In contrast to historical forms of organisation and to all-out and greedy institutions, which are oriented towards a total commitment of their members which does not permit them to have competing memberships, the access of organisations to their members via the function-specific orientation of formal membership roles is only partial and, furthermore, imbued with increased uncertainties and limited possibilities for control (Coser 2015; Goffman 1973).

In order for them to function as smoothly as possible, organisations have a great interest in influencing their members and adapting them to the organisation's intended processes. Although organisations make great efforts to instruct their members about formal requirements and to align their behaviour with the purpose of the organisation, organisations are only partially successful, so that a more or less large area of behaviour is subject to little or no control. For example, Luhmann notes:

> Every organization consists of actions. But no one can act without being a part of it. He brings himself, his personality, to the workplace. However, the organization only demands specific services from him. His feelings and his self-representational interests are hardly ever called upon. They lounge around functionlessly during work and cause damage if they are not kept under control. (Luhmann 2016, p. 43; my translation)

As a result, uncertainties of expectation arise which cannot be formally dealt with and which, among other things, are outsourced to the informal sector. But "even such informal roles must be brought into a relationship with system membership; they are only taken over because of membership of the formal system" (Luhmann 1976, p. 49). In contrast to formal roles, however, the influence and control

[1] This leads to the fact that, in addition to the organisational purpose, the hierarchy and membership are also shown as fundamental characteristics of an organization (Kühl 2011; for further characteristics Endruweit 2004; Preisendörfer 2005).

possibilities relating to informal roles are much more difficult. In addition, organisations function inadequately without their informal areas, so that although the structural uncertainties are not easy to deal with, organisations still need to establish a link between the informal and formal areas (Bensman and Gerver 1963; Luhmann 1976; Tacke 2015).

While organisations face various uncertainty problems as a result of the inadequate standardisation of their members' behaviour, visitors pose even greater challenges to them, as visitors are usually not members of the organisations they visit and the organisations therefore cannot apply membership-based role requirements and rules of conduct and authority.[2] In particular, art organisations that exhibit or perform art and make it accessible to visitors are faced with the problem of how with the limited means at their disposal they can exert maximum influence on the behaviour of visitors so that disturbances are minimised. In addition to house rules, formal visiting and opening hours, and access to rooms, art organisations can regulate the behaviour of visitors through architecture, the manner in which art and objects are presented, pricing,[3] supervision by supervisory staff, guided tours, and art educational interventions.

In order to examine how art organisations manage to regulate the behaviour of their visitors so that disturbances are minimised, the results of empirical studies that investigate the behaviour of visitors are presented below.

6.2 How Do Visitors Behave in Art Organisations?

The empirical basis for the following material are studies from the year 2002 and follow-up studies from the year 2011 at various theatres and at the opera in Frankfurt am Main as well as various research from museums in Oldenburg and in Frankfurt am Main between 2017 and 2018.[4] In addition to ethnographic

[2] Even if visitors are members of the organisation visited, for example because they are in the organization as employees, visitors cannot be formally addressed via their membership-based roles, but only via informal expectations.

[3] For example, Johann Wolfgang von Goethe endeavoured to impose disciplinary measures on visitors through the pricing of admission fees and not to make too many places available to students. Goethe remarks in a letter: "Because if we don't take this precaution, we will, before we know it have filled up the upper gallery with students. We have also recently witnessed some bad behaviour again. The students, especially on the right side, made fun of the guard and kept taking their hats on and off, they also began to drum, which one absolutely should not have to put up with" (Goethe 1988, p. 332).

[4] For the sociological theatre studies, *see also* Steuerwald 2012.

observations and questionnaires, which included questions on sociodemographic characteristics, interest, competence and frequency of visits, various crisis experiments were also carried out, which revealed through people's irritation certain cultural norms.

In general, the behaviour of visitors to museums and theatres can be described as a "withdrawal of the body" (Schulze 1992; Steuerwald 2012): Concentrated listening, quiet observation, sitting immersed in thought. The experience is contemplative, and the body is at rest. For example, hardly any visitor moves particularly fast or hectically. Rather, their movement indicates "having time", which is the complete opposite to the diagnosed social acceleration of modernity and the subjectively connected feeling of a shortage of time and a professional life characterised by a shortage of time; it recalls the bourgeois practice of promenading and strolling in the 19th century (König 1996; Rosa 2010). The amount of noise is also reduced. Even in museums, where people sometimes move through the building in small groups, the volume of conversation in the exhibition rooms is comparatively low. A reduced volume is also noticeable in the foyers of theatres before the performance and during the breaks. During performances in theatres and in the opera, conversations can hardly be detected. The few observable conversations are clearly reduced in volume due to the reduced distance between speaker and listener. In addition, many elements of visitors' interaction are non-verbal, for example indicating where to look by means of a gesture, not saying which way one should go but rather one person going ahead and the others following, or visitors waiting until vantage points occupied by other visitors, which allow a good view of the exhibit, are vacated (for similar results, *see also* vom Lehn 2006). In addition to the "withdrawal of the body", these studies show that visitors almost exclusively interact with those persons who visit the performance or exhibition with them. As a rule, these groups of people are closer together and there is usually a spatial gap between them and other persons present, so that they form a common close-up space or a common "territory of the self" (Goffman 1982). Others present are acknowledged but politely ignored (Goffman 1982, 1986; Steuerwald 2012).

Differences in behaviour can be observed primarily in relation to age. In younger people (under 20 years of age) the "withdrawal of the body" is less pronounced. Younger visitors usually move faster, are louder and the time intensity of their experience is shorter. Time studies show that younger visitors not only on average view fewer exhibits in a time-intensive manner (by which I mean more than 20 s per exhibit) but, in addition, the time taken to view a particular exhibition object is also shorter. It can also be observed that for younger visitors who attend exhibitions or performances in larger groups, such as school classes, changes in their group structures may occur. For example, not only the group sizes change, but also the group dynamics are time and again likely to change. As a result, groups can become louder or quieter, attention spans can change or the speed of movement sequences can change.

In order to supplement the results of ethnographic observations and to obtain further evidence of how people interact and their associated behaviour in art organisations, various crisis experiments were carried out by Harold Garfinkel (2003). In terms of methodology, the "breaching", as the crisis experiments are described by Garfinkel, endeavours to break up everyday (and usually unconscious) norms and cultural standard practices, for example where acts are deliberately misunderstood or placed in a different context. Following on from Alfred Schütz and Talcott Parsons, Garfinkel assumes that trouble-free interactions are only possible because the actor has access to shared cultural classifications and standardisations which guide the actor and are usually not communicated. In addition to references to the functioning of interaction orders and the cultural self-understandings, it is also possible to methodically work out a standardisation of behaviours which are seldom explicitly communicated, but which guide action, such as tact and politeness.[5] Below are three brief examples from the studies which show the extent to which behaviour in art organisations is socially regulated and sanctioned.

6.2.1 Interaction Order 1: Applause at the Wrong Passages

In general, applause can be understood as a regulated symbolic practice of acknowledgement or thanksgiving that is expressed through clapping hands (Steuerwald 2012). However, applause is not a spontaneous expression of consent and reaction or an uncontrolled emotional expression of an experience, but a regulated system of behaviour which indicates when and how people applaud. Historical studies, for example, document that in Roman antiquity, numerous situations required the applause of those present, such as coronations, sacrifices, the investiture of high-status individuals or plays, which can also be distinguished, depending on the occasion, by different variations of clapping, such as clapping

[5] For example, Goffman also writes: "It makes sense to investigate these forms of misconduct because they draw our attention to some aspects of good behaviour that we usually take for granted" (Goffman 1986, p. 89).

with the flat or hollow hand or with the forearms (Jenniches 1969). In order to reduce potential uncertainty among those present and to discipline applause, symbolic indicators can usually be observed showing when one should clap and when not. In addition to the claqueurs who were paid in the 19th century to applaud in certain places in order to guide applause desired by actors, directors or artistic directors, theatres also use short breaks, closing or letting fall the stage curtain and darkening the hall lights (Dreßler 1993; Jenniches 1969).

The investigation of clapping in theatres shows that in plays (as opposed to musical productions) there is usually no clapping during the performance (Bab 1931; Jenniches 1969). Applause could only be observed in the ethnographic studies before the breaks and at the end of the performance. The applause before the break is usually much shorter. Apart from the longer duration, the applause at the end of the performance is also characterised by other potential actions which symbolically express a special acknowledgement of the performance, such as clapping while standing, shouting or whistling. In addition, the final sequence is characterised by a special communicative interaction between actors and audience. Thereby the actors enter the stage individually, one after the other, with the actors in the supporting roles usually appearing first, followed by the actors in the main roles. In all cases observed, the applause for the leading roles lasted longer, so that the strength of the applause is not only dependent on the actor's performance, but also on the importance of the role.

In contrast to spoken theatre productions, in musical theatre productions applause could also be observed during the performance.[6] As a rule, however, this is limited to outstanding performances in the leading roles or "significant" minor roles but does not apply to performances in "insignificant" minor roles, so that the applause there also depends on the individual's position within the status system of the performers. In addition, no applause at all could be observed during the performance at group performances such as the singing of a choir. Furthermore, observations in the theatres and the opera show that some visitors are active and others passive while clapping. It is possible that these differences can be attributed to different cultural competences in terms of judgement, so that passive visitors orient themselves towards the active clappers, to whom a higher competence is attributed. This uncertainty of judgement could also explain why the applause was always longer in the leading roles than in the supporting roles, since the actors are attributed greater competence as a result of their leading role.

---

[6] The ban on clapping during performances at the Richard Wagner Festival in Bayreuth is in contrast to this (see Gebhardt and Zingerle 1998, p. 70).

The crisis experiments carried out in the spoken theatres, which provided for clapping in the wrong places, also point to a regulated system of behaviour that is secured by normative expectations. None of the visitors could be motivated to clap their hands in applause during the performance. Nevertheless shouts from the audience like “Please be quiet!”, “Calm down” or “I want to see and hear the performance!”, which could only be heard after several false pieces of applause, indicate sanctions as a result of the disappointment of expectations. In addition to clapping during the performance, a further experiment will show how, as a result of increased uncertainty, visitors can orient themselves to others and what measures are available in the event of misconduct. The following is a brief excerpt from the examination protocol.[7]

> Normally at the end of the second act the hall light goes out and a short pause of 2–3 seconds follows. Then an actress speaks a few sentences in the dark and afterwards the break begins. After the darkening of the hall light I wait a second and start clapping loudly. First a few clap along and then probably the majority of those present. When the clapping ends, the actress speaks her lines. A short pause of 2 to 3 seconds follows and the applause resumes. Shortly afterwards, the lights come on and the applause ends. Observable reactions: Amused looks, smiles and a few blushing cheeks.

The uncertainty as to when one should clap is usually shown by the fact that, as the end of a performance is not always apparent, symbolic indicators such as the darkening of the hall lights are used. However, it can be observed that, as a rule, because the darkening of the hall lights is not clearly reliable information which allows one to act with complete confidence, a few seconds elapse as visitors wait to see what happens. In such cases, the initial clapping amounts to a decision-making aid in cases of uncertainty. However, the experiment not only points out that orientation towards others is used as a problem-solving strategy for actions under conditions of uncertainty, but also how unintentional incorrect conduct leads to corrective compensatory action (Goffman 1986). Smiles and amused glances are therefore symbolic indicators of one’s own incorrect conduct, aimed at ensuring that in the situation in question the presentation of Self, personal status and corresponding recognition as a person is not damaged.

[7] The experiment was carried out on 20.02.2002 at the TAT (Theater am Turm) in Frankfurt am Main during the performance of Europa.

6.2.2 Interaction Order 2: Singing in the Opera House

As historical studies have shown, the interactional orders in opera and the demands placed on visitors have changed considerably since the 19th century (Dreßler 1993; Müller 2006). Müller, for example, writes: "The participants in musical performances gradually transformed themselves into listeners who concentrated on following the course of the music, moderately restricting applause and displeasure, and completely stopping communication with the artists and their neighbors during the performance" (Müller 2006, p. 178). One consequence of the shift from participants to concentrating listeners, which went hand in hand with the more or less simultaneous introduction of seating on the ground floor of theatres and the blocking of access to the stage for the audience, is that, in contrast to rock and pop concerts for example, and following the demand for a "withdrawal of the body" in the opera, nobody dances to or sings along with the music (Gebhardt and Zingerle 1998).[8]

The various conversations with opera-goers that took place during the ethnographic observations document the fact that non-singing in the opera is so self-evident that it is not even the subject of discussion by visitors. Furthermore replies to specific questions such as "What a bizarre idea!" or "Who would want to sing along!" point to a cultural impossibility. Just how strong this impossibility is can be seen above all in the following attempt to sing along in the opera during a performance. Here is an excerpt from the examination protocol[9]:

> The hall lights finally go out and the actors come on stage. I wonder what the best part would be to sing along to. Somehow I'm a little nervous. ... After about 20 minutes I try to sing along. Quietly and without being heard, I hum along. I'm trying to get a little louder. But the voice fails me and it comes out only in a short caw. I'll stop trying for now. After the break I start a second attempt. My nervousness increases. Again I start humming and somehow hit on the melody. I try to get a little louder again until someone in my near environment hisses softly: "Shh!". I shut up immediately. Somehow this is embarrassing for me. About 10 minutes later I start another attempt. After about 5 seconds a man in the row in front of me turns around and says quietly but firmly: "Can you be quiet now! That's completely disrespectful".

[8] At pop and rock concerts there is usually no seating or only seating for some visitors. The ban on access to the stage is also repeatedly lifted through individual visitors coming back to the stage again and again and singing along or diving into the audience from the stage.

[9] The experiment was carried out on 3 April 2002 in the performance of "Das verratene Meer" at the Frankfurt Opera.

Although the singing along in the opera did not really work, the experiment is revealing in that it points to normalisations of behaviour in the opera, which are expected by the visitors and are sanctioned at a certain point in the case of rule violations. On the other hand, the experiment shows to what extent norms can be internalised and inscribed in the bodies, so that norm violations lead to physical reactions and trigger emotions (*see also* Neckel 1991). The fact that some norms can be so emotionally anchored is demonstrated by this test both in terms of the embarrassment connected with it and, for example, in the nervousness that can be observed in the action planning and the intended rule breaking in the opera, as well as in the failure of the voice.[10]

6.2.3 Interaction Order 3: Disturbing the Acoustic Warning Barriers

In contrast to the two previous experiments, the last experiment does not focus on the behaviour of visitors, but on the behaviour of members of the art organisation. The task was to trigger the warning barriers, which acoustically secure the space immediately in front of and next to paintings, several times in succession.

> I slowly move towards a painting and at first look at it at a distance of 2 meters. Slowly I go closer. At a distance of about 30–50 cm I stop and walk with my head closer to the painting. Suddenly an alarm signal sounds and after about 20 seconds there is a person from the supervisory staff looking at me. I step back from the work of art and wait until the alarm stops. After about 30 seconds I approach until the warning signal sounds again. The supervising person reappears immediately. I step back again and as the alarm is again silenced, I trigger the warning barrier again. The supervising person comes directly to me and politely says: "Excuse me Sir. It's forbidden to get so close to the painting". I apologize. I look at the artwork for a while and go to the next painting in the room that I know is secured. There, too, I directly trigger the alarm. Again, the supervising person comes directly to me. I shout to her: "Wait a moment" and quickly move towards the first painting and cross the warning barrier to trigger the sound and shout: "Look, it's the same sound!". The supervising person is right next to me and politely, but explicitly points out to me that if I trigger a safety barrier again, she will accompany me to the exit and I will have to leave the museum.

[10] Garfinkel (2003) also reports that the breach of everyday rules can lead to problems, so that some experiments had to be stopped prematurely.

As well as the supervisory person's references to formal rules of action for visitors and sanctions in the event of transgressions, this examination protocol also demonstrates what is expected of visitors. For this reason, it is only on the third occasion that the supervisor explicitly refers to the standards expected and to the rules of conduct. The first triggering of the warning barrier is thus usually classified as an unintentional transgression, which can happen because the warning barrier is not visible. Only if it is obvious that the norm is not known or does not give an indication of how one should behave is the instruction as to how to behave given. Finally, the examination protocol also shows behaviour which is designed to acknowledge and protect the status and integrity of the persons present. Thus, the conversation should not only be understood as a compensatory action in which, as Goffman (1986) pointed out, the transgression is corrected. The introduction to the conversation about an apology is also an indication of a recognition of the person that takes place via tact and politeness norms (Goffman 1982, 1986).

6.3 Why Do Visitors Behave Similarly in Art Organisations?

The results of the crisis experiments indicate that at least some of the highly cultural behaviour patterns in art organisations are standardised and that transgressions are sanctioned by other visitors as well as by members of the organisation or security personnel commissioned for this purpose (cf. Trondsen 1976).[11] In any case, patterns of expectation and irritation as a result of transgressions can be identified. The fact that there are behavioural norms in art organisations is also confirmed by surveys of theatres. For example, 91.2% of those surveyed at the Mousonturm in Frankfurt and 96.5% of those surveyed at Schauspiel Frankfurt stated that rules of decency apply in the theatre. This includes clapping at the end of the performance, being quiet and avoiding excessive alcohol consumption. It is revealing, however, that from what was said during the ethnographic discussions in the theatres and museums, almost no one knew the formal house rules. As a rule, they only assumed that there would be house rules.

[11] Following Erving Goffman (1986), these cultural behaviours can also be understood as ritualisations. See also in this context: Metzger 1998; Rösing and Barber-Kersovan 1994.

However, the existing possibilities of theatres and museums such as the way in which objects are presented, opening hours or the control and supervision by staff are not sufficient to explain why visitors behave similarly and orient themselves towards the "withdrawal of the body" and highly cultural standards of decency. For example, the organisation cannot insist on applause or sanction transgressions. Historical studies also show that the influence of art organisations on the behaviour of visitors is very limited (e.g. Dreßler 1993; Höhne 2016; Müller 2006). For example, the numerous police regulations in the 18th and early nineteenth centuries, which prohibited disapproval and expressions of displeasure by theatregoers during performances in the presence of police officers, show that art organisations alone are hardly in a position to regulate behaviour adequately (Dreßler 1993). If, as a result of the limited tools at their disposal, art organisations cannot adequately regulate the behaviour of their visitors, there must be other reasons for why their behaviour is the way it is. Accordingly, I next elaborate on the thesis that art organisations are dependent on the performance of other social sectors such as school and family in order to minimise disturbances and unexpected behaviour.

The first indicators of this can be derived from the previous audience research as well as the statistical evaluation of the quantitative survey. For example, sociological analyses document that the art public is characterised above all by a high level of formal education, is on average older, and that women are usually disproportionately represented (cf. e.g. Reuband 2018; Rössel et al. 2005; Schulze 1992; Steuerwald 2012).[12] The results of the visitor survey in the museums in Oldenburg confirm these findings. There, 50.14% of the interviewees had a university degree. When one includes general higher education entrance qualifications and technical college entrance qualifications, about two thirds (67.79%) of the surveyed visitors had a high degree. About two thirds of the visitors are female, the average age is 49.9 years and 28% are retired or pensioned. It is also interesting to note that one third of those surveyed said that their interest in art did not develop until they became older, and that just under half of those surveyed did rather well in their last school year in art classes, compared with 14% who did rather badly. As Bourdieu (e.g. 1999) has also shown, the high importance of formal education as well as the importance of art education allow the conclusion to be drawn that how people get on in school has an effect on the likelihood of their visiting art organisations. Taking into account studies in school sociology which show that pupils acquire not only knowledge but also behaviours at

[12] See for international studies about visitors Falk 2015.

school which are connected to norms and which, among other things, are related to modes of experience and perception, one can identify a first possible explanation of the homogenisation of behaviours in theatres and museums (e.g. Dreeben 1980; Parsons 1968). The "withdrawal of the body" as an adequate way of experiencing and behaving in art organisations could thus be understood as a side effect of school learning.

However, statistical evaluation also shows that the frequency of one's own visits to a museum depends on the extent to which one was in the museum with one's parents as a child. Put simply, this means that people whose parents used to go to museums with them are more likely to go to museums. Accordingly, family has a special significance. If you go to a museum with your parents during your childhood, you acquire certain behavioural, experiential and perception skills relatively early on and you develop an increased interest (measured by the frequency of visits).[13]

Finally, historical studies investigating behavioural patterns in theatres document that from the end of the 18th century onwards an increasing disciplining of visitors' behavioural patterns could be seen in theatres, which harmonised the diversity of behavioural patterns and led to a change from participants to listeners and spectators (e.g. Dreßler 1993; Höhne 2016; Müller 2006). While the numerous efforts of the theatres to educate their audiences were only partially successful, for example by prohibiting access to the stage, placing seating in the stalls or darkening the auditorium, from the 19th century onwards an increasing disciplining and standardisation of the visitors to the auditorium could be observed, which, just like the efforts of the theatres, was a consequence of overall social processes of civilisation and disciplining, modernisation and bourgeoisiation (Elias 1976, 1998; Foucault 1994; Steuerwald 2016). Höhne (2016), for example, uses books on decency and manners in the form of encyclopaedias of conversations, which cite behaviour patterns that are adequate, situation-specific and regarded as correct and appropriate, to demonstrate that the demands and expectations placed on behaviour patterns in theatres have changed greatly since the 19th century and increasingly converge with bourgeois standards of decency and behaviour. For example, Spemanns *goldenes Buch der Sitte*, which is about reasonable behaviour in theatres says: "Laugh and weep not in a conspicuous and exaggerated way. A man who is civilised is expected to be able to control his passions and feelings. This means that he knows how to be moderate in joy and grief,

[13] Whether a higher level of artistic competence, understood as a broad knowledge of art history and contemporary art, can be demonstrated in this context cannot yet be answered.

and that he controls his irascibleness and his other feelings" (Baudissin 1901, quoted in Höhne 2016, p. 155). In addition to these references to the "withdrawal of the body", which, as Norbert Elias (1998) vividly demonstrates through historical sources, gradually develops into a status-high characteristic in courtly society from the early modern period onwards and spreads throughout society with a time lag, the books on decency and manners also document above all modes of behaviour which are intended to make possible an undisturbed and contemplative way of experiencing what is offered. Accordingly, there are prohibitions on talking or eating during the performance and, for example, a request not to use the opera glasses to "admire the ladies' cleavages" (Smolka 1957, quoted in Höhne 2016, p. 160; my translation).

6.4 Concluding Remarks

The aim of the present study was to examine the problem of how art organisations, despite limited access as a result of non-memberships, succeed in influencing the behaviour of visitors in order to minimize possible disruptions of the interaction order. As the empirical investigations of the norms of interaction in theatres and museums show, behaviour patterns can be observed which, depending on age, are more or less oriented towards a "withdrawal of the body" as well as towards highly cultural rules of behaviour, so that behaviour patterns are largely adapted to the requirements of the art organization and disturbances rarely occur. In addition, the empirical studies indicate that the behaviours which can be observed are mediated through socialization processes at school and in the family, which are to be understood as the consequences of processes of civilization and discipline in society as a whole, which ultimately, from the end of the 18th century onwards, brought about an approximation of behaviours and status differences and homogenized the audience. This led, among other things, to the fact that, as the crisis experiments documented, sanctioned behavioural expectations developed which are sanctioned not only by the organisation but also by other visitors. Overall, the findings indicate that the existing tools at the disposal of art organisations are not sufficient to achieve this, so that in order to ensure adequate functioning of art exhibitions and productions, they must take advantage of what has been brought about through other social areas such as family and school. In this respect, art organisations are parasitic organisations.

References

Bab, Julius. 1931. Das Theater im Lichte der Soziologie. Leipzig: Hirschfeld.

Behnke, Christoph. 2012. Gründe für den Besuch von Ausstellungen und Fragen der Kunstvermittlung. In Das Kunstfeld. Eine Studie über Akteure und Institutionen der zeitgenössischen Kunst, Eds. Munder, H. and Wuggening, U. Zürich: jrp|ringier.

Bensman, Joseph and Gerver, Israsel. 1963. Crime and Punishment in the Factory. The Function of Deviancy in Maintaining the Social System. American Sociological Review 28, pp. 588–598

Bourdieu, Pierre. 1974. Elemente zu einer soziologischen Theorie der Kunstwahrnehmung. In Zur Soziologie der symbolischen Formen. Ed. Bourdieu, P. Frankfurt: Suhrkamp.

Bourdieu, Pierre. 1999. Die feinen Unterschiede. Kritik der gesellschaftlichen Urteilskraft. Frankfurt: Suhrkamp.

Bourdieu, Pierre and Alain Darbel. 2006. Die Liebe zur Kunst. Europäische Kunstmuseen und ihre Besucher. Konstanz: UVK.

Coser, Lewis A. 2015. Gierige Institutionen. Soziologische Studien über totales Engagement. Berlin: Suhrkamp.

Daniel, Ute. 1995. Hoftheater. Zur Geschichte des Theaters und der Höfe im 18. und 19. Jahrhundert. Stuttgart: Klett-Cotta.

Dreeben, Robert. 1980. Was wir in der Schule lernen. Frankfurt: Suhrkamp.

Dreßler, Roland. 1993. Von der Schaubühne zur Sittenschule. Das Theaterpublikum vor der dritten Wand. Berlin: Henschel.

Elias, Norbert. 1976. Über den Prozeß der Zivilisation. Soziogenetische und psychogenetische Untersuchungen. Zweiter Band: Wandlungen der Gesellschaft. Entwurf zu einer Theorie der Zivilisation. Frankfurt: Suhrkamp.

Elias, Norbert. 1998. Über den Prozeß der Zivilisation. Soziogenetische und psychogenetische Untersuchungen. Erster Band: Wandlungen des Verhaltens in den weltlichen Oberschichten des Abendlandes. Frankfurt: Suhrkamp.

Endruweit, Günter. 2004. Organisationssoziologie. Stuttgart: Lucius & Lucius.

Falk, John. 2015. Visitor Studies. Encyclopedia of Science Education, pp. 1097–1101.

Foucault, Michel (1994). Überwachen und Strafen. Die Geburt des Gefängnisses. Frankfurt: Suhrkamp.

Garfinkel, Harold. 2003. Studies in Ethnomethodology. Cambridge: Polity Press.

Gebhardt, Winfried and Arnold Zingerle. 1998. Pilgerfahrt ins Ich. Die Bayreuther Richard Wagner-Festspiele und ihr Publikum. Eine kultursoziologische Untersuchung. Konstanz: UVK.

Glogner-Pilz, Patrick and Patrick Föhl (Eds). 2016: Handbuch Kulturpublikum. Forschungsfragen und -befunde. Wiesbaden: Springer VS.

Goethe, Johann Wolfgang von. 1988. Briefe Bd. 2 (1786–1805) (Hamburger Ausgabe in 6 Bänden). 3. Aufl. München: Beck.

Goffman, Erving. 1973. Asyle. Über die soziale Situation psychiatrischer Patienten und anderer Insassen. Frankfurt: Suhrkamp.

Goffman, Erving. 1982. Das Individuum im öffentlichen Austausch. Mikrostudien zur öffentlichen Ordnung. Frankfurt: Suhrkamp.

Goffman, Erving. 1986. Interaktionsrituale. Über Verhalten in direkter Kommunikation. Frankfurt: Suhrkamp.

Graf, Bernhard and Heiner Treinen. 1983. Besucher im Technischen Museum. Zum Besucherverhalten im Deutschen Museum München. Berlin: Gebr. Mann.

Höhne, Steffen. 2016. Kulturpublikum. Zur Genese und Transformation des bürgerlichen Konzert- und Theaterpublikums seit der Aufklärung. In Handbuch Kulturpublikum. Eds. Glogner-Pilz, P. and Föhl, P. Wiesbaden: Springer VS.

Jenniches, Karl M. 1969. Der Beifall als Kommunikationsmuster im Theater. Kölner Zeitschrift für Soziologie und Sozialpsychologie 21, pp. 569–584.

Kirchberg, Volker. 2005. Gesellschaftliche Funktionen von Museen. Makro-, meso- und mikrosoziologische Perspektiven. Wiesbaden: Verlag für Sozialwissenschaften.

Klein, Hans-Joachim. 1990. Der gläserne Besucher. Publikumsstrukturen einer Museumslandschaft. Berlin: Staatliche Museen Preußischer Kulturbesitz.

König, Gudrun M. 1996. Eine Kulturgeschichte des Spazierganges. Spuren einer bürgerlichen Praktik 1780–1850. Wien et al: Böhlau.

Kühl, Stefan. (2011). Organisationen. Eine sehr kurze Einführung. Wiesbaden: Springer VS.

Luhmann, Niklas. 1976. Funktionen und Folgen formaler Organisation. Berlin: Duncker & Humblot.

Luhmann, Niklas. 2016. Spontane Ordnungsbildung. In Der neue Chef. Ed. Kaube, S. Frankfurt: Suhrkamp.

Martin, Uta. 1998. Typologisierung des Theaterpublikums. Das Erkenntnispotential der verhaltensorientierten Marktsegmentierung für das Marketing öffentlich-rechtlicher Theater. Dresden: Technische Universität Dresden.

Mayntz, Renate and Rolf Ziegler. 1977. Soziologie der Organisation. In Handbuch der empirischen Sozialforschung. Ed. König, R. Band 9. Stuttgart: Enke.

Metzger, Heinz-Klaus. 1998. Rituelle Aspekte des bürgerlichen Musiklebens. In Musik und Ritual. Eds. Barthelmes, B and La Motte-Haber, H de. Mainz: Schott.

Müller, Sven Oliver. 2006. Distinktion, Demonstration und Disziplinierung: Veränderungen im Publikumsverhalten in Londoner und Berliner Opernhäusern im 19. Jahrhundert. International Review of the Aesthetics and Sociology of Music 37, pp. 167–187.

Neckel, Sighard. 1991. Status und Scham. Zur symbolischen Reproduktion sozialer Ungleichheit. Frankfurt. New York: campus.

Parsons, Talcott. 1968. Die Schulklasse als soziales System. Einige ihrer Funktionen in der amerikanischen Gesellschaft. In: Sozialstruktur und Persönlichkeit. Ed. Parsons, T. Frankfurt: Europäische Verlagsanstalt.

Powell, Walter W. and Paul J. DiMaggio (Eds). 1991. The New Institutionalism in Organizational Analysis. Chicago: Chicago University Press.

Preisendörfer, Peter. 2005. Organisationssoziologie. Grundlagen, Theorien und Problemstellungen. Wiesbaden: VS.

Renz, Thomas. 2016. Nicht-Besucherforschung. Die Förderung kultureller Teilhabe durch Audience Development. Bielefeld: transcript.

Reuband, Karl-Heinz. 2018. Das Kulturpublikum im städtischen Kontext. Wie sich das Publikum der Oper von anderen Kulturpublika unterscheidet. In Oper, Publikum und Gesellschaft. Ed. Reuband, K-H. Wiesbaden: Springer VS.

Rosa, Hartmut. 2010. Beschleunigung. Die Veränderung der Zeitstrukturen in der Moderne. Frankfurt: Suhrkamp.

Rösing, Helmut and Alenka Barber-Kersovan. 1994. Konzertbezogene Verhaltensrituale. In Handbuch der Musikpsychologie. Ed. Bruhn, H München, Wien, Baltimore: Urban & Schwarzenberg.

Rössel, Jörg, Hackenbroch, Rolf and Angela Göllnitz. 2002. Die soziale und kulturelle Differenzierung des Hochkulturpublikums. Sociologia Internationalis 40, pp. 191–212.

Rössel, Jörg, Rolf Hackenbroch, and Angela Göllnitz. 2005. Soziale Differenzierung und Strukturwandel des Hochkulturpublikums. In Jahrbuch für Kulturpolitik 2005 – Thema: Kulturpublikum. Ed. Institut für Kulturpolitik der Kulturpolitischen Gesellschaft. Essen: Klartext

Schulze, Gerhard. 1992. Die Erlebnisgesellschaft. Kultursoziologie der Gegenwart. Frankfurt. New York: Campus.

Robinson, Edward Stevens, Sherman, Irene Case and Lois E. Curry. 1928. The behaviour of the museum vistor. Washington D.C.: American Association of Museums.

Steuerwald, Christian. 2012. Mit Goffman im Theater. Inszenierungsweisen von Kunstorten und ihren Besuchern. In Perspektiven der Kunstsoziologie. Praxis, System, Werk. Eds. Steuerwald,C and Schröder, F. Wiesbaden: Springer VS.

Steuerwald, Christian. 2016. Die Sozialstruktur Deutschland im internationalen Vergleich. Wiesbaden: Springer VS.

Tacke, Veronika. 2015. Formalität und Informalität. In Formalität und Informalität in Organisationen. Eds. von Groddeck, V and Wilz, S M. Wiesbaden: Springer VS.

Tröndle, Martin. Greenwood, Steven , Kirchberg, Volker and Wolfgang Tschacher 2014. An integrative and comprehensive methodology for studying aesthetic experience in the field: Merging movement tracking, physiology, and psychological data. Environment and Behavior 46, pp. 102–135.

Trondsen, Norman. 1976. Social Control in the Art Museum. Urban Life 5, pp. 105–119.

vom Lehn, Dirk. 2006. Die Kunst der Kunstbetrachtung. Aspekte einer pragmatischen Aesthetik in Kunstausstellungen. Soziale Welt 57, pp. 83–99.

Zahner, Nina Tessa. 2015. Das Laienpublikum als Herausforderung für die Feldanalyse. In Kunst und Öffentlichkeit. Ed Danko, D et al. Wiesbaden: Springer VS.

Part II
The Arts and the Power of Social Structures

Equality, History, Tradition: Gender-Political Issues in the Oxford Collegiate Choral Scene

7

Sigrun Lilja Einarsdottir

Abstract

The focus of this chapter is the status and experience of female singers within the historically male-dominated collegiate choral tradition in Oxford. The collegiate tradition is directly linked and related to the cathedral choral tradition within the Church of England, which up until the early 1990s, was not accessible to girls and women. Until 1850, the University of Oxford and its colleges were primarily monastery-like religious institutions where fellows (all men) were forbidden to marry and expected to live in celibacy. Since the late 1870s, women had permission to read and attend classes at Oxford, but the University excluded women as fellows and students until 1920. Until the early 1990s, when girls were first admitted as choristers, the musical life at Oxford colleges and halls was historically and exclusively male-oriented, with men in the back rows (tenors, basses and counter tenors); and a boy treble line in the front rows, mirroring the cathedral choral tradition within the Church of England. During the last decades, the Oxford collegiate choirs have undergone drastic changes, with the establishment of mixed voice choirs. This text will present findings of a music-sociological study on the Oxford collegiate choral tradition, where the focus is on three main pillars. Firstly, the issue of aesthetics and preservation of the all-male voice choirs as a cultural phenomenon; secondly, the contradictory position of female sopranos and altos versus boy trebles and counter tenors; and thirdly, the institutional-based gender discrimination and lack of performance, training and career opportunities for women within the collegiate and cathedral choral traditions.

S. L. Einarsdottir (✉)
Bifröst University, Borgarnes, Iceland
e-mail: sigrunlilja@bifrost.is

L. Gaupp et al. (eds.), *Arts and Power*, Kunst und Gesellschaft,
https://doi.org/10.1007/978-3-658-37429-7_7

Keywords

Collegiate choirs · Gender politics · Oxford · Choral tradition · Gendered musicality

7.1 Introduction

Oxford is not only known for its magnificent architecture and academic excellence; it is also a city with a thriving cultural life and a particularly rich music tradition. The history of Oxford is also very much a history of political and socio-economic power in the United Kingdom (UK) (see Brennan 2004; Trow 2007). Although the University of Oxford has a strategy in place for increasing diversity among its students and staff, historically it was first and foremost an elite educational institution, fostering the education of the ruling class (and still criticised as such), not only in the field of politics but also in the field of culture. Today, the University of Oxford and its colleges and halls consist of people of all genders; however, this institution (the oldest university in the English-speaking world) is historically a male-oriented dominion. Music plays a predominant role in the Anglican collegiate chapels, where most of the college choirs can perform up to seven to eight times a week.[1]

The presence of women in Oxford colleges and halls is a relatively recent phenomenon, dating back to the late 1870s, when women had permission to read and attend classes at Oxford—or, more correctly, stand in the hallway listening to the lectures through closed doors, as it was feared that their presence would cause "unacceptable distractions" to their male counterparts. The University excluded women as fellows and students until 1920 and they were not allowed to obtain degrees until 1941. To secure the segregation of genders, women's colleges were established in Oxford. In addition, there was a systematic strategy within the University of keeping the women colleges poor and thus maintaining, as newspaper headlines indicated, a "sex war at Oxford". One of the main reasons given was the scarcity of resources, library spaces and limited lodgings (Harrison 1994, pp. 346–359).

A graduate from Lady Margareth Hall (in 1958) commented that "in my days going to Oxford as a female was like being on the 'side-lines' of a giant male

[1] This varies between colleges.

public school" (Harrison 1994, p. 355). The great Virginia Wolff concluded that the poverty of the women's colleges was a result of "the reprehensive poverty of our sex" (Harrison 1994, p. 355). The gender ratio did not start to change until the first men's colleges accepted women in 1974 (Harrison 1994, p. 355) and from 2008, all colleges and halls (apart from Blackfriars) have been mixed. This environment of segregation was reflected in the collegiate choirs at Oxford (Harrison et al. 2012a, b).

As within other cathedral choirs in the UK, choristerships[2] in collegiate choirs were only available to boys (aged 6–13). It was not until in 1991 that the Salisbury Cathedral took the initiative and recruited girl choristers for the first time ever. This following a heated debate in terms of preservation and aesthetic politics on the "destruction of a tradition and an elimination of a particular sound of the all-male voice choirs" (Mould 2007, pp. 267–268). The same development occurred within the collegiate choirs in Oxford with all choirs being mixed, except the three choral foundations of Oxford: Magdalen College Choir, Christ Church Cathedral Choir, and the Choir of New College.

The arguments for the continued segregation and gender-based discrimination in the abovementioned choral foundations indicates a historical, yet conservative, conservation of a tradition derived from a centuries' old, male-oriented power structure. Power is an inevitable dimension in this debate. Power tends to depend on the permanent power structures that have the tendency to defend the existing status quo (see Foucault 1995/1975). The cathedral and collegiate choral tradition in England are examples of such a structure, which determines the access and opportunities (or lack thereof) for women as choristers and choral scholars.

Furthermore, the unique sound of the boy trebles and counter tenors is, by many, considered a cultural phenomenon, based on the assumption that the sound of girls and boys voices are supposedly different (which is based on the anatomy of the voice) as well as the sound of a female alto vs counter tenors. These arguments have been criticised in the public debate which claims that the unique sound of the boy trebles is a result of intensive musical training and if boys and girls are put together these differences are not noticeable (see Howard et al. 2002; Sergeant et al. 2005). A longitudinal research conducted by Graham Welch, David Howard, and other scholars on girl choristers at Wells Cathedral

[2]A position for a boy aged 6–13, attending a cathedral school or a school attached to an Oxbridge college (boarding school or day school), receiving intensive musical training in the form of daily rehearsals, music lessons and singing at chapel / cathedral services up to six to seven times a week during term time (see Mould 2007).

demonstrates that girls (when provided with the same vocal training as boys) do not only sing like girls; they can also imitate the boys (see Howard et al. 2002; Welch et al. 2002; Welch 2017).

Given the fact that the existing tradition is historically all-male choral, the main question is how it affects the female singers in Oxford collegiate choirs. Nationwide, many choral singers and conductors that the researcher has spoken to throughout the years, constantly address their worries about the decline in the number of choral singers—and particularly men in mixed-voice amateur choirs. Ashley (2009) suggested that the presence of girls would "disrupt" the boys in mixed children's choirs, a similar argument that is used to argue against mixed school education.

The context of girls singing within the English choral tradition is a relatively recent topic and the gender-based exclusion of women and girls in the choral foundations is still a topic of political debate. Despite this fact, these issues have received limited scholarly attention and no academic book or paper is known of at the time of this publication that deals with gender issues within the collegiate choral context in the UK.

Present literature on Oxford choirs seems to be very scarce and primarily of historical nature (see Mould 2007; Burgess and Heale 2008; W. Tuckwell 1900; L. S. Tuckwell 1913; Judge 1986). Further historical writings are doubtless to be found in biographies and autobiographies of Oxford alumni, as well in obituaries. Boys and men singing in choirs is a topic that has received some scholarly attention (see Ashley 2009; Welch 2017; Freer 2009a, b; Elorriaga 2011; Harrison et al. 2012a, b; Spiegel et al. 1997) but very little exists on male choirs in the collegiate choral scene.

This chapter aims at observing the perceptions of choral members on gender-political issues, exclusiveness, and systematic gender discrimination in terms of performance and the training opportunities in choirs associated with Oxford colleges and halls and linked to the University of Oxford. The research question is the following: Is there an institutional and systematic gender-discrimination ingrained in the Oxford collegiate choral scene, and how is it perceived by the choral singers?

7.2 Methods and Participants

The methodological approach of this study was based on grounded theory, which consists of systematic collection and analysis of empirical data to construct theories "grounded" in the data themselves—hence, the analysis of the data generates the concepts constructed. Instead of defining and writing a massive literature review, the researcher starts by going out in the field to collect data (Charmaz 2006; Glaser and Strauss 1967).

Data collection consisted of in-depth, semi-structured individual interviews with ten choir members from seven choirs within Oxford colleges and halls; five focus group interviews with choir members (six to ten individuals in each group) and a paper-based survey (Gable 1994; Schwarz et al. 1998) in which 16 choirs with 30 members took part. Of 344 members of the choirs, 233 responded (68 % participation rate).

In terms of gender, 125 (or 54 %) identified as female, 103 (or 44 %) identified as male and four (or 2 %) identified as other. This gender ratio was in line with the University of Oxford Admission Statistics (2010–2016). The average age was 23,56 years and 80 % of the participants were from the UK (compared to 64 % student ratio at the UO), 8 % from other EU countries and 11 % from other non-EU countries.[3] Looking at the gender differences, there was a higher ratio of UK men than women, whereas 26 % of women come from outside the UK but only 14 % of the men come from outside the UK.

Little over half of respondents were academically affiliated with the Humanities division of the University of Oxford. Inquiring about general choral participation, 63 % of respondents only sing in one choir and 36 % sing in other choirs (8 % in another college choir, 19 % in another non-college choir and 10 % in both a college choir and a choir not linked to a college).

The overwhelming majority of participants have learned to play musical instruments, or 93 % and 77 % of them mentioned piano as one of the instruments they have learned to play and 29 % mentioned the violin. Around one-third (33 %) had private singing lessons in secondary school and one-third sings in more than one choir (one soprano sung in 5 choirs).

Overall, 29 % (N=233) of respondents had a chorister's background. More men than women had chorister background and 58 % of choristers had independent school education whereas 36 % were state school educated and 6 % were educated in other types of educational institutions. Cross tabulation between school types and gender indicates a slightly higher ratio of independently educated males.

Data collection took place from October 2016 until April 2017. Participation in the study was anonymous and in accordance with the Data Protection Act (Data Protection Act 2018).

---

[3]Average ratio 2010–2016—University of Oxford Academic Administrative Division—Student Statistics 2017.

7.3 Aesthetics and Equal Access versus Preservation of a Tradition

The three choral foundations (Christ Church Cathedral Choir, Choir of New College and Choir of Magdalen College—male voice choirs with boys in soprano and alto voice pars) maintain the tradition of the all-male voice choir and were frequently mentioned and discussed in the interviews in relation to gender politics. Furthermore, these three choirs were also referred to as the most prestigious choirs in the Oxford collegiate context (and even on a national level) and were not accessible for women or girls in 2016–2017.

In the survey, participants were asked to express their views about the current situation regarding gender politics. Fig. 7.1 demonstrates results from the survey about respondent's views on four situational statements regarding the status and role of the three choral foundations.

Apparently, there is a strong disagreement among respondents regarding preferences in terms of the sound of an all-male voice choir vs a mixed-voice choir. However, the respondent's perspectives towards the status of choral foundations were rather mixed but most respondents agreed on equal opportunities for women.

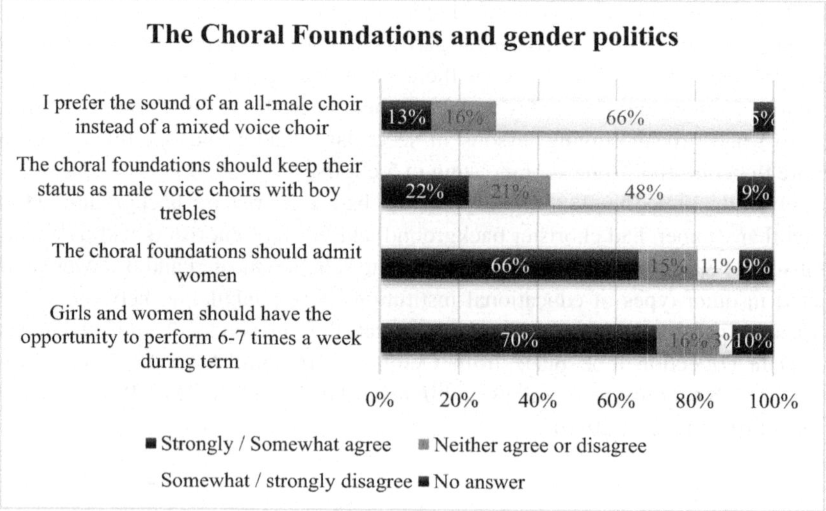

Fig. 7.1 The Choral Foundations and gender politics

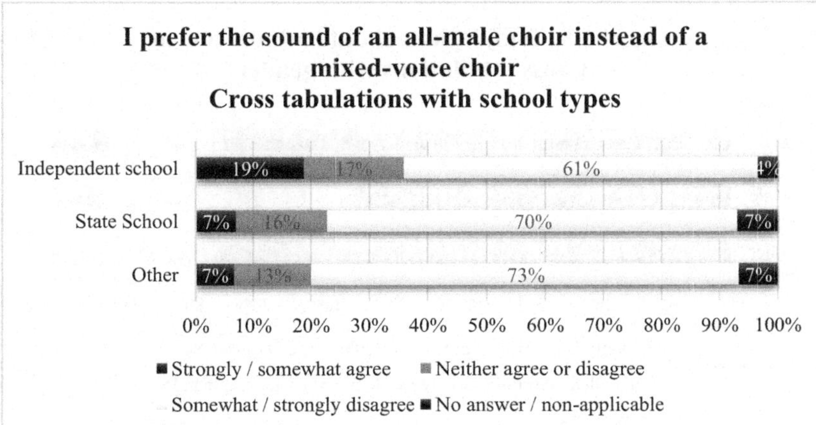

Fig. 7.2 I prefer the sound of an all-male choir instead of a mixed-voice choir (Cross tabulations with school types)

Cross-referencing the situational statement with school types yielded a significantly stronger opposition to the sound of the male voice choir sound among respondents that came from state schools (before attending university) than independent schools (see Fig. 7.2), and a significantly higher ratio of women agreed that the choral foundations should admit women (see Fig. 7.3).

Figure 7.4 presents the situational statements in terms of the controversial position of female adult sopranos versus the traditional boy trebles. Looking at Pearson's correlation between these statements, a significant positive correlation ($r=.777$) was found between the statements on the treatment of sopranos like boy trebles and that sopranos may tend to emulate boy trebles. No other medium or significant correlation was found between these statements and other statements in the survey, indicating how controversial this issue is. The first statement on Fig. 7.4 deals with the allegation of whether young women with a particular sound in their voice would be more likely to achieve a choral scholarship. Twenty percent strongly or somewhat agreed, but over 50 % did not take a stand. However, it must be noted that almost one third of the respondents neither agreed nor disagreed with this assumption.

The existence of the counter tenors is based on the allocated positions in cathedral and collegiate choirs that are earmarked for men; where they might be chosen over a female alto, regardless of vocal quality and musical talents.

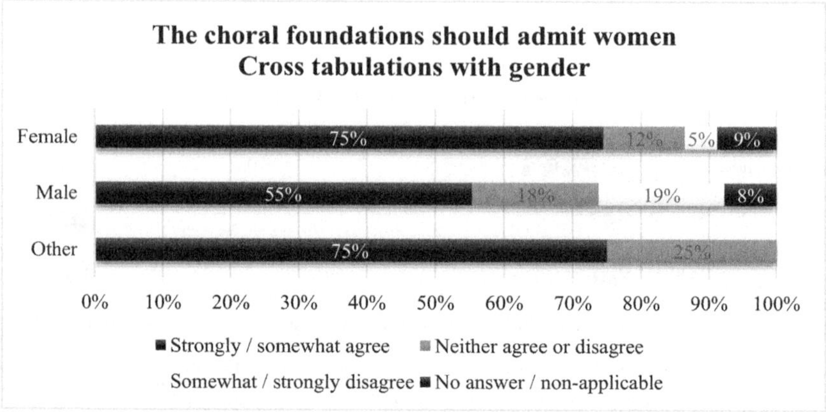

Fig. 7.3 The choral foundations should admit women (Cross tabulations with gender)

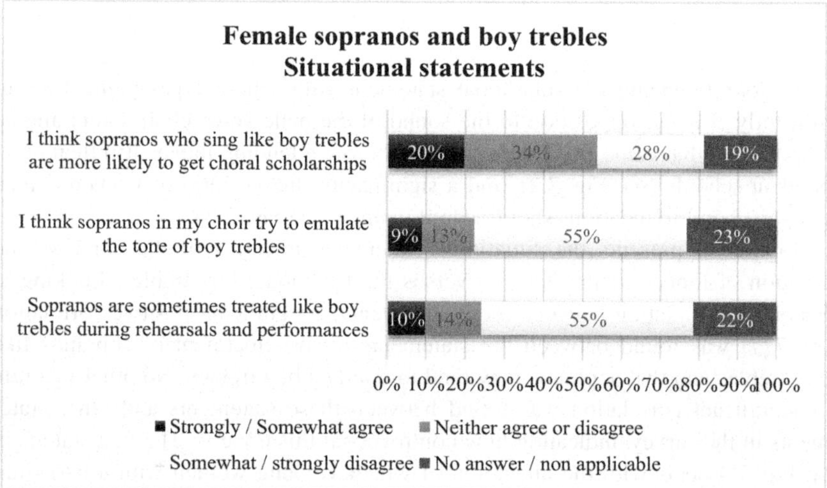

Fig. 7.4 Female sopranos and boy trebles—situational statements

This theme was one of the most predominant aspects of the discussion in the interviews in terms of gender politics.

Figure 7.5 represents findings of situational statements regarding the interesting conflicts and clashes in terms of the historic references of the counter tenor

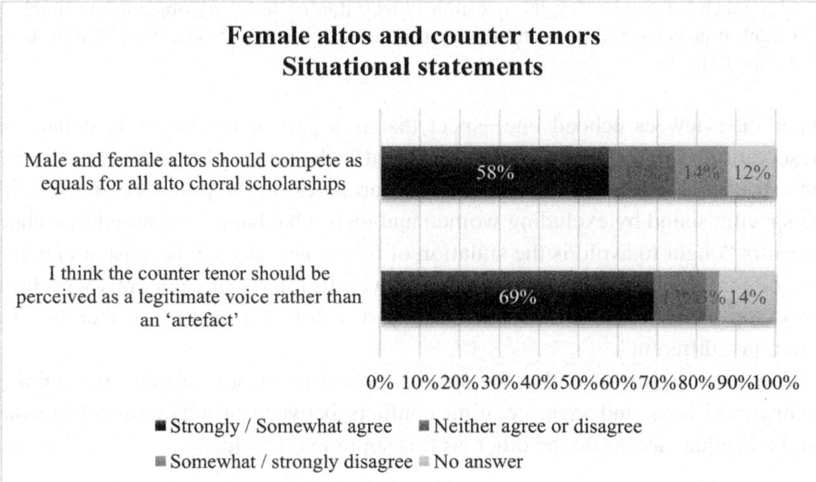

Fig. 7.5 Male / female altos—situational statements

and the presence (and rights) of the female alto voice. Apparently, respondents were rather mixed in their views towards the equal rights of male and female altos, 58 % agreed that they should compete for all alto choral scholarships on equal basis. However, 69 % of respondents agreed that the counter tenor should be viewed as a legitimate voice.

Some scholars (such as Ashley 2009) have claimed that the presence of girls may have a negative effect on boys in choirs, lead to a significant shortage of men in mixed choirs, and that the girls would take over. This discussion was also evident among some of the music directors. Inquiring on this matter, one of the counter tenors who had previous experience as a chorister commented on this issue, that boys might have difficulty in singing with girls and then the girls would take their roles when the boy's voices break (Counter tenor).

Throughout the data collection, many of the participants—and other people connected to the university—claimed to be conflicted about access to choristerships versus the historic gender-discriminating tradition:

> I don't necessarily resent the tradition for not allowing me, because the tradition is a beautiful thing aesthetically, the way it has been set up. I don't have any objections to the inequality of it, partially because I'm a conservative but at the same time I'm

> very much a feminist. So, the question is why don't I have any objection to that? I think it does have to do with the way it is set up—it's so aesthetically pleasing. It works. (Alto 1)[4]

Other interviewees echoed one aspect that is a part of the on-going debate on preservation vs equal opportunities—and that is the aspect of the sound of the all-male voice choir. There are two different narratives on the politics of preserving the specific sound by excluding women and girls. One bass (1) claimed that choir directors "ought to avoid is the situation of blending too much because it can create quite a weird sound. It tends to [blend weirdly] in lot of cathedrals who have boys' and girls'"—but bass 2 said: "It's just a different quality of sound— not better, just different".

The other argument is that the specific sound is a result of intensive training exclusive to boys and men, creating conflicts between preservation on the one hand and equal access on the other, as this soprano refers to:

> I feel like there could be a conflict of interest for me of the whole idea of boys and men's choir because it is a problem of equal opportunities, like centuries of sexism but the choirs are so beautiful … the impression I get on the institutional level is like there's much less effort, less training, less of tendency to put quality directors into the mixed choirs. (Soprano 1)

The boy trebles attract visitors and tourists to Oxford, and some of the choral singers interviewed in this study (particularly in the auditioned choirs) stated that this tradition is worth preserving and is important for the community and the music scene as well, so the choral foundations should stay as they are. The politics of preservation seem to cause this inner conflict in the minds of the participants, between equal access versus tradition:

> I think that a part of that whole psyche that men don't do arts, they do sciences and sport and the men that do arts … I think choirs are one of the few places where it's preserved—it's a masculine thing to do. At the same time, on the outside it's not. (Tenor 1)

> I guess it's hard to avoid when the foundations have been around for hundreds of years and in 1550 it would have been ridiculous to propose that women should sing in the choir. (Soprano 2)

[4] All quotes from singers are from personal communication and interviews, conducted from October 2016—March 2017.

Despite the presumable reluctance of changing this tradition within the choral foundations in Oxford, girls are being admitted into the cathedrals:

> When you have boys' and a girls' choir that automate—everyone benefits. The boys get a better musical training since they have more time to rehearse and actually get that kind of experience ... I find that there is a lot of even people that I wouldn't expect, have this kind of a prejudice against women singing in chapel choirs because of this idea of that they've never had chorister's training … their reading is not that good. For a start, a lot of men didn't either … and this idea that you can only become a good sight reader from being a chorister. (Tenor 2)

The tenor's perspective above is a point often mentioned in informal conversations. The tremendous workload of singing every day, up to two to three times a day may be considered (by some) to be quite a lot for a young voice, while others consider this intensive rhythm in training to be hugely beneficial in the long run.

Since choristerships are increasingly becoming accessible to girls, one of the music directors claimed that as more and more women become music directors in the near future, the women who used to be choristers are increasingly gaining choral scholarships and pursuing long-term music education on a university level—that "it's about to happen" (music director). As a result, not only is it important to have more women within the most influential choirs as singers (and therefore attempting to correct the almost thousand-year-old gender-based discrimination) but also as music directors, conductors and professional musicians in general. However, as previously addressed, the Anglican choral tradition (particularly within the cathedral choirs on national level) still offer less possibilities for girls and women to seek a music career within professional, cathedral choirs. This narrative is quite well-known in the fight for gender equality—that women must be "patient" and things will "happen automatically". In the Oxford context, participants had their doubts, whereas the hindrances for women and girls are still quite tangible within the collegiate choral scene.

7.4 Gendered Musicality and Institutional Obstacles

In discussions with music directors in the choral foundations, they emphasized the importance of role models for the boys—as they look up to the undergraduates and lay clerks and in many cases boy choristers continue on to become choral scholars later—and some of the interviewees partially agreed with this viewpoint.

Therefore, the aesthetic statement regarding the particular sound of the boy's voices seems to be somewhat irrelevant, so why the reluctance? This brings up another controversy:

> Why not ask [other] choirs … to do more services? They can, but they don't want to—or there's no demand. (Tenor 3)

The lack of institutional interest seems to be on the top of interviewees' minds in terms of opportunities for women. One of the interviewees pointed out that the colleges that host the choral foundations are among the richest colleges in Oxford, so the issue of funding (and access to it) plays a role in the decision-making on how much should be spent on choral activities. Furthermore, the Oxford context has a widespread image of institutions that are not overtly keen on radical changes and tend to cling to their history and traditions (see Trow 2007), "fostering a past tradition that is at odd with the general trend" (Tenor 2) and not "a place that is renowned for its course for change" (Tenor 3). There seems to be little to no studies or research on the admittance of girls into cathedral and Oxford collegiate choirs; their circumstances, opportunities, and the scope of gender discrimination are still prevalent, despite the presence of girl choristers since the early 1990s. Little is known about the general, institutional attitudes towards the gender equality aspect in terms of opportunities:

> It seems to me to be an issue which is sort of oh we'd like to have girls in the top line but that would require changes in the school and its sort of circular problem. (Bass 1)
>
> You cannot escape to notice that almost every other cathedral in the country now has girls in the front row and therefore I think it is a conscious policy to keep the boys only—for whatever reason. (Tenor 3)

These arguments are frequently heard. Construction costs and difficulties in putting up girl's toilets and facilities are used as the main arguments for the reluctance to change and grant equal access to all children willing to obtain intensive musical training, regardless of gender. However, in conversations with fellows, students and MDs in Oxford, colleges seem to be under political pressure to increase equality, not only in terms of gender, but also of socio-economic and ethnic backgrounds. Interviewees frequently mentioned the gender-based lack of performance opportunities and access to training. Tenor 3 claimed that the choirs needed to be mixed if they intend to be inclusive or to make it possible for girls to become choristers at the choral foundations and to be able to access the level of musical training needed. It is far easier for men to get an excellent vocal education in a professional choir at Oxford than for women (Bass 1), and singing every day, like in the choral foundations, is not available for women—the biggest commitment they can get is singing three to four times a week in chapel ceremonies (Female soprano 2):

> Even though I knew I was a competent singer and that I had experience doing that type of singing, I would have actively avoided applying to the big three foundation colleges; not because I didn't think I was up to it. It was simply because I wanted also to have a life outside singing. (Counter tenor 1)

In short, the choices are there if you are a man; but not if you are a woman. A higher ratio of female participants sings in more than one choir and they need to do so if they want to have a similar amount of performance opportunities as their male counterparts in the choral foundation. This kind of intensive training is associated with a higher prestige and having access to such a workload is considered the pathway to more high-level musical training. One of the sopranos claimed that she was singing in five choirs (thereof two college chapel choirs) and alternating seems to be quite common, where students substitute for one another. One of the music directors said that from one week to another, it was rare to get the exact same choir as the week before. Consequently attitudes towards the workload were somewhat controversial, some of the interviewees did not want to sing more than three times a week (Female alto 2):

> … but sometimes I do think I'm not too sad about the fact that I'm not expected to sing 6–7 services a week, because sometimes I do love singing all the time but sometimes, I just don't fancy singing every day. I quite like to do other sorts of singing, whether with different choirs or operas or consort singing that is different. (Female soprano 2)

Within the Anglican choral tradition there is still a significant lack of female role models for women and girls. One of the participants in the focus group, who sings under the direction of a female MD addressed this issue:

> There's one thing I noticed when I came to Oxford and that is the lack of female conductors and it wasn't until I got here when I got here "oh, that's actually the first female conductor I've had in UK". That's disturbing (laughter) because in every other place I've been, I've never thought about it. (Female alto 3)

In addition, in 2019 adult female altos are still a rare sight in cathedral choirs, and opportunities for adult women to become lay clerks in cathedrals are still scarce. The Choir of St Paul's cathedral appointed its first female alto in its one-thousand-year history in February 2017. One alto claimed the following:

> If I had my way, I'd have been a boy treble … I don't have any objections to be scorned away into a mixed voice choir, but I just wish I could be a boy treble and grow up to be a lay clerk or something because I just think that aesthetic is so beautiful. (Female alto 1)

This institutional-based discrimination is therefore a hurdle that participants frequently discussed:

> I don't really see the appeal of trebles in that everyone seems to love the sort of pure sound, but I think it's often set up in contrast to the sort of image of female singers that often it's very exaggerated and not exactly the case of how women in choirs sing. … I can't see the appeal working with children who might have beautiful voices but would inherently have lower concentration and won't have as many years of musical experience. (Female soprano 3)

> I don't know if adding women would necessarily ruin it [the tradition] … does it actually need fixing or just creating exactly the same situation for women? (Female alto 1)

Some of their male counterparts agreed that female sopranos would be more competent in terms of musical skills than an eight-year-old boy. Female participants somewhat experienced a reminiscence of the all-male tradition; the female sopranos "in the role of the boy trebles"—and some of their male counterparts addressed these issues as well:

> Sopranos have problems with the system because I think almost every mixed choir in Oxford has twice as many sopranos as the other parts … I think there is a fundamental issue in treating sopranos like children … I have heard conductors ask why sopranos are not as confident and not as good as the rest of the parts. I think it's because we are not trusted in the same way as the other parts are. (Female soprano 4)

With many of the mixed-voice choirs, the repertoire often consists of renaissance choral music that was originally (in many cases) written for all-male choirs with boy trebles. Hence, the interpretation of such music tends to a similar direction sound wise, emphasising this pure sound; directly or indirectly influencing the development of the adult sopranos and female altos, and thus affecting the overall sound of the choir:

> It is interesting to note that when lots of girls are auditioning, the quality of their voices and the style in which they sing is often catered towards the English choral tradition which requires them to act like trebles … I find a lot of girls, especially in Oxford, are trained in this very light, heady style of production. (Bass 3)

> When I was singing [outside UK], all the women were aiming for getting rid of all vibrato and sounding like boys. (Female alto 3)

When the membership numbers of the choirs who participated in the study are observed they consist of 111 sopranos in total and 87 altos, and the tendency is usually to have more sopranos than altos. More than half of the eighteen permanently appointed music directors interviewed for this study were choristers themselves as children. Furthermore:

> You hear claims like "oh female sopranos can't sight-read" and then "women shouldn't be in the choir" but then again there are lots of men who sight read very badly—and no one says "oh, the cannot have men in the choir". (Tenor 3)

Despite most cathedrals and collegiate choirs opening their doors to girls, the access may have been established but equality in terms of opportunities may not be quite the same for girls and boys:

> In those cathedrals [in UK] who have girls, the girls do not get the same opportunities as the boys—they may sing once a week whereas the boys sing perhaps once a day. It's not actually equal. (Female soprano 4)

> I was one of the first [female] members; we only sang once a week, we didn't get paid, but the boys would. It really got to my head when I got older when I realized they were getting paid, but we weren't—specifically when we were doing services with them. They were bringing in older girls in to help the boys out but [the girls] wouldn't get paid, but they [the boys] would and we were replacing them at concerts. We asked, "why aren't we being paid" and they said, "oh it would be so little, it wouldn't impact you". But if you'd pull all that money together, then the girls could go on a trip to the cinema. It doesn't have to be a big thing but there's an equality aspect to it. (Female alto 2)

This testimony is somewhat striking, the girls were expected to help the boys even though the boys got paid and the girls did not. This reflects an even deeper reality of gender discrimination, where the girls still have to do the work, but the boys reap the benefits. In terms of access and equality, the economic aspect emerged repeatedly in the data:

> I quite like singing and I get paid for it in various places, but if I were a man I could apply as a lay clerk in Christ Church and if I got it, I get a house and a salary, so I wouldn't have to do another job. (Female soprano 4)

> The problem of not having any adult women in the choir at all is that economic perspective ... If you are a man, there are 40 cathedrals in the country and they all offer stable jobs, more than often accommodation and an income and it's a very safe thing ... which adult women don't get at all. (Tenor 3)

7.5 Conclusions

The initiative of Salisbury Cathedral started an on-going debate on the issue of aesthetics and preservation; the gender discrimination is based on a prestigious cultural tradition that has survived wars, plagues, dictators and such, and is still ongoing with its unique characteristics. Recent studies by Welch, Howard and other scholars have demonstrated that girls seem to fare quite well within the cathedral choral scene, without having much of an impact on the specific boy treble sound (see Howard et al. 2002; Sergeant et al. 2005). Hence, the aesthetic justification of gender-based discrimination does not hold water in this context. Visually, boy trebles seem to attract visitors and the all-male choir is therefore a certain cultural landmark in terms of the English choral tradition, particularly in the cathedral and Oxford collegiate choral context.

In short, the findings of this study indicate a deeply ingrained, systematic gender discrimination in the choral culture at Oxford colleges. Not only are opportunities for training and singing (in terms of frequency and financial gain) limited for women; they also seem to be judged more harshly than their male counterparts, based on the myth that since they are less likely to receive proper chorister training. Lay clerkships are available for tenors, counter tenors, and basses in many cathedrals in the UK but are almost non-existent for women. The traditional power structures are well in place, and how to disestablish to secure more equality and inclusion seems to be a complicated matter. The fact that tradition has numerous characteristics that makes it somewhat elitist does not make it any easier.

Therefore, the driving force for women to dedicate themselves to a career in choral singing is very limited and the institutional restrictions are still a factor in the 21st Century. Hence, it could be a cultural element or normalised attitudes and behavioural patterns that may be ingrained into the choral culture at Oxford colleges.

Further studies are suggested. Little is known about the girl chorister pioneers that started their careers as girl choristers in Salisbury Cathedral in the early 1990s and in other cathedrals that followed suit. How did they experience entering a thousand-year-old male-dominated tradition, did they receive equal training and performance opportunities as their male counterparts, and how have their potential careers as musicians and/or musical directors fared? In addition, the perspectives of female alumni from Oxford collegiate choirs are unknown as well; this study focused on present students who were in “the eye of the storm” as such. It would be interesting to inquire about their experience in the future as they look back in time, in context with their progress within the music industry, both in the

UK and abroad. Making suggestions for a long-term plan to make the tradition more inclusive, with such a study in place, will be more fruitful.

Acknowledgements This research was funded by the Marie Sklodowska-Curie Action within the Horizon 2020 (H2020-MSCA-IF-2015, EF 707827 OXFORDCHOIRS). I would like to thank the participants of this research for their valuable contribution. Furthermore, I would specifically like to thank Professor Tia DeNora at University of Exeter and Professor Eric Clarke at University of Oxford for their mentoring and support.

References

Ashley, Martin. 2009. How High Should Boys Sing? Gender, Authenticity and Credibility in the Young Male Voice. Ashgate Publishing, Ltd. http://www.google.com/books?hl=en&lr=&id=NP7-1xa37-sC&oi=fnd&pg=PA5&dq=Ashley+2009+sing&ots=JBQv7xfb6k&sig=D7KoqEjU-opPPYXKc_xD0hKtOts.

Brennan, John. 2004. The Social Role of the Contemporary University: Contradictions, Boundaries and Change. Ten Years on: Changing Education in a Changing World, 22–26.

Burgess, Clive and Martin Heale. 2008. The Late Medieval English College and Its Context. Boydell & Brewer Ltd.

Charmaz, Kathy. 2006. Constructing Grounded Theory: A Practical Guide through Qualitative Analysis. Pine Forge Press. http://www.google.com/books?hl=en&lr=&id=w2sDdv-S7PgC&oi=fnd&pg=PP1&dq=Charmaz+2006+grounded+theory&ots=p-kj7REF6s&sig=YGrjY110k0OqXSbbMsCkJrPlrs0.

Data Protection Act 2018. n. d. http://www.legislation.gov.uk/ukpga/2018/12/contents/enacted. Accessed: 2 January 2020.

Elorriaga, Alfonso. 2011. The Construction of Male Gender Identity through Choir Singing at a Spanish Secondary School. International Journal of Music Education. October. doi:https://doi.org/10.1177/0255761411421091.

Foucault, Michel. 1995/1975. Discipline & Punish: The Birth of the Prison. Translated by Alan Sheridan. New York: Vintage Books.

Freer, Patrick K. 2009a. Boys' Descriptions of Their Experiences in Choral Music. Research Studies in Music Education 31(2): 142–60. doi:https://doi.org/10.1177/1321103X09344382.

Freer, Patrick K. 2009b. "I'll Sing with My Buddies" — Fostering the Possible Selves of Male Choral Singers. International Journal of Music Education 27(4): 341–55. doi:https://doi.org/10.1177/0255761409345918.

Gable, Guy G. 1994. Integrating Case Study and Survey Research Methods: An Example in Information Systems. European Journal of Information Systems 3(2): 112–126.

Glaser, Barney and Anselm Strauss. 1967. Grounded Theory: The Discovery of Grounded Theory. Sociology: The Journal of The British Sociological Association 12: 27–49.

Harrison, Brian. 1994. The History of the University of Oxford: Volume VIII: The Twentieth Century—Google Books. UK: University of Oxford.

Harrison, Scott D., Graham F. Welch, and Adam Adler. 2012a. Men, Boys and Singing. In Perspectives on Males and Singing, 3–12. Springer. http://link.springer.com/chapter/10.1007/978-94-007-2660-4_1.

Harrison, Scott D., Graham F. Welch, and Adam Adler. 2012b. Perspectives on Males and Singing. Springer Science & Business Media.

Howard, David M., John Szymanski, and Graham F. Welch. 2002. Listeners' Perception of English Cathedral Girl and Boy Choristers'. Music Perception: An Interdisciplinary Journal 20 (1): 35–49.

Judge, Roy. 1986. May Morning and Magdalen College, Oxford: Folklore, 97(i), 15–40.

Mould, Alan. 2007. The English Chorister: A History. Hambledon Continuum.

Schwarz, Norbert, Fiske, Susan T., and Gardner, Lindzey. 1998. Survey Methods. In The Handbook of Social Psychology, edited by Susan T. Fiske and Gardner Lindzey, 1:143–79. New York: McGraw-Hill.

Sergeant, Desmond C., Peta J. Sjölander, and Graham F. Welch. 2005. Listeners' Identification of Gender Differences in Children's Singing. Research Studies in Music Education 24 (1): 28–39.

Spiegel, Joseph R., Robert Thayer Sataloff, and Kate A. Emerich. 1997. The Young Adult Voice. Journal of Voice 11 (2): 138–43. https://doi.org/10.1016/S0892-1997(97)80069-0.

Trow, Martin. 2007. Reflection on the Transition from Elite to Mass to Universal Access: Forms and Phases of Higher Education in Modern Societies since WWII'. In International Handbook of Higher Education, edited by James J. F. Forest and Philip G. Altbach, 243–80. Springer International Handbooks of Education. Dordrecht: Springer Netherlands. https://doi.org/10.1007/978-1-4020-4012-2_13.

Tuckwell, Lewis Stacey. 1913. Old Magdalen Days, 1847–1877. Blackwell.

Tuckwell, William. 1900. Reminiscences of Oxford. WENTWORTH Press.

Welch, Graham F. 2017. Singer Identities and Educational Environments. In Oxford Handbook of Musical Identities, edited by Raymond MacDonald, David J. Hargreaves and Dorothy Miell, 543–565. Oxford University Press.

Welch, Graham F. and David M. Howard. 2002. Gendered Voice in the Cathedral Choir. Psychology of Music 30 (1): 102–120.

"When She Talks, I Hear the Revolution": Pop Feminism Between Hegemony and Subalternity

8

Magdalena Protte

Abstract

Can you hear the revolution? When talking about pop feminism, the first associations that pop into our minds might look something like this: loud, rhetorical, shiny, pink, depoliticized, devoid of meaning, shallow. Especially from an academic point of view, pop feminsim might seem not critical enough and thus unnecessary for, or even harmful to feminist struggles. Using Antonio Gramsci's political theory, this article argues against these accusations. Pop feminism is understood to be part of the force relations within the State. It therefore plays an active part in the struggle for hegemony within civil society and can tie in with the common sense of subaltern groups. It can therefore develop subversive character—and at the very least builds necessary bridges between theory and praxis.

Keywords

Pop feminism · Pop culture · Hegemony · Civil society · Subaltern · Integral State · Common sense · Theory · Praxis · Gramsci

---

M. Protte (✉)
Philipps-Universität Marburg, Marburg, Germany
e-mail: magdalena.protte@posteo.de

L. Gaupp et al. (eds.), *Arts and Power*, Kunst und Gesellschaft,
https://doi.org/10.1007/978-3-658-37429-7_8

> We teach girls to shrink themselves, to make themselves smaller. We say to girls: You can have ambition, but not too much. You should aim to be successful, but not too successful, otherwise you will threaten the man. ... Because I am female I'm expected to aspire to marriage. I'm expected to make my life choices always keeping in mind that marriage is the most important. A marriage can be a good thing; it can be a source of joy and love and mutual support. But why do we teach girls to aspire to marriage and we don't teach boys the same? ... We raise girls to see each other as competitors. Not for jobs or accomplishments, which I think can be a good thing, but for the attention of men. We teach girls that they cannot be sexual beings, in the way that boys are. … Feminist: The person who believes in the social, political, and economic equality of the sexes. (Adichie 2011, 12:59, 14:07, 16:36, 28:43, quoted in Knowles 2013, 1:26)

The quote above has become quite famous over the course of a couple of years. It has been repeated over and over, taken apart, and put together again. Originally, it was coined by the feminist Nigerian writer Chimamanda Ngozie Adichie in 2011. Back then, Adichie presented these words in a TEDx Talk, addressing a rather niche audience. It was only when pop star Beyoncé Knowles sampled the quote into her song *"****Flawless"* in 2013, that it got noticed by a broad audience[1]. When interviewed by the Dutch newspaper *deVolkskrant*, Adichie said about this development:

> Literally every major newspaper in the world wanted to speak with me about Beyoncé. I felt such a resentment (laughs loudly). I thought: are books really that unimportant to you? Another thing I hated was that I read everywhere: now people finally know her, thanks to Beyoncé, or: she must be very grateful. I found that disappointing. (Adichie 2016)

To me, the discomfort Adichie seems to feel when confronted with Beyoncé's spin on her words is paradigmatic for the way most academics seem to think about pop feminism: It is feminism gone famous—people finally know what it is about, yet it is de-politicized, somehow emptied of its once radical meaning. It is not in a book therefore it cannot be critical thought. Pop feminism seems stuck in between two worlds: The world of "serious feminism" on the one hand, a world where people read complicated books by Butler and de Beauvoir, and talk about being made a woman, about the undoing of gender, and about the heteronorma-

[1] "Broad audience" is an understatement. Literally everybody knows who Beyoncé is. I admit to not having scientific proof of this claim, but just ask around. There is not getting past Queen B!

tive matrix. And the world of pop culture on the other hand, a place of an industry designed to deceive society and turn people into passive consumers; a dire place for Adorno, yet a fun and very shiny one for the (undoubtedly manipulated) rest of the world.

On the following pages, I want to show that pop feminism—luckily—is part of both worlds and can thus build necessary bridges between theory and praxis. Using Antonio Gramsci's political theory, I will argue that pop feminism plays an important part in *civil society's* (and therefore the *Subaltern's*) struggle for hegemony. This text is a contribution to a theory of the state, rather than to pop theory. A theory of the state, to be precise, that considers pop culture to be part of the force relations within the State. As such, pop culture can—but does not necessarily have to—develop subversive character. I will start with a very short paragraph on pop culture, followed by a brief summary of what I understand to be the baselines of Marxist Theory of the State, insofar as they are crucial for an understanding of Gramsci's *Integral State*. I will attempt to weave my understanding of Gramsci's thoughts on hegemony and culture organically into these assumptions, so as to—eventually—explain how pop feminism fits into the picture.

Although it might not be necessary for this article to make sense, I would like to make my own *conception of the world* transparent: I have an intersectional[2] approach to feminism. To me, feminist struggles must include Women of Colour, trans people, queer people, the working class (for my Marx fanboys out there!), disabled folks, sex workers, and everyone else who fights against, or is affected by, oppression. As I will explain in more detail later, all these people are part of what Gramsci calls the *Subaltern*. Additionally, this article tries to break open the traditional form of academic text by occasionally using colloquial language as well as pop cultural references. In doing so, I want to try and make this text a bit more accessible to a non-academic audience, while at the same time being fully aware that I will probably fail spectacularly to do so at times.[3] And now: POP!

---

[2] The term *intersectionality* was introduced by professor of law Kimberlé Crenshaw. She describes how, in antidiscrimination struggles, Black women have the choice to either position themselves along the axis of gender (focussing on *white* women), or along the axis of race (focussing on Black men), and are therefore erased in their specific position as Black women. Crenshaw argues instead for an approach that understands the intersections and overlappings of different forms of oppression (cf. Crenshaw 1989).

[3] Another fun perk of changing to a more colloquial language once in a while: It exposes academics to a feeling they might have forgotten: What it is like to be confronted with scientific text and not understand the half of it.

8.1 "Hot Topic": Pop Culture

The word "pop" was not always considered a self-sufficient noun, least of all a noun connected to art and culture. Although the verb "to pop" had already existed in the English language for a long time, it was first used in an art context in 1947, in a collage by Eduardo Paolozzi, a representative of what would later become known as *Pop Art* (cf. Behrens 2010, p. 17). There, at the muzzle of a gun that is aimed at the head of a pin-up girl, we see the word "POP!", written in bright red letters (cf. Paolozzi 1947). Pop then went rather quickly from being an art form in the 1960s, to becoming the expression of alternative youth culture in the 1970s and 1980s, to being universalised in the 1990s as a "dummy-term" (Diedrichsen 1998, p. 14) that "expresses the gigantic and mass-character of pop culture per se" (ibid.; my translation). It is part of this development that pop does not require any specific form of education or knowledge. It is "easily accessible, does not require a lot of effort and follows fixed rules, patterns, and formulas" (Aydın and Henning 2010, p. 11; my translation). Pop, in other words, is "an argument where anyone can join in" (Behrens 2010, p. 23). As such, it can be described in Gramscian terms as part of the *common sense* and therefore plays an important role in *subaltern* struggles, as I will show on the following pages. In order to do so, I will start where any self-respecting SJW[4] would: with Marx.

8.2 "All About that Base": Structure and Superstructure

Marx and Engels never actually managed to write a whole book on the state, so there is really not very much to go on when studying Marxist Theory of the State. Nevertheless, there are a few essential assumptions we can gather from their work. According to them "[i]t is not the consciousness of men that determines their being, but on the contrary it is their social being that determines their consciousness" (Marx 1976, p. 3). "Being", in this context, meaning the relations of production and productive forces, or in other words: the economic base. This economic base influences (some people prefer to stick with Marx's exact word: determines) our social environment, which, in turn, influences the way we

[4] Social Justice Warrior (cf. Merriam-Webster Dictionary 2021). As this term is nowadays mostly used in a derogatory way, I am ironically reclaiming it here.

develop our conscious political thoughts and actions and therefore the state. The state, in turn, provides a necessary frame for said relations of production in the shape of a legal relationship (cf. Marx and Engels 1887). The state apparatus in its modern form has grown together with the capitalist mode of production, that is to say: neither can exist without the other. Or as Marx puts it:

> The totality of these relations of production constitutes the economic structure of society, the real foundation on which there arises a legal and political superstructure and to which there correspond definite forms of social consciousness. The mode of production of material life conditions the social, political and intellectual life-process in general. (Marx 1976, p. 3)

This quote unfortunately often serves as an argument for the presumption of an economic determination of human life. It is indeed very tempting to assume a single and direct line of influence from the economic base to the superstructure of the state—but here is where Gramsci comes in. He broadens the idea of what the state is, by fixing

> … two major superstructural "levels": the one that can be called "civil society", that is the ensemble of organisms commonly called "private", and that of "political society" or "the State". These two levels correspond on the one hand to the function of 'hegemony' which the dominant group exercises throughout society and on the other hand to that of "direct domination" or command exercised through the State and "juridical" government. (Gramsci 1999, p. 145, Q 12, § 1)[5]

Both levels are part of the *Integral State*. The crucial point here is that the state not only consists of the state apparatus, which in turn not only serves as an instrument in the hand of the ruling class. Instead, civil society gives a frame for the continuous struggles for hegemony that eventually manifest on the level of political society. "State = … hegemony protected by the armour of coercion" (Gramsci 1999, p. 532, Q 6, § 88). Every group within civil society takes part in these struggles for political, cultural, and economic hegemony. Struggles, that eventually translate into the state apparatus. The latter is no longer understood as a fixed and all-powerful construct. Instead, the *Integral State* consists of individ-

[5] I will cite the prison notebooks as follows: (Gramsci vvvv, p. xx, Q yy, § zz), v being the year and thus the edition as listed in the bibliography, x being the page number of said edition, y being the number of the notebook (Q for "quaderno"), and z being the paragraph in question.

ual people who are shaped by it, but who in turn also shape the state with their actions. We can now begin to see why—with this emancipatory approach—we might have to include more than just an elite of academics if we want to change hegemony within this *Integral State*.

Traditionally, the struggle for hegemony is seen as one between classes. Marx and Engels themselves thought that "all struggles within the State … are merely the illusory forms in which the real struggles of the different classes are fought out among one another" (Marx and Engels 1932/1968, p. 12). I do, however, think that—as we take a closer look at Gramsci's conception of the (economic) base—we will come to see that the groups fighting for hegemony do not always align along the axis of class. Marxist feminists have long argued for a more intersectional approach, saying for instance that with care economy, the heteronormative matrix comes into the equation of force relations on a structural level. For Gramsci as well, the structure must be considered "historically, as the ensemble of social relations, in which the real people move and operate" (Gramsci 1929–1935, Q 10, § 8; my translation).[6] We can interpret those social relations not only as class relations, but also—for instance—as gender and race relations. Although most Marxists have accepted the fact that more than one underlying structure of society exists, there have been many discussions about main and side contradictions within this structure. There is an interesting part in the prison notebooks, where Gramsci argues as follows:

> In fact every real historical phase leaves traces of itself in succeeding phases, which then become in a sense the best document of its existence. The process of historical development is a unity in time through which the present contains the whole of the past and in the present is realised that part of the past which is "essential"— with no residue of any "unknowable" representing the true "essence". (Gramsci 1999, p. 754, Q 7, § 24)

The way I am reading this, it means that when talking about the structure of society, we should not fixate on any chicken or egg dilemma. Whether they existed from the beginning or not, those structures have manifested throughout history.

[6] To my utter dismay, I could not find a full translation of the prison notebooks in English. I therefore had to translate some of the quotes in this article myself. Since my Italian sadly never reached the point of being able to read Gramsci, I translated from German (Gramsci 1991–2002), cross referencing with the Italian original (Gramsci 1929–1935). I apologize in advance.

They are entangled, interwoven and eventually inscribed in the state. With every change in hegemony they have changed as well and have left traces in history, and they will continue to do so. In other words: Maybe it does not really matter whether racism or sexism was there first. Maybe it is not important which one was the original underlying cause for the mess we're in. These questions are futile. Both racism and sexism have manifested over time to be mutually dependent. They form an historically grown net that shapes our reality. More important question might now be: Where can we see this net in the world that surrounds us and how do we work together to untangle it?

8.3 "A Walking Contradiction"

To Gramsci, "all men are 'philosophers' …" (Gramsci 1999, p. 626, Q 11, § 12).[7] Philosophers, not in the sense of an academic (let alone: Ivy League) education, but philosophers in the sense of a spontaneous philosophy. This kind of philosophy is a product of the aforementioned historical structure and thus "it is a conception [of the world; M.P.] which, even in the brain of one individual, is fragmentary, incoherent and inconsequential, in conformity with the social and cultural position of those masses whose philosophy it is" (Gramsci 1999, p. 769)[8]. Gramsci calls this philosophy *senso commune—common sense*. A philosophy of the masses, *common sense* is "mechanically imposed by the external environment, i. e. by one of the many social groups in which everyone is automatically involved from the moment of his entry into the conscious world" (Gramsci 1999, pp. 626–627, Q 11, § 12).

In orthodox Marxist understanding these groups are mostly classes and class fractions. However, with the abovementioned interpretation of Gramsci, we now understand social relations to be a complex and multilevel structure. We can see how diverse axes, like gender, race, or sexuality influence the formation of groups within civil society. Groups that are strongly affected by these relations and have not managed to become hegemonic, are what Gramsci calls the *subaltern*. Resulting from the complex structure, *subaltern* groups are manifold and are alike only in being different from the hegemonic bloc (cf. Gramsci, Q 3, § 46). *Subalterns*

[7] Obviously not only all men as in "male", but as in "people". Even language is affected by the patriarchy. Who would have thought.

[8] So basically your philosophy is random af, Horatio!

struggle to push their own philosophies, but their struggles are hard to unite, as they are under constant threat of the hegemonic groups (cf. Gramsci, Q 25, § 2). We can see this dynamic playing out even within feminist struggles: To this day, feminism is more often than not associated with white cis women in pussy hats who fight for equal pay. Black women or trans women, however, might have very different concerns as they face different and more substantial threats. At the same time they are often (consciously or not) excluded by white feminists. Only this year a famous author Who-Must-Not-Be-Named, and who claims to be feminist, has outed herself to be a TERF[9] in a series of very unfortunate tweets. She is thus part of a group of *subalterns* who——while undoubtedly struggling against hegemony as "feminists" themselves—are at the same time complicit in the oppression of other women who they refuse to unite with.

When looking at Gramsci's conception of the state, several things become apparent: Firstly, the economic base is covered by a net of multiple social relations that have grown over time and throughout history. Secondly, those relations shape our *conception of the world* and thus our *common sense*. They also shape the various groups struggling for hegemony within civil society, who in turn try to shape the state and thus the current historic phase that is going to become another layer in the complex net of social relations. Lastly, because of this complexity, it is hard for the *subaltern* groups to unite their struggles.

8.4 "Strong Enough": Universalising Good Sense

So how do these assumptions connect to pop culture? What is the deal with common sense? How do we enforce our own feminist position? Obviously, just lah-di-da-ing our way through life, oblivious about our own position in the world, and about how messed up our *common sense* really is, was not Gramsci's idea of critical thought. He rather wanted everyone to question their own *common sense*, understand its randomness and develop it into a *coherent* unity (cf. Gramsci 1991–2002, Q 11, § 12). In other words: The beginning of critical thought "is 'knowing thyself' as a product of the historical process to date which has deposited in you an infinity of traces" (Gramsci 1999, p. 628, Q 11, § 12). Understanding how our eclectic *conception of the world* came to be, what different kinds of

[9] Short for *Trans Exclusive Radical Feminist*, although I personally would not call them either radicals or feminists.

groups we are part of and how that affects our everyday life, is part of the process of developing *good sense—senso bueno*. In understanding our own history as part of a continuous narrative, we can recognise injustice not as individual ill fate or unfortunate law of nature, but as structural conditions that need to be tackled.

The good news: because our own common sense is "not critical and coherent but disjointed and episodic" (Gramsci 1999, p. 627, Q 11, § 12) and because it is the product of random coincidence, we already carry various contradicting *conceptions of the world* in us. Thus, the potential for sparking our *senso bueno* already lies within every single one of us. And this is indeed good news, since individuals are not only mechanically and passively influenced by their groups. They are part of, but also play an active part in, the universalisation of their own philosophy: Everybody "participates in a particular conception of the world, has a conscious line of moral conduct, and therefore contributes to sustain a conception of the world or to modify it, that is, to bring into being new modes of thought" (Gramsci 1999, p. 141, Q 12, § 3). As mentioned before, this is a very crucial notion that implies an emancipatory form of politics. It means that every person is a political being and that—with our very existence in it—we shape the world and take part in the struggle *against* (as well as *for)* hegemony that later translates into political society. As feminists, we are part of a *subaltern* group. We fight against manifestations of injustices, as we strive to change the hegemony and universalise our own *senso bueno.*

8.5 "Smells like Teen Spirit": Pop Culture in Civil Society

Yet, although every individual can contribute to this fight, in order to change hegemony we need help.

> An historical act can only be performed by "collective man", and this presupposes the attainment of a "cultural-social" unity through which a multiplicity of dispersed wills, with heterogeneous aims, are welded together with a single aim, on the basis of an equal and common conception of the world, both general and particular, operating in transitory bursts (in emotional ways) or permanently (where the intellectual base is so well rooted, assimilated and experienced that it becomes passion. (Gramsci 1999, p. 665, Q 10.II, § 44, 223)

In other words, there is not much use in a small elite of feminist academics finally seeing the world for the twisted, messed up, and patriarchal hell it actually is.

We need to form a cultural-social unity in the struggle for hegemony. I shall now attempt to show how this unity can be achieved.[10]

The "entire system of beliefs, superstitions, opinions, ways of seeing things and of acting … surfaces collectively under the name of 'folklore'" (Gramsci 1999: 626; Q 11 § 12). Folklore is the cultural manifestation of *common sense*, an "implicit 'conception of the world and of life' of specific … social classes", that is opposed to "'official' conceptions of the world (or in a broader sense to those of the historically determined educated parts of society)" (Gramsci 1929–1935, Q 27, § 1; my translation). Like *common sense*, folklore is something we have to overcome in order to form a coherent *senso bueno*, a critical *conception of the world*. At the same time, however, in order to universalise this critical *conception of the world*, it is crucial to take folklore seriously. A *subaltern* group, in order to become a philosophical movement—or in other words: in order to form a historic bloc—needs to tie in with *common sense* and reach those who Gramsci calls "semplice"—ordinary people[11].

> Creating a new culture does not only mean one's own individual "original" discoveries. It also, and most particularly, means the diffusion in a critical form of truths already discovered, their "socialisation" as it were, and even making them the basis of vital action, an element of co-ordination and intellectual and moral order. (Gramsci 1999, p. 629, Q 11, § 12)

This works best by "taking roots in the humus of the culture of the people" (cf. Gramsci 1991–2002, p. 1777, Q 15, § 58), i. e. in folklore. Folklore can be something as old as religion or morals (cf. ibid.), yet it can also be art, popular literature, and—as I would suggest—nowadays: pop culture. Once we understand pop culture as a form of folklore, the "gigantic and mass-character of pop culture" (Diedrichsen 1998) becomes comprehensible: In pop culture, we see the manifestation of a hegemonic *conception of the world*, the socialisation of *common sense*. It seems random, fragmented into oh-so-many forms of art, music, television, podcasts, memes, social media trends etc. Pop is incoherent, yet it does not require a specific education or academic knowledge. This is the very reason why pop culture is "an argument where anyone can join in" (Behrens 2010, p. 23).

---

[10] Spoiler alert: through pop culture. Obviously.

[11] This term—albeit normatively neutral—is in no way unproblematic even in Italian, and the nuances get lost even more in translation. The English word that first springs to mind—"simple"—is not necessarily the best in my opinion. "Semplici" can mean a variety of things, including "ordinary, naïve", but also "modest", and in German "gutbürgerlich".

However (and this is crucial), in letting anyone join, pop culture can create points of contact for individual and divergent *conceptions of the world* and thus can create and universalise the frictions we need to develop our *senso bueno*. Of course, pop feminism cannot always depict all the developments in feminist theory. After all, artists can only create an image, they can only capture "'that which is' in a specific moment" (Gramsci 1929–1935, Q 15, § 58; my translation), whereas theorists can see the becoming of a movement and what a person should be like in order to attain a specific goal (cf. ibid.). Under these circumstances, one could easily perceive pop culture as anachronistic, being not critical and subversive enough: Pop feminism, in this context, is only the manifestation of feminist *conceptions of the world* in pop culture. A very brief example of this can be given in the development of queer TV representation in the last 20 years: Back in 2001, a kiss between two *white* cis women in "Buffy the Vampire Slayer" (cf. Whedon 2001) was considered groundbreaking. Later that decade, shows like "The L Word" (2004–2009) pushed the boundaries of socially acceptable queer content by focusing on Lesbian subculture and featuring at least one Black woman. Nowadays, the 2018 TV series "Pose", a show about Ballroom culture in the 1980s, stars mostly queer and trans People of Color. I would argue that TV series like these create frictions in our *common sense*, thus universalizing specific *conceptions of the world*.[12] If civil society is the ground on which struggles for hegemony take place, pop feminism is a growing part of that ground.

8.6 "This is the End": Pop Feminism Can Build Bridges

Gramsci's conception of the state is emancipatory at its core: People within the state are no longer viewed as a mass to be governed. They are instead political beings who fight for their positions, and they are subjects, capable of critical thought, who bear in themselves a collection of various *conceptions of the world* and thus the possibility for a spark of resistance. It is only in being confronted with other *conceptions of the world*, in understanding and contextualizing our own position in history, and in making our own consciousness coherent, that we are able to develop our *senso bueno*. And finally, as Gramsci puts it:

[12] Having said this, I want to stress that not all is well just because people are finally represented in pop culture. There is still a long way to go!

> For a mass of people to be led to think coherently and in the same coherent fashion about the real present world, is a "philosophical" event far more important and "original" than the discovery by some philosophical "genius" of a truth which remains the property of small groups of intellectuals. (Gramsci 1999, p. 630, Q 11, § 12)

Undoubtedly, there are plenty of feminist problems that cannot be solved by pop feminism. It is important to keep that in mind when talking about pop culture. Nevertheless, it is also important to make feminist theory accessible to everyone, not just academics, so that—as pop feminist heroine Buffy Summers puts it—"every girl who could have the power will have the power" (Whedon 2002). Universalising certain ideas and building bridges between theory and praxis, must be an integral part of feminism. Pop feminism can build those bridges. In the end, pop feminism, although part of *common sense*, might just be necessary to get to a place of *good sense*, "to work out consciously and critically one's own conception of the world and thus … take an active part in the creation of the history of the world" (Gramsci 1999, p. 627, Q 11, § 12).

References

Adichie, Chimamanda Ngozi. 2011. We should all be feminists. Ted Talk: TEDxEuston. TEDx Talks. YouTube video, 30:15, December 4 2013. https://www.youtube.com/watch?v=hg3umXU_qWc. Accessed: 29 October 2019.

Adichie, Chimamanda Ngozi. 2016. Beyoncé's feminism isn't my feminism. Chimamanda Ngozi Adichie on the use of her Ted talk in Beyoncé's ****Flawless. Interviewer Aimée Kiene. deVolkskrant, October 7. https://www.volkskrant.nl/boeken/ngozi-adichie-beyonce-sfeminism-isn-t-my-feminism~a4390684/. Accessed: 15 April 2019.

Aydın, Yaşar, and Astrid Henning. 2010. Zum Verhältnis von Gesellschaft, Populärkultur und Wissenschaft im globalen Kapitalismus. In Pop Kultur Diskurs. Zum Verhältnis von Gesellschaft, Kulturindustrie und Wissenschaft, ed. Adam, Holger, Yaşar Aydın, Zülfukar Cetin, Mustafa Doymus, Jonas Engelmann, Astrid Henning, and Sonja Witte, 9–15. Mainz: Ventil Verlag.

Behrens, Roger. 2010. Traditionelle und kritische Poptheorie. Anmerkungen zur fröhlichsten Wissenschaft, heute. In Pop Kultur Diskurs. Zum Verhältnis von Gesellschaft, Kulturindustrie und Wissenschaft, ed. Adam, Holger, Yaşar Aydın, Zülfukar Cetin, Mustafa Doymus, Jonas Engelmann, Astrid Henning, and Sonja Witte, 17–39. Mainz: Ventil Verlag.

Crenshaw, Kimberlé. 1989. Demarginalizing the Intersection of Race and Sex: A Black Feminist Critique of Antidiscrimination Doctrine, Feminist Theory and Antiracist Politics. University of Chicago Legal Forum. http://chicagounbound.uchicago.edu/uclf/vol1989/iss1/8. Accessed: 30 June 2020.

Diedrichsen, Diedrich. 1998. Alles ist Pop. Wanderer. Was bleibt von der Gegenkultur? Süddeutsche Zeitung, August 08.

Gramsci, Antonio. 1999. Selections from the Prison Notebooks of Antonio Gramsci. Trans. & ed. Quentin Hoare and Geoffrey Nowell Smith. Essential Classics in Politics: Antonio Gramsci. London: ElecBook.

Gramsci, Antonio. 1991–2002. Gefängnishefte. Kritische Gesamtausgabe, ed. Bochmann, Klaus, Wolfgang F. Haug, and Peter Jehle. Hamburg & Berlin: Argument.

Gramsci, Antonio. 1929–1935. Quaderni del Carcere, ed. Alberto Soave, Private Online Archive. https://quadernidelcarcere.wordpress.com/indici-e-sommario/indici/. Accessed: 24 May 2019.

Knowles, Beyoncé 2013. ***Flawless. Parkwood Entertainment, Columbia Records. Beyoncé. YouTube video, 4:12, November 24 2014. https://www.youtube.com/watch?v=IyuUWOnS9BY. Accessed: 29 October 2019.

Marx, Karl. 1976. Preface and Introduction to A Contribution To The Critique Of Political Economy. Peking: Foreign Languages Press.

Marx, Karl, and Frederick Engels. 1887. Capital. A critique of political economy. Marxists Internet Archive. https://www.marxists.org/archive/marx/works/1867-c1/ch02.htm. Accessed: 11 May 2019.

Marx, Karl, and Frederick Engels. 1968 [1932]. Idealism and Materialsism. In The German Ideology. Marxists Internet Archive. https://www.marxists.org/archive/marx/works/download/Marx_The_German_Ideology.pdf. Accessed: 11 May 2019.

Merriam-Webster Dictionary. 2021. Social Justice Warrior. https://www.merriam-webster.com/words-at-play/what-does-social-justice-warrior-sjw-mean. Accessed: 15 December 2021.

Paolozzi, Eduardo. 1947. I was a rich man's plaything. Collage. Printed papers on card. http://www.tate.org.uk/art/artworks/paolozzi-i-was-a-rich-mans-plaything-t01462 Accessed: 25 May 2019.

Whedon, Joss. 2001. Buffy the Vampire Slayer, Season 5, Episode 16: The Body.

Whedon, Joss. 2002. Buffy the Vampire Slayer, Season 7, Episode 22: Chosen.

Female Bodybuilding and Patriarchal Civilization. The Intrusion of a Practice in Sport into Artistic Fields and Visual Culture

9

Isabel Fontbona Mola and Ulf Wuggenig

Abstract

This essay focuses on the practice of female bodybuilding (in the form we know it today) from the perspectives of domination and empowerment, social influence and power. The primary theoretical frame of reference is Bourdieu's field-habitus-capital approach, though this essay also discusses some limitations of applications of this theory, since Bourdieu linked bodybuilding with the working class only and did not discuss it as a middle class and feminism connotated female practice as well. Using a critical queer approach, the practice of female bodybuilding is discussed in terms of its similarities and fundamental difference with male bodybuilding. The essay furthermore looks at the practice on a systemic level, analyzing its expansion from sport and entertainment into several cultural and visual fields, including eroticism and the arts, in a process of social differentiation; and its globalization beyond its origins in the United States and the United Kingdom, which was also facilitated by the internet—psycho power—and which redefined the practice's visibility after the iconic turn in an overall economy of appearance and attention. Of special relevance, this essay analyzes some of the intra- and intergender struggles within the field in light of the male dominance on the institutional level,

I. F. Mola (✉)
University of Girona, Girona, Spain
e-mail: isabel.fontbona@udg.edu

U. Wuggenig
Leuphana University of Lüneburg, Lüneburg, Germany
e-mail: uwuggeni@leuphana.de

L. Gaupp et al. (eds.), *Arts and Power*, Kunst und Gesellschaft,
https://doi.org/10.1007/978-3-658-37429-7_9

which is enforced by heteronomous orientations (demand as well as male gaze) that support heavily sexualized, essentialist expectations of "femininity" for performing bodybuilders. These expectations are similarly reinforced by the demands of marketing interests and are supported by the complicity of the negatively privileged, which is characteristic for the functioning of symbolic power in Bourdieu's sense. The ambiguity involved in the affirmative reproduction of the dominant gender order, as well as the critical intentions and dissident interventions in the patriarchal order, are discussed using the theoretical frame provided by French constructivist structuralism and further approaches to social power.

Keywords

Female bodybuilding · Body · Arts and power · Field-theory · Third wave feminism

9.1 Introduction

In the last decades, bodybuilding, which Wacquant (1995, p. 168) narrowly characterized as a practice people perform "with the ultimate goal of sculpting their bodies for ritualized competitive display, public shows at which they parade their musculature on stage," has become more and more popular beyond the field of sport. In this essay, we focus primarily on the practice of female bodybuilding, investigating it from the perspectives of domination and empowerment, social influence and power.

Russell famously defined power in general terms, referring to an impulse that could be applied to human agents as well as to things: "Power may be defined as the production of intended effects .... A has more power than B, if A achieves many intended effects and B only a few" (Russell 1938, p. 23). French and Raven's more specific definition of power, however, has more widespread applications because it is not restricted to intentional behavior. "Social influence," according to their agency-centered definition can be understood as "change in the belief, attitude, or behavior of a person (the target of influence), which results from the action of another person (an influencing agent)" (Raven 2008, p. 1). In this framework, social power is relationally "defined as the *potential* for such influence, the ability of the agent or power figure to bring about such change, using resources available to him or her" (ibid.).

Pierre Bourdieu's field theory offers a constructivist-structuralist approach to power. On the level of agency, it emphasizes the relevance of general *and* field-specific capital—whether that capital is economic, cultural, social, or symbolic, all different species of power from Bourdieu's perspective—of individual or collective agents (like organizations, regions, etc.) who are involved. In Bourdieu's formulation, these agents hold different forms of power positions, which are reflected in their dispositions, strategies and practices (Bourdieu and Wacquant 1992, pp. 67, 76).

Female bodybuilding might be regarded as a practice that shows the effects and reflections of "patriarchal civilization" (Marcuse 1974), or of "masculine domination," the term Bourdieu (2001) chose. On the other hand, many feminists especially of the third wave describe the practice as a behavior that demonstrates female empowerment, in that it attempts to change the attitudes, dispositions, perceptions, beliefs, and behaviors agents, female as well as male or nonbinary (Richards et al. 2017). On the systemic level female bodybuilding has been interpreted, as we will show, as an intervention that is either affirmative or critically dissident, or a hybrid combination of these two impulses.

Influence and power are at least partially a function of the volume (i.e., the number) of the agents and institutions who are not only involved with, but also impacted by effects. This is a question of scope that ranges from the micro- to the macro-level in social, as well as geographical, terms. Thus, visibility and its mediation, whether it is through photography, film, video, specialized or mass media, or the internet, are important aspects of various agents' influence and power, especially after societies' pictorial turn. Attention in contemporary societies might be said to function as capital (Franck 1999). "Psycho-power," being based on attentional technologies, has become of great importance:

> Psycho-power is the systematic organization of the capture of attention made possible by the psycho-technologies that have developed with the radio (1920), with television (1950) and with digital technologies (1990), spreading all over the planet through various forms of networks, and resulting in a constant industrial canalization of attention. (Stiegler 2012, p. 104)

Additionally, it seems relevant to note, that we look at the extension of the field beyond the spheres of sport and entertainment, where it originated. That means considering other social universes as well. In addition to framing practices in their own ways, some of these universes with their renowned agents and through their high amount of "psycho-power," are mediators for enhancing the practice's visibility. They have the ability to reach audiences with different class connotations than bodybuilding is usually able to as a mere sport or cultural industry entertainment sphere.

In terms of the media-supported social extension of bodybuilding—first in its male version and later in its female iteration—the activities of single agents and their network in the 1970s and early 1980s proved to be decisive. As it is in art, science, or literature, where singular actors are able to change the whole game in a field—a phenomenon Bourdieu refers to when he asks, "what would a sample of writers of the 1950s be without Sartre?" (Bourdieu 1998, p. 60)—within the sport field of bodybuilding, it was the charismatic figure of Arnold Schwarzenegger who gathered and wielded the cultural, symbolic, and social capital that expanded and transformed the field. Schwarzenegger's strategy involved using another field, as well as some of that field's agents with high symbolic capital, to enhance his own visibility and that of the practice as a whole. By involving an elitist institutional context, he succeeded in gaining legitimacy and much more visibility for the practice. In 1976, for example, in a performance that would become legendary, he performed *The Body as Art* at one of the worldwide leading museums for contemporary art, the Whitney Museum of American Art in New York City.[1] Performing in front of 2,400 visitors—many who belonged to the center of New York's internationally dominant art world—Schwarzenegger declared his muscular body to be an "art medium," which affirmatively demonstrated an ideal of virile masculinity that was familiar from certain kinds of images from art history. Schwarzenegger, however, did not replicate the dominant art historical image of male representation, which has connotations with the upper class. In that context, the ideal man (i.e., the *gentle-man*) does not typically showcase his muscles—which in the upper classes are conventionally seen as vulgar, meaty indicators of labor—because he does no physical work. Instead, it is the lack of showcased muscles that are conventionally the sign of real power (Kunitz 2017). Sociologist Klein (2007) thus proposed linking male bodybuilding less with art historical references, but with the exaggerated comic book depictions of masculinity.

A primary reason for the successful transformation of bodybuilding, as well as the extension of the field since the 1970s, was the cooperation Schwarzenegger initiated with highly visible artists, some who were on their way toward consecration and some who had already attained it. In the 1970s, Schwarzenegger collaborated with Robert Mapplethorpe, an artist famous for his formalist artistic approach to an array of spectacular subjects, including celebrity portraits,

[1] YouTube [1976] 2012.

male and female nudes, self-portraits, and the homosexual male BDSM subculture of New York City. In the 1980s it was Andy Warhol, who found interest in Schwarzenegger.

In 1977, the docudrama *Pumping Iron* [Film] 1977 appeared, which signaled a moment when the practice expanded into the field of popular visual culture. *Pumping Iron* centered on the competition for the title of Mr. Olympia, bodybuilding's most famous award. The film gained widespread popularity after its initial release because of Schwarzenegger's rise to Hollywood superstardom, which reached its peak in the 1980s. "Bodybuilding was brought to the broader public consciousness," Kunitz (2017) writes, "by Schwarzenegger and the 1977 film *Pumping Iron*. It offered an ideal of exaggerated musculature that was obviously unattainable, no matter how much effort one put in, without technological enhancement in the form of steroids and specialized weight machines." Schwarzenegger's body and performances seemed edgy, something highly acclaimed within the field of contemporary art.

Bourdieu, who together with Foucault reintroduced the body into social theory, dealt briefly with the practice of bodybuilding in *La Distinction* (1979), in its English translation—*Distinction* (1984)—one of sociology's most influential books of the twentieth century. Bourdieu, however, did not show an awareness of Schwarzenegger or of Mapplethorpe's artistic appropriation of bodybuilding. Instead, he considered bodybuilding solely as a type of sport and, furthermore, as a genuinely masculine practice. Modern female bodybuilding, however, had just originated in Southern California, breeding ground for exotic lifestyles and subcultures, in the second half of the 1970s (Ian 2001, p. 69).

In 1985, *Pumping Iron II: The Women* was released. In tandem with the reception of female bodybuilding in cultural industry media the film decisively enhanced the visibility of the practice and also gave it new connotations:

> The new image of strong, independent women is exemplified in the film *Pumping Iron II: The Women* and by cover stories on women bodybuilders in *Life* and *Time* magazines. The mildly sensational and congratulatory tone of the articles panders to the world of women and fashion …. (Klein 1985, p. 68)

Female bodybuilding is fundamentally different from the male practice. It does not affirmatively represent a conventional ideal of femininity in the way that most male bodybuilders do when they realize a certain ideal of lower- and middle-class masculine stereotypes including their caricatures. As Aoki notes, a female bodybuilder instead "looks like a woman who fails to look like a man who fails to look like a woman—thus performing a failed impersonation of a failed impersonation" (Aoki 1996, p. 61).

The higher visibility female bodybuilding gained after the release of *Pumping Iron II* was largely due to the strategies of a single woman. Lisa Lyon, a former Californian art student and a bodybuilder who had initially trained with Schwarzenegger, became the first worldwide female bodybuilding champion when she won Women's World Pro Bodybuilding Championships in Los Angeles in June of 1979.

Lyon acted similarly to Schwarzenegger, though she used the artistic field even more skillfully than he did to enhance her own symbolic capital as well as that of her field. Between 1980 and 1983, she collaborated closely with Robert Mapplethorpe.[2] Along with punk celebrity Patti Smith, Lyon became the "main muse" of this artist, who produced art history's "most dangerous pictures". At the same time, Lyon also collaborated with another well-known photographer, Helmut Newton. Thus, Lyon is remembered on the one hand for representing queerness in an exemplary way. Mapplethorpe's portraits of her are not only delicate studies of the interplay between strength and vulnerability as well as disguise and identity, but also—because Lyon's muscular body defied traditional conceptions of what a female body "should" look like—between masculinity and femininity. On the other hand, Lyon was a kind of one-woman media-relations activist on behalf of her sport and of herself. Apart from opening the doors to fashion, she did not hesitate to pave the way between female bodybuilding and soft pornography (followed later, by hard-core pornography with the rise of the internet).[3] She not only posed for Newton, who was known for his objectification of women by photographing fashion models of upper (middle) class taste in a visually sophisticated form, but she was also the first female bodybuilder to appear in *Playboy* magazine in October 1980.

In the 1980s and 1990s, the practice of bodybuilding continued to be perceived and visually appropriated in the artistic field. Pop art icon Andy Warhol—who together with Xenakis, Boulez, and *Tel Quel* symbolically represented the taste of artistic producers in the left upper region of Bourdieu's (1984,

[2] After meeting in 1980, Lyon and Mapplethorpe collaborated over the next few years on several artistic projects, including a film, a series of portraits, and a famous catalogue, *Lady, Lisa Lyon,* which also included contributions from British writer Bruce Chatwin. The book, which originally appeared in the early 1980s (Mapplethorpe and Chatwin 1983) was published for a second time in 1991. The reissue was aimed at a broader audience and was able, via internet distribution, to reach many more spectators (YouTube 2016).

[3] On a porn website like *Letmejerk*, there are hundreds of videos with muscled women and female bodybuilders, both commercial and noncommercial—but as a whole they are presented in a commercial context.

p. 128) famous visual representation of social space in 1960s France—depicted Schwarzenegger on the cover of his *Interview* Magazine in 1985. Bodybuilding also gained more and more appreciation in other fields of visual culture like fashion, beauty, and eroticism. As female bodybuilding expanded into fields beyond sport, the practice also began to be discussed not only as a strategy for overcoming self-ascribed victim status (Heywood 1998, p. 160), but also as a queer approach that transgressed conventional heteronormative gender stereotypes, including the celebration of nonbinary identity models. It was also discussed as a practice of enhancement of attractiveness, of narcissism, autoeroticism, and muscle worship (cf. Thirer and Greer 1978; Klein 1994; Parsi 1997; Richardson 2004; Schippert 2007; Gardiner 2012; Richards et al. 2017; Beutler 2017).

Isabel Fontbona Mola, an art historian and performative artist who contributed to this essay, is active in one of the two main subfields of bodybuilding as a competitive sport. She performs in "natural," (i.e., "drug free") bodybuilding, which is be distinguished from the subfield that uses "performance- and image-enhancing drugs" (PIEDs) (cf. Garratt 2015; Liokaftos 2019). This essay, which is based on the practical experiences of an insider as well as on sociological reflections, also will raise thc question about how attacks on the reproduction of the gender binary and of "masculine domination" (Bourdieu) might be judged, not least from a political perspective.

9.2 Some Aspects of the Practice of Bodybuilding

Competitive bodybuilding can broadly be described as "a systematic training regime used to develop and sculpt the body for the purpose of increasing muscle size and definition, and body symmetry through the use of weights" (Bell 2008, p. 43). This form of construction of the corporeal self "involves the labor-intensive process of attaining the imagined 'perfect' muscular body, but the product, in effect the climax of a fundamentally aesthetic movement, is extremely fleeting" (Locks and Richardson 2012, p. 3).

In fact, the competitive practice, whether it is drug-free or based on PIEDs, requires a high level of involvement and discipline. It involves much more than simply moving weights at the gym, the use of tools, the application of methods, the following of rules, routines, orders and rituals. Everything has to be structured according to each individual's specific goals: the competition category (which up until now has been binary—male or female), the requirements of the shape of the body, the starting point from which to build the specific shape, an account for any limitations, and metabolism. Klein (2007) provides detailed descriptions of the daily routines of bodybuilders, including workouts, strategies to modify appearance (e.g., tanning, shaving, poses), body depilation (i.e., hair removal) and food practices (including severe caloric restriction).

Coaches often assume a supportive and controlling role in training, determining the necessary patterns for carrying out the preparation and presentation. Training with a coach also involves interpersonal interaction rituals that are based on power relations. The coach, male or female, might use informational, expert, reward, and—depending on the type of social relation—identification power, an aspect of "symbolic power" in Bourdieu's frame. In one of his definitions of "symbolic power" in *Masculine Domination* Bourdieu writes, "the dominated, often unwittingly, sometimes unwillingly, contribute to their own domination by tacitly accepting the limits imposed" and that symbolic power often takes "the form of bodily emotions—shame, humiliation, timidity, anxiety, guilt—or passions and sentiments—love, admiration, respect" (Bourdieu 2001, p. 38). The entire interaction between bodybuilder and coach is likewise structured by the more or less hierarchically ordered distribution of capital, either symmetric or asymmetric, as well as by their general social position, which includes gender, ethnicity, and age.

The process of bodily transformation, often compared to a metamorphosis, can last several months or even years, when bodybuilders have to go through different phases of preparation to build up their "specific body capital." They aim either to build more muscle or to lose weight.

When gaining muscle mass is prioritized—"bulking"—bodybuilders eat more, mainly carbohydrates, in order to increase energy and to speed recovery. In doing so, they can move more weight, which stimulates the muscles and makes them grow, constructing the "pumped phallicized body of the woman bodybuilder" (Jagodzinski 2003, p. 35).

When bodybuilders work on the "definition phase," where losing fat is the main goal, they reduce the intake of calories and increase the cardiovascular elements of their workouts and their duration. In this phase, training is focused on showing a shape that is as ripped,[4] cut, striated, shredded,[5] or lean as possible.

---

[4] The term "ripped" describes "a person who has more separation in the muscles than someone who is merely 'cut.' The separation of muscles is more defined in that some cross-striations are visible. This condition occurs when a bodybuilder has low levels of body fat and excellent muscle separation. Sometimes referred to as 'slide'" (Lowe 1998, p. 180).

[5] Being "shredded," means: "The highest level of muscle separation. There is no subcutaneous water or fat between the skin and the muscles, to the extent that muscle fibers (cross-striations) can be seen easily (there are ridges in the muscles)" (Lowe 1998, p. 180).

In the physical appearance in this phase, gender differences according to the "two-sex model" (Laqueur 1990, p. 16), are also expressed.

> A male bodybuilder who is shredded is the ultimate because he has the quality muscle and has effectively purged his body of excess water and fat. However, female bodybuilders are judged down for getting "shredded" because it is assumed, they can get this way only through the use of illegal substances and diuretics. (Lowe 1998, p. 180)

What bodybuilders are aiming for during this leaning period is a way to make the cross striation of the muscles, fibers, and the veins more visible. Making the veins remarkable—showing "vascularity"—is particularly evaluated in contests.

Usually, when bodybuilders are trying to push themselves to this extreme point—aiming to be maximally ripped and shredded—they have to develop strategies, which look quite unhealthy and deviant, to make their body react. One example involves manipulating their water intake. About one to two weeks before a contest, a bodybuilder will drink around seven to eight liters of water per day. A few days before the competition, this step is then followed by a process known as water cutting, which aims to eliminate the subcutaneous layer of water that exists between the skin and the muscle, "dehydration."

Bodybuilders also aim at a "deload of food". This process starts in the weeks leading up to a contest and often continues until the day before its start. It is followed by "carbohydrate loading," where the intake of carbohydrates increases. They are expected to eat carbohydrates and fats, but not to drink anything. This mixture of procedures, among others, results in a body covered by muscles, without any fat layer. It brings the agents to the optimal bulging point, which is labeled as being "pumped." (Fig. 9.1)

Thus, the aim of the practice as sport might be described as building an ephemeral sculpture, showing it on stage, and exalting it through flexing poses. This lasts only a few seconds and largely ignores the physical costs the preparation of such a process might entail. "We could call this (flexing pose) the 'penile pump' as the bodybuilder tries to display as much tumescent muscle as possible; the skin must be well-tanned and oiled, the physique rock-hard, showing striations and bulging veins—in other words to look as much as a giant erection as possible" (Jagodzinski 2003, p. 24).

Most competitive bodybuilders are organized in and tend to participate in "non-natural" federations, using drugs to help accomplish the goal of perfection: building their "best version." This process might be interpreted as a specific "technology of the self," subsumed by Foucault under "technologies of individual

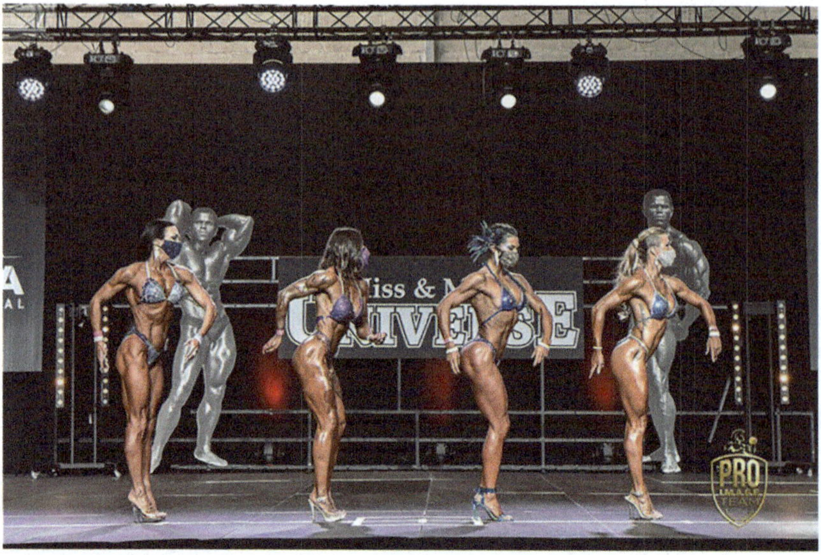

Fig. 9.1 Ms. and Mr. Universe contest 2020. Miss shape category, Isabel Fontbona third from left. Organized by World Amateur Body Building Association (WABBA), Estoril (Lisbon), December 7th 2020. *Photo*: PRO Image Team

domination"; that is, when individuals act upon themselves. These technologies "permit individuals to effect by their own means or with the help of others a certain number of operations on their own bodies and souls, thoughts, conduct, and way of being, so as to transform themselves in order to attain a certain state of happiness, purity, wisdom, perfection, or immortality" (Foucault 1988, p. 18).

9.3 Pierre Bourdieu and Loïc Wacquant on Bodybuilding and Sports

In *Distinction,* Bourdieu presented a photo and a portrait of a man showcasing and talking about his muscles and strength: "I was no weakling for my age when I started, but all the same I've put 5 inches on my shoulders, 3 inches on my chest and 1 1/2 inches on my arms, and all that in just three months. It's beyond my wildest hopes. My muscles are several inches bigger." (Bourdieu 1984, p. 210).

Other authors emphasize that in "bodybuilding the development of the muscular physique (it means, 'getting bigger') is the final goal" (Locks and Richardson 2012,

p. 21). For this reason, this process of bodily transformation from the perspective of traditional field-habitus-capital theory, has been largely related to the masculine domain, focusing on male action and the demonstration of power: "Muscles are *the* distinctive symbol of masculinity" (Wacquant 1995, p. 171). Obviously, this symbolic difference is constructed to naturalize a social hierarchy, because apart from specific muscles that are associated with male and female genitalia, muscle anatomy does not differ between men and women (Scott 2011, p. 71). As Bourdieu emphasized, "The particular strength of the masculine sociodicy comes from the fact that it combines and condenses two operations: it legitimates a relationship of domination by embedding it in a biological nature that is itself a naturalized social construction" (Bourdieu 2001, pp. 22–23).

Marcuse similarly pointed to the naturalization of difference after a "long historical process in which the social, mental and even physiological characteristics of women developed as different from and contrasting with those of men" (Marcuse 1974, p. 280). In response to the question about whether female characteristics are in any sense "natural", he noted, "over and above the obviously physiological differences between male and female, the feminine characteristics are socially conditioned. However, the long process of thousands of years of social conditioning means that they may become 'second nature,' which is not changed automatically by the establishment of new social institutions" (Marcuse 1974, p. 280). Thomas Laqueur (1990) described the discursive making and unmaking of sex over the centuries in detail, from the ancients to the moderns in a precise account of developments in reproductive anatomy and physiology, the plots into which they have been woven by texts, images and rhetoric of scientists, theorists, literary and artistic figures, and theorists.

Images and stereotypes conventionally associated with women include, among other things, passivity, receptivity, sensitivity, nonviolence, tenderness, the promotion of the aesthetics of delicacy and vulnerability, occupying the least possible space, or the creation of weak and skinny bodies. A thinner body ideal and anorectic appearance are socially much more accepted in women. This became particularly true after feminism turned away from motherhood and its bodily connotations, in visualizations including large breasts and the wider, broadening hips that often come with childbirth (Jagodzinski 2003, p. 29).

Bourdieu, in his work on masculine domination, refers to the role education and its rituals—"rites of passage" and "rites of institution"—play in underlining the somatized construction of sexual difference and the mechanisms of the social exclusion of women—based on male domination—in practices of sports. Implicitly, he subsumes bodybuilding in this formulation:

> Education exercises a psychosomatic action leading to the somatization of sexual difference, that is, of masculine domination. One particularly important domain of application of this work of psychosomatic inculcation is the embodied construction of social differences between the sexes. It operates according to several modalities. The first consists in rites of institution (rather than rites of passage: Bourdieu, 1982/1990), such as circumcision, which mark the opposition not between a "before" and an "after," "youth" and "adulthood," but between those who participate in the rite—men—and those who do not—women (historically, sports have played this critical role in our societies). (Bourdieu 1996–7, p. 195)

In *Distinction*, Bourdieu differentiates between the different benefits of sports, and he stresses the class connotations, that are linked with different kinds of sports. He explicitly mentions bodybuilding within the context of enumerating benefits labeled as "physical," "social," "economic," and "symbolic." In his words, the choice of sports links the following:

a) "immediate or deferred 'physical' benefits (health, beauty, strength, whether obviously visible through body-building or discreetly visible through 'keep-fit' exercises)";
b) "economic and social benefits (upward mobility etc.) … including gains in distinction … regarding the access to highly selective groups which some of these sports give (golf, polo etc.)";
c) "immediate or deferred symbolic benefits linked to the distributional or positional value of each of the sports considered (i.e., … greater or lesser rarity, and its more or less clear association with a class, with boxing, football, rugby or body-building evoking the working classes, tennis and skiing the bourgeoisie and golf the upper bourgeoisie)." (Bourdieu 1984, p. 20)

In his model of social space, Bourdieu makes a tripartite class differentiation, furthermore distinguishing among factions. These three groups are labeled as working class, petit bourgeois, and bourgeois respectively. In this context, he links bodybuilding again with the working class. In his approach, demands—and habitus—related to bodily practices are closely associated with social positions. First, in his three-dimensional model, the classes and class fractions agents belong to are defined by their cultural and economic capital. Their volume (the sum of the species of capital) and their structure (the relations of the species of capital) are as relevant as the trajectory of the agents in social space (i.e., the space of classes) over time (intra- and intergenerational mobility, transversality). Second, habitus and (bodily) practices are associated with gender, age, ethnicity, region (country of origin, urban center/capital vs. province/periphery), as well as with social capital, properties that are often more hidden. They interact more or less strongly with the primary factors of the division of habitus, practices, and lifestyles, which are the volume and structure of capital (*see* Bourdieu 1984, pp. 104, 107–108, 112), not to forget field specific capital (as artist, as bodybuilder etc.).

From this perspective, the way gender is enacted is dependent to a high degree on one's position in the class structure, which is defined by the volume and structure of capital one has inherited or achieved. Bourdieu's specific version of what later was called the gender intersectionality thesis, mentioned but not fully reconstructed by Fowler (2003), prioritizing class over what Anglo-Saxon social theory does not shy to call "race", reads:

> Sexual properties are as inseparable from class properties as the yellowness of a lemon is from its acidity: a class is defined in an essential respect by the place and value it gives to the two sexes and to their socially constituted dispositions. This is why there are as many ways of realizing femininity as there are classes and class fractions, and the division of labor between the sexes takes quite different forms, both in practices and in representations, in the different social classes. (Bourdieu 1984, pp. 107–108)

Bodily class (fraction) demands are characterized by Bourdieu in the following way: "one is practically never entitled to assume that the different classes expect the same thing from the same practice. For example, gymnastics may be asked—this is the popular demand, satisfied by body-building—to produce a strong body, bearing the external signs of its strength" (Bourdieu 1984, p. 211). A "healthy body" on the other hand reflects bourgeois demand, "satisfied by 'keep-fit' exercises or 'slimnastics.'" And for specific fractions of the middle and upper classes, he underlines the demands "characteristic of women in the new fractions of the bourgeoisie and petite bourgeoisie," which emphasize the "'new gymnastics,' a 'liberated' body" (Bourdieu 1984, p. 211). Lisa Lyon, with her art school background, and her social capital, which gave her access to the upper regions of the fields of art, media, and fashion, could serve as a celebrity example.

Klein (1985) did not refer to Bourdieu, but discussed the class connotations of bodybuilding as well, based on his ethnographic research in Gold, California's—and perhaps the world's—most famous gym in the 1980s. The results were illuminating regarding the statistical interaction of gender with class: Among his sample of forty men, 79% had come from a blue-collar, working-class background. Only 16% had graduated. In contrast, in the sample of thirty-five Gold's females, 69% came from professional or semiprofessional homes. Their educational background reflects this difference as well: 80% attended college and 40%

graduated. The jobs that women at the gym hold included a high proportion of skilled occupations such as teaching, editing, business ownership, and performing arts.

9.4 Bodybuilding as a Kind of Sport vs. Bodybuilding in the Artistic Field and in Popular Visual Culture

Bodybuilding cannot be considered solely as a kind of sport, as authors like Bourdieu and Wacquant, who brought the body back into sociology might have assumed. It is true that during preparation, its quality as sport is quite manifest (for example, planned diets are similar to those of professional athletes). On the other hand, there is much higher average retirement age than in professional sport. Klein, who discusses differences and similarities at length, concludes that "bodybuilding rests precipitously between sport and spectacle" (Klein 2001, p. 415), without considering bodybuilding's connections e.g. with the arts.

Bodybuilding has been considered as a practice closely related to sculptural art since the 1970s and early 1980s, when connections to the fields of art, fashion and media were being established. These links enhanced its visibility and acceptance beyond the lower regions of social space where especially male bodybuilding had previously been encapsulated. Female bodybuilding already in the late 1970s had a clear middleclass connotation.

Soon after the pioneering impulses of Mapplethorpe and Lyon, female bodybuilding was appropriated on a broader scale not only in the media but as an edgy type of presentation of the self and the body in the field of contemporary art. Numerous renowned contemporary artists explored it, including female artists like Nicole Eisenman, Annie Leibovitz, Cindy Sherman, Nancy Spero, Valérie Belin, and male artists like Matthew Barney and Andres Serrano, as well as an alternative comics artist figure like Robert Crumb (*see* Frueh et al. 2000; Crumb 1996).

We will offer two illustrative examples only, both trying to transcend aspects of the centuries old two-sex story. British performance and video artist Francesca Steele (2009) intentionally tries to challenge expectations about femininity by reconfiguring her body within codes that are usually ascribed to men, in other words, through muscles. In her own words, for her performance art,[6] for which

[6] *See* Steele's performance, *Routine*, from 2010 at The National Review of Live Art (The National Review of Live Art database n.d. Francesca Steele. Routine.).

she is also uses hormones, she is going "through the early stages of a female to male gender transition." This involves "a deeper voice, increased facial hair; increased body hair, including a new line of growth from my pubic hair to my navel; and enlarged clitoris; and altered personal odour..." (Steele 2019, p. 51).

Similarly, Cassils, a transgender performance artist, attacks binary identity conceptions of patriarchal civilization by using their body as a "social sculpture,"[7] playing with and, at the same time, denouncing the fluidity of such identities. In performing, Cassils also intentionally transcends conventional bodybuilding sports regulations. Duality is attacked, and even melted away, opening up new possibilities that overcome seemingly immutable categorizations (*see* Wickstrom 2014; Steinbock 2014).

This malleability, which is characteristic of a body modified through discipline—technology of the self—allows Cassils to redraw the deeply embedded line between the masculine-feminine binary. Via their interventions, the artist attempts to play with the fluidity of gender, taking flight from dualism: "I use my exaggerated physique to intervene in various contexts in order to interrogate systems of power and control" (Cassils quoted in McTavish 2015, p. 80). Cassils creates a transgressive body, often considered as monstrous, asexual, and sterile as long as it is linked to a "hyper-masculine and transgendered body" (Getsy and Steinmetz 2015, p. 20).

Looking at artistic practices of a less edgy kind, Heywood not only emphasizes the role that "third wave feminism" played for legitimizing female bodybuilding, but also hints at the functions of visualization, especially of the production and social use of a certain type of "art photography" for expanding the practice and increasing its visibility. In this context, she compares the work of two photographers, Bill Dobbins and Bill Lowenburg. Dobbins (1994, 1999), who also collaborated with Schwarzenegger, glorifies and naturalizes female bodybuilders, trying to reach a general public by marketing bodybuilders' sex appeal and erotic capital, including through the delivery of pornographic pictures. Lowenburg, on the other hand, tends to focus on bodybuilding's problematic, bizarre, and monstrous qualities, striving "for an alienation effect that makes the viewer reflect critically on the subject" (Heywood 1998, p. 93). Of course, there are noteworthy female art photographers as well, e.g. Sarah van Ouwerkerk, who

[7] Cassils, as is common in the queer community, uses plural, nonbinary pronouns like they, them, and their to refer to themself.

participated in the exhibition *Picturing the Modern Amazon* at the New Museum in New York in 2000,[8] or Valérie Belin.[9]

While Heywood still classified bodybuilding as a subspecies of a sports practice—though she was also aware of its integration in the (visual) cultural fields of artistic photography and eroticism/pornography[10]—the artistic sculptural aspect of bodybuilding is especially highlighted by Jörg Scheller, who curated a big bodybuilding exhibition in Kunsthalle Zürich in 2015 and 2016.[11] He coined the phrase, "*Sculpto, ergo* sum"—"I sculpt, therefore I am" (Scheller 2017, p. 42). The argument that bodybuilding produces sculptures is of course an older one, as demonstrated in *The New Encyclopedia of Bodybuilding*.[12]

Bordo spoke of "plasticity as postmodern paradigm" (1990). This idea implies the spreading of empowerment convictions that there are no limits to the possibilities for shaping our bodies. This belief is heavily supported by the rise of the constructivist paradigm in academia and constructivist feminist theory since the 1980s (*see* Heywood 2006).

The concept of bodybuilding and the voluntaristic perceptions of the body as a potential sculpture that can be shaped freely while ignoring biological, social, or cultural boundaries, seems also to be based on this paradigm. Another influence is "technological optimism," which implies that the body might be constructed through the use of pharmacological and medical means. Likewise, ideas about the ways the body can be modified through the use of various forms of prostheses have gained popularity since the 1970s. New digital technologies replace or augment the functioning of inner organs, body surfaces, senses, and limbs (*see*

[8] *See* the documentation in the Galleries section of van Ouwerkerk's website (van Ouwerkerk, Sarah. n.d. Sarah van Ouwerkerk Photography) and in the catalog of this exhibition (Frueh et al. 2000).

[9] Valérie Belin's work on male and female bodybuilders aims at depicting the "plasticity of the body" and "the way it can be manipulated, transformed and even deformed" (*see* van Ouwerkerk, Sarah. 1999. Sarah van Ouwerkerk Photography. Bodybuilders II).

[10] On the relations of bodybuilding to soft—as well as hard-core pornography, *see* Heywood 1998 and Kimmel 2005, pp. 87–89. *See* also "female muscle worship" pages on the internet.

[11] *See* Kunsthalle Zürich 2015–2016. Exhibition: Building Modern Bodies. The Art of Bodybuilding Curated by Jörg Scheller.

[12] "The bodybuilder's physique is a carefully balanced combination of many factors, including shape, proportion, and symmetry. Bodybuilding has been compared with sculpture, with the bodybuilder creating and shaping a physique the way the artist sculpts a statue from marble or granite" (Schwarzenegger and Dobbins 1998, p. 200).

Featherstone 1999). "Much of bodybuilding's imprint," Liokaftos furthermore notes, "owes to its association with the larger phenomenon of human enhancement through pharmacological means. Shifting and competing notions of 'nature,' 'development,' and 'health' central to this phenomenon are forcefully inscribed on the materiality and imagination of the physical self in bodybuilding" (Liokaftos 2019, p. 754).

The artistic aspects of bodybuilding need not necessarily be based on its potential sculptural dimensions. Bodybuilding might also be justified via the "institutional definition of art." To classify some practices of bodybuilding in this way would not only be compatible with the definition of art which was suggested by philosopher Dickie, but also with the labeling approach represented by Becker's art world theory (2008) and a field-theoretical approach like Bourdieu's (1993), all of which have been reconstructed well by van Maanen (2009) (Fig. 9.2).

In contrast to sociologists of sport like Klein or Liokaftos, Aranyosi argues from a phenomenological perspective: "bodybuilding should not qualify as a sport, given that at the competition stage it lacks an essential feature of sports, namely, skillful activity" (Aranyosi 2018, p. 401). While sports are about manifestations of *Leib*, he argues, the purpose of bodybuilding competitions is to present the *Körper*. This quality would align it more closely with an activity like beauty competitions or classical sculpture. Other attempts to deny bodybuilding's classification as a sport emphasize its aesthetic aspects, how the activity is tied to visual beauty. The primary criterion would be how the body looks. In contrast to a sport, where the training one does is similar to the competition one engages in, the rigors of training are not what matters. Instead, it is only about what the performer presents in the competition itself: "in bodybuilding training and competition differ" (Heywood 1998, p. 96).

It does not seem necessary, however, to subsume bodybuilding into only one or two social fields. From field-theoretical point of view, bodybuilding can be classified according to all the fields it is actually integrated in. These fields might include entertainment, sports, beauty competitions, performative or performance art, visual art, photography, fashion, and eroticism and pornography among others, but it could also include bodybuilding's status as an activity, agents simply pursue as a noncompetitive lifestyle practice. This is our position, which contrasts with Bourdieu's treatment of bodybuilding and Heywood's approach, which is not inspired by sociological differentiation theory.

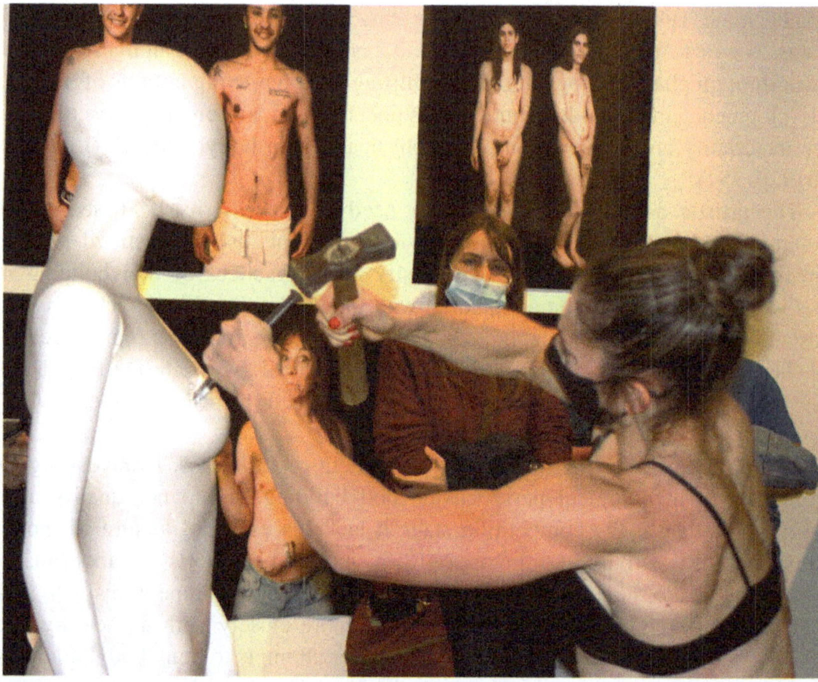

Fig. 9.2 Artistic performance of Isabel Fontbona 'Corporeal Turbulences. Between Absurdity and Resistance'. Panoràmic Festival 2020 'Extimacy'. Exhibition: 'The other ubiquitous. Fragmented body migrations'. Arts Santa Monica's Contemporary Cultural Center, Barcelona, October 20th 2020. *Photo*: Eugeni Prieto

9.5 Aspects of the History of (Female) Bodybuilding in the Second Half of the Twentieth Century: Primary Social Antagonisms and Gender Tensions

During the 1950s, Muscle Beach at Venice in Southern California became the congregation par excellence for most bodybuilding adherents. It was the perfect setting for something as visually exotic as this professional subculture (Klein 2001, p. 41; Locks and Richardson 2012, p. 10).

The Muscle Beach scene also produced Pudgy Stockton, a professional strongwoman who quickly became a favorite of the media. Already by the end of the 1940s, she had been featured in dozens of magazines, including *Life*. She was considered to be a turning point in the subsequent development of female bodybuilding because she presented a figure "with miraculous curves, and amazing strength … competent, feminine, strong and sexy" (Todd 2000, p. 56). She paved the way for female bodybuilding becoming popular in a first step, without having the art and fashion model appeal of a Lisa Lyon or a media woman charisma of a Rachel McLish in the 1980s.

Bodybuilding developed quickly after World War II, mainly in the US and the UK. The IFFB emerged as dominant organization, founded in the mid-1940s by Joe and Ben Weider. They introduced the Mr. Olympia competition in 1965, which was to become the ultimate professional contest, "Big O." In the late 1960s, bodybuilding experienced significant changes due to the increased use of performance enhancing drugs. The use and abuse of these substances allowed more extreme results in terms of muscle definition and massiveness.

The 1970s were a time when the practice continued to grow in popularity. Audiences grew, media interest increased, and competitions drew more participants. Arnold Schwarzenegger, bodybuilding's biggest star, in this decade won the Mr. Olympia competitions—which as a sign of internationalization took place in Paris and New York, but also in Essen, Germany and Pretoria, South Africa—six times.

It was not until the second half of the 1970s, after the rise of the "second wave" feminist movement and the proliferation of female powerlifting events, that women were judged as capable of competing in their own bodybuilding competitions. It was a man, however, Henry McGhee, who in 1977 founded the United States Women's Physique Association (USWPA). He was convinced that women should be expected to display their body in a similar way that men did. The first official female bodybuilding competition, was held on a regional level in Canton, Ohio, in November 1977. The first US National Women's Physique Championship followed in the next year. McGhee assured the competitors that they would be judged "like the men," with an emphasis on muscular development, symmetry, and physique presentation.

The Best in the World contest of August 1979 was the first IFBB-sanctioned bodybuilding event for women that awarded prize money to the top finishers. It was also the forerunner for the Ms. Olympia competitions of the following decades. In a field strongly supporting as well as demanding the visualization of appearances they offered the highest degree of media visibility, i.e. popular success. Although these events were sanctioned as bodybuilding contests, the participants were required to appear on stage in high heels. Thus, early in the history of institutionalized female bodybuilding, the foundations for the tensions characterizing the field, the oppositions between size and aesthetics, between muscles and femininity, between pure and heteronomous practices were implemented.

Lyon was the first female bodybuilders of high visibility to combine the showcasing of muscles with male defined "femininity." She is the figure who, seen from the perspective of her famous peer Schwarzenegger, "essentially invented the kind of combination of muscle-posing and dancelike movements that characterizes presentation in women's contests to this day" (Schwarzenegger and Dobbins 1998, p. 45). Nevertheless, from the point of view of a well-known feminist sociologist of sport, her appearance on the stage of female bodybuilding could be perceived as an important change, since female bodies now convincingly "argued that muscles are not the exclusive domain of men" (Bolin 2012, p. 29).

The 1970s finally also saw the first bodybuilding organization run exclusively for women and by women. Doris Barrilleaux founded the Superior Physique Association (SPA) in 1978. In April 1979, SPA held Florida's first official women's contest in which thirteen persons competed. That same year, the IFBB formed the IFBB Women's Committee and appointed a woman, Christine Zane, as the organization's first chairperson.

The first half of the 1980s was another era when female bodybuilding "took off" further. But it was also the period that saw the internal antagonism of the field along the autonomy-heteronomy axis enforced: size and muscle vs. aesthetics, femininity, and more and more eroticism. And it was also the time when women's bodybuilding was officially recognized as a sport discipline, which happened at the 1982 IFBB Congress in Brugge, Belgium.

In 1980, the first Ms. Olympia contest, which was to become the most prestigious competition, was held. Rachel McLish, the competition's first winner, was one of the main figures during the eighties. In addition to being a successful athlete, she turned out to be a very "promotable" person as well, a cheerleader type beauty, receiving a large amount of media attention. She inspired many women to start training and competing, not least through publishing a series of popular books on bodybuilding with titles like *Superflex*, *Flex Appeal* or *Perfect Parts* (e.g. McLish and Reynolds 1984; McLish and Vedral 1987). Cory Everson, who dominated the field of female bodybuilding in the second half of the eighties, continued in this tradition by winning six consecutive Ms. Olympia titles from 1984 to 1989.

The field-specific antagonism during this era is easy to reconstruct due to the fact that it was taken up in visual culture and thus stored in cultural memory. In 1985, the documentary *Pumping Iron II: The Women* was released. This movie documented the preparation of several women for the 1983 Caesars Palace World Cup Championship in Las Vegas. Among the competitors prominently featured in this film, whose main theme was nothing else than "What is femininity?" were Bev Francis and Rachel McLish. The movie pitted them against one another in a symbolic way. McLish was represented as a "sultry and curvaceous woman," while Francis, the main champion of this period was represented as "almost manly, super-muscular" (Pumping Iron [Film] 1985), in Bourdieu's frame of reference, an opposition between high erotic combined with low athletic capitals vs. the inverse structure of capital.

This "rivalry" and its visual narrative exposes the roots of many controversies in the field. "Pure" bodybuilders in the relatively autonomous subfield—those who emphasize body size and muscles—are competing with rivals who, in a heteronomous way, are being supported by the demands of a public who wants to see a "show," as well as by competition regulations (developed by men) that emphasize "femininity," which in this context can be read as a euphemism for "sex appeal" and "erotic capital" (Hakim 2011).

The history of bodybuilding in the 1980s is also remarkable for the fact that representations of built sculptures of flesh in magazines began to more commonly objectify and sexualize female bodies. In media from that time, it is not difficult to find headlines that link female bodybuilders with terms that are more closely aligned to their "hotness" rather than to the muscular improvements they have achieved. This evolution did not happen without the complicity of several successful bodybuilders. Like Lisa Lyon, four-time Olympia competitor Anita Gandol posed for *Playboy* (in 1984), and Erika Mes, a two-time Olympia pro-bodybuilder and the 1980 NABBA Universe champion from the Netherlands, posed for the Belgian edition of *Playboy* in 1987. Because the IFBB's interpretation of femininity was less sexualized, both women earned a one-year suspension from the federation as a result of their soft porn magazine presentations.

The debate about aesthetics, femininity, and beauty vs. size and muscles took another direction in the 1980s, when successful bodybuilding athletes like Bev Francis and Kim Chizevsky-Nicholls emerged (. These women emphasized strength and power via muscle development, which broke with the traditional stereotypes that had been linked to the female body. At this point, the issue was no longer about empowering the female body, but instead about breaking the image of the "weaker sex" and seeing the extent to which the limits of "femininity" could be pushed through the muscle.

In bodybuilding, as it is it any field of cultural production, when changes to the fields' game are of interest, it is the players with a high level of field-specific symbolic capital whose moves and effects have to be considered. In the early 1990s, Linda Murray succeeded Cory Everson as the next major dominant figure in the bodybuilding sport field. Like Bev Francis, she displayed an increased body size. In reaction to the larger body size Murray and Francis displayed as previous Ms. Olympias the IFBB made efforts to "feminize" the sport again. For example, the requirement that competitors should not be "too big" was an attempt to retain a higher level of "femininity."

In 1995, a new subcategory of female bodybuilding, the "physique contest," was introduced with the aim of fostering conventional femininity. In 2000, a further category, "figure" was introduced. In 1996, however, Kim Chizevsky-Nicholls, who represented a "pure" bodybuilding model, became the next great champion by winning both the Ms. International and Ms. Olympia contests (Dobbins 1999). According to Bolin (2012), Chizevsky-Nicholls broke with the common type of female embodiment that had prevailed since 1992, a prototype that was stylish and smooth—influenced by the appearance of British fashion model Kate Moss—heavily promoted by the fashion industry. After winning the 1999 Ms. Olympia title, Chizevsky-Nicholls decided to retire, a response to the gender guidelines, set up by the men who dominated the IFBB, that advocated for less "muscularity" of women.

9.6 The Twenty-First Century: The Evolution vs. Decline of Female Bodybuilding Disputes

The main dynamics of female bodybuilding in the twenty-first century before the spread of the corona pandemics, occurred on the structural level. The changes are characterized by the field's internationalization and globalization as well as its internal differentiation. These tendencies amounted to an expansion of the field as a whole, though from the perspective of several observers, the practice's "golden era" had already passed. Depending on the observer, the end of the golden era had either already taken place in the 1990s or in the first half of the 2000s. In view of what had happened in previous years—including the abolishment of some of the activity's most iconic contests—some defenders of the view of decline even spoke about the "death" of female bodybuilding. Others, however, saw an "evolution", with symptoms of crisis—especially in the field's center, the United States—in "pure" bodybuilding only.

Bolin (2012) hints at four watershed events. The effects were caused by initiatives of the IFBB, "a near-monopoly at the professional level in the US and also international," the dominant governing body of both the professional field and its corporate players (Liokaftos 2019, p. 759).

The first change impacted the Ms. Olympia contest. In 2000, the IFBB no longer held this event as a separate contest, but instead turned it into part of the Olympia Weekend in Las Vegas; it was scheduled to take place the day before the men's show. The second event occurred when heavyweight and lightweight classes were abolished in 2005. The third change involved the introduction of new judging guidelines. Jim Manion, the chairman of the IFBB's Professional Judges Committee, stated in 2000 that women would be judged "on healthy appearance, face, makeup, and skin tone." The criteria given also included the (in) famous statement, "symmetry, presentation, separations, and muscularity BUT NOT TO THE EXTREME!".

This move could be interpreted as a reaction against successful bodybuilders, like Chizevsky-Nicholls, who, unlike former media stars like McLish, Everson, Murray, and Lyon, were not conventionally pretty, sexy, or marketable and did not represent what Rae Connell in his typology of female appearances had termed "emphasized femininity":

> One form [of femininity] is defined around compliance with this subordination and is oriented to accommodating the interests and desires of men. I will call this "emphasized femininity". Others are defined centrally by strategies of resistance or forms of non-compliance. Others again are defined by complex strategic combinations of compliance, resistance and co-operation. (Connell 1987, pp. 184–185)

The fourth transformation happened when the IFBB amateur rules for bodybuilding were changed in 2001:

> First and foremost, the judge must bear in mind that this is a women's bodybuilding competition, and that the goal is to find an ideal female physique. Therefore, the most important aspect is shape—a muscular yet feminine shape. The other aspects are similar to those described for assessing the male physique, but muscular development must not be carried to such an excess that it resembles the massive muscularity of the male physique. Competitors shall also be assessed on whether or not they carry themselves in a graceful manner while walking to and from their position onstage. (IFBB Amateur Rules for Bodybuilding 2001)

Regarding the internal differentiation of the field, in addition to using "bodybuilding" as a category, "fitness" competitions (originally pageants) were introduced in 1992. In 2001, "figure" competitions, also named "figure fitness", followed as a classification. This development was accompanied by its own changes as well. In the "aesthetic vs. size" antagonism, also known as the opposition between "beauty or the beast," the dominance clearly shifted more and more to the heteronomous pole of the field's axis.

In 2004, the (in)famous “20% rule” was introduced by IFBB Chairman Jim Manion: “For aesthetics and health reasons, the IFBB Professional Division requests that female athletes in Bodybuilding, Fitness and Figure decrease the amount of muscularity by a factor of 20%.”

These top-down changes were supported by the expected demands of fans and consumers; that is, the audiences of competitions, images and videos. In line with the complicity thesis, many competitors also adjusted to these expectations, though they did so in different ways. For instance, Chizevsky-Nicholls decided to move from a more muscular, “monstrous” appearance in order to accentuate more “feminine” and sweet lines. Others, like Monica Brant and Timea Majorova, took advantage of the increasing curiosity about and attention toward the muscular bodies of women and became “superstars” of fitness. More and more participants, like Denise Masino, even entered the sexual field of pornography, supported by the new visual culture of photos and videos with worldwide outreach based on the internet. Others, like Iris Kyle, refused to participate and remained in the subfield of “pure bodybuilding,” which had been labeled by its opponents as “freak culture” and a field of “monster beauties,” represented on the internet as well.

Years ago, it was virtually impossible to get in touch with elite athletes or artists, when not being an insider. The internet, however, has allowed to connect with them through social networks, by “following” them and by commenting on their photos and videos. But besides the most conventional methods of connecting, what other options exist for female bodybuilders? The ability to establish video calls or live videos has opened a new market. “Muscle worship,” to give an extreme example, is closely linked to the eroticism of the muscle (“erotic muscles,” “dirty muscles”). Whether it is of the narcissistic autoerotic type, or consumer-driven, muscle worship is also embedded in the advertising of prostitution and pornographic material on the internet. Richardson interprets aspects of the latter orientation “as a form of sexual fetishism … Muscle worshippers gain sexual gratification from actions such as ‘oiling up and massaging muscles’ and ‘watching a bodybuilder flexing (especially seeing the muscle bulge and swell) and displaying feats of strength’” (Richardson 2008, p. 290).

In terms of the differentiated field, “figure” soon developed as the fastest growing sector of the three categories of competition. This growth was not only enforced by corporate players and functionaries, but it also corresponded to the incorporated preferences of a certain subgroup of women as well, since the category is a far less physically demanding than fitness or bodybuilding in its pure sense and it is attractive for commercial reasons as well:

> Like fitness, figure competitions attract women who want to succeed in an industry where beauty and an embodied athletic aesthetic provide career/economic opportunities such as modeling, promoting, endorsing, acting, and other activities related to selling fitness products (i.e. clothing, supplements, weight-loss programs, equipment, etc.). (Bolin 2012, p. 46)

Figure competitors are judged in terms of their overall athletic appearance, as well for properties, which correspond to stereotypes of how women should look:

> the hair and facial beauty; the overall athletic development of the musculature, the presentation of a balanced, symmetrically developed physique, the condition of the skin and the skin tone; and the athlete's ability to present herself with confidence, poise and grace. The physique should neither be excessively muscular nor excessively lean and should be free from deep muscle separation and/or striations. Physiques that are considered too muscular or too lean must be marked down. (IFBB 2009)

A romanticized gender discourse penetrated the field, and the effect was an explosive growth of figure competition events and fitness pageant competitors. In this way, even transgressive bodies were tamed via self-objectivation and by exposing their sexuality for the male gaze (Szymanski et al. 2011). On the other hand, no matter how much the grand bodies of women bodybuilders are adorned with insignias of femininity, their transgressive bodies are regarded as less marketable than the bodies of the fitness and figure physique competitors.

While the two most important awards (Ms. Olympia and Ms. International) were abolished in 2013 and 2014, this was not really the end of female bodybuilding.[13] At the same time that the erasure of female bodybuilding was diagnosed—based on a study, which unfortunately referred to representations in old media only (Bolin 2012) and ignored bodybuilding's broad visual representation on the internet—one could observe a growing penetration of body practices, aesthetics, and philosophies that originated in bodybuilding.

A basic feature of bodybuilding's development in the period before in 2020 the coronavirus pandemic changed so much, was the proliferation of national and international organizations and competitive events in different parts of the world. Another important aspect was the internationalization strategy of this period. This process has included the formation of affiliations with existing organizations,

[13] In 2019, the documentary *Too Big for the World* was released, presenting reasons why female bodybuilding in its pure version has been marginalized over the last decade (Too Big for the World. Irene Anderson 2019).

the creation of branches in new countries and regions, and the increased number of international events. Bodybuilding was rapidly becoming globalized, both as sport and body culture (Liokaftos 2019).

Detractors and skeptics claim that female bodybuilding is on the decline or has even undergone erasure, but others argue that these diagnoses fail to recognize what was actually happening in the field after its hype in the Anglo-Saxon world in the 1980s and 1990s.[14] The main structural changes seem to involve a large differentiation of forms and styles under the overall term of female bodybuilding, which is only partly used anymore, and the fact that since the mid-2000s, hyper-muscular size based presentations were appearing less frequently on competition stages. While the number of contests focusing on larger physiques as well as the term female bodybuilding in book publications has certainly decreased, that does not mean, as some argue, that it has faded away as a practice completely. The number of women who want to develop their bodies has increased. For example, through a quick internet search, one can see that the muscular female form is more visible now than at any other point in human history. In July 2020 there were e.g. not less than 27 k images of female bodybuilders and muscled women presented in photo and video sharing social networking Instagram only (Instagram #femalebodybuilders n.d.). There likewise exists a huge number of fans, partly driven by the erotization of the practice represented via images and videos on the internet.

Bodybuilding can no longer just be called *one* sport, cultural activity or subculture. New categories have evolved beyond "pure" bodybuilding—fitness, figure, physique, and also bikini.[15] And there are also some other reasons for that, beyond corporate commercial interests and hegemonic masculine interventions. Women, in fact, are different in their habitus as well as in their body shape and corporal dispositions. The figure category, for example, seems appropriate for those who are not able to perform acrobatics on stage and who are larger. Thus,

---

[14] A decline of "female bodybuilding" in English as an N Gram after the turn of the millennium can be observed in the time series curves of Google Books N Gram Viewer, however, this tool's units are alphabetic signs and not images (Google Books N Gram Viewer n.d.).

[15] Bikini Fitness, "the softer look," is currently the most popular category of the bodybuilding subfield on its heteronomous pole. Competitors wear "bling-bling" bikinis, high heels, and jewellery. They are judged on their overall physical appearance, including their complexion and presentation. One of the most relevant aspects for evaluation is the way women move on stage; they have to perform model walks and hit front, rear, and over-the-shoulder poses.

it seems to be a choice for women who are unable to do the gymnastic moves required by the fitness category and who are not muscular enough for the bodybuilding category (*see* Femuscleblog 2017).

The divisions according to weight and height, and the differentiations between the bikini, figure, physique, fitness, and pure bodybuilding categories, give women of multiple body types the chance to participate on stage. Women of ectomorphic types would find fitness and figure more to their preference, while larger women of the mesomorphic or endomorphic type would find the bodybuilding category more suitable. What should be clear, however, is that female bodybuilding is not a beauty pageant—these types of competitions differ greatly from a pageant.

Another argument beyond diversity of body size and shape refers to the fact that there is also a diversity of ideas among fans, or among the bodybuilding subculture in general, in terms of which body type best represents the ideal—a sleeker model, a mid-range size, or a body with larger musculature.

With these arguments in mind, it seems clear female bodybuilding has not died in the twenty-first century or been replaced by a beauty pageant format. Instead, it is evolving along internal differentiation, developing factions that account for greater diversity and that reflect different aesthetic ideals. Female bodybuilding has not died, but changed form.

With the construction of these new categories of competition, women are also able to switch between them if they desire and if they feel a certain category would provide a better fit. To have such possibilities for transversality is an advantage, and there is more flexibility in the practice than ever before.

Some critics argue that evidence of the practice's decline can also be found in the fact that during the golden age, ticket sales to the Ms. Olympia contests were high, but that they declined in the twenty-first century; likewise, even during the golden era, women's pay in terms of prize money often was much lower compared to their male counterparts (the difference was sometimes as high as a factor of ten).

The rise of the internet and new media, however, has also given women the opportunity to start their own websites rather than waiting for an organization or a fitness magazine to promote them. Pay websites and session wrestling, for example, have become new methods to finance their athletic endeavors, and hardcore fans, known as "schmoes," have served as a financial lifeline in an industry that tends to ignore its athletes. If this new business model seems to be partly or wholly inappropriate, it can be argued that it is not more inappropriate than the ways that fitness magazines have previously presented women in an objectified, sexualized way.

There has been an additional development in the field that may not have been intentional. Female bodybuilding has spread across the world, from the United States and Europe to Asia, Africa, Europe, and Latin America, a fact that is often ignored when detractors insist that female bodybuilding is dead. These critics overlook the fact that the practice has spread widely over numerous countries outside the West. If there has been a decline in the practice, it has primarily occurred in the United States and in some parts of the West. There are many other countries, however, including India, Brazil, South Korea, and Uganda, where it is easy to see that female bodybuilding has expanded.

9.7 Views on Female Bodybuilding—Gender Transgression and Parody. Subversive Acts, or the Reproduction of Ideals of Femininity?

The ideal female body historically has often (though certainly not always) been associated with the maternalistic ideals like "curves, voluptuousness and softness" (Shilling and Bunsell 2009, p. 146) or "softness, passivity and non-aggression" (Ian 1991). For this reason, the female bodybuilder has frequently been considered as a "gender outlaw" (Shilling and Bunsell 2009), as freaky, grotesque, bizarre, deviant, someone considered unfeminine who is causing "gender trouble" by entering into a masculine terrain and breaking with some ideals of maleness and with masculine domination in general.

Morton-Brown lucidly expresses what kinds of expectations female bodybuilders are challenging and disrupting:

> [F]emale bodybuilding is a performance of gender bending-women who construct hypermuscular bodies, disrupt social norms of gender, performing a kind of self-styled hermaphroditism that begs the question, "why?" … [F]emale body builders challenge identity labels of sex and gender because of the fact that their gender transgressions occur at the physiological level of the body … Female bodybuilders parody dominant labeling philosophies of sex, gender, and sexual orientation, challenging our sex/gender paradigm in very unique ways. (Morton-Brown 2004, p. 27)

From the perspective of some "third wave" feminist positions, female bodybuilding might seem to hold enormous potential for challenging traditional concepts of femininity, but several points might be made in response to the claim that it can serve as a form of politically effective practice or radical feminist empowerment. For example, even as female bodybuilding may represent a specific type of resistance to contemporary Western stereotypes surrounding femininity and what it means to be female, it has been argued, on the other hand, that the field of bodybuilding in the Western world, since its inception in the second half of the twentieth century, has been dominated by male actors and normatively regulated by male influence.

Bell (2008) points to three ambivalent or paradoxical effects of women being engaged in bodybuilding: First, female bodybuilding offers another ideal of bodily perfection to pursue obsessively; second, bodybuilding objectifies women; and third, bodybuilding allows for the idealization of muscular bodies in an intrinsically androcentric way.

Scott similarly argues that acts of subversion still operate under male-dominated conventions—the normative expectations and regulations governing the public presentation of female bodybuilders in contests and competitions—that further perpetuate a problematic form of ideal femininity: "bodybuilding may represent aspects of social resistance to the gendered norms of femininity, (but in front of this threat) the competitive regulations of this sport enforce compliance to societal expectations of womanhood" (Scott 2011, pp. 70–71). She sees the ideal of femininity upheld even when "gender-bending" occurs in bodybuilding.

> While these women may manipulate their bodies and challenge social norms of femininity, they also comply with those norms in competition when they are forced to minimize their musculature, submit to feminizing aesthetics and modify their posing routines to diminish their gender-bending physique ... [T]he physical body is an object of resistance in that it deviates from the socially acceptable norms of womanhood and demonstrates the power of the female body. In contrast to this physical resistance, the attitude of bodybuilding represents compliance to socialized norms of womanhood. (Scott 2011, pp. 81–82)

From the perspective of constructivist theories of gender performativity (Butler 1999; Mikkola n.d.), there are no essential properties; gender is an illusion maintained by prevalent power structures. From an action-theoretical perspective, gender only comes into being through gendering acts. Activities and practices *make* one gendered as a man or woman. Gender is not something one is, it is something one does. It is instituted through a *stylized repetition of acts*; that is, through wearing certain gender-coded clothing, walking, posing, dancing, realizing interactions in certain gender-coded ways, styling one's hair or face in a gender-coded manner, and so on. Gender only comes into being through these gendering acts, and genders are true and real only to the extent that they are performed.

From a related, discourse-theoretical perspective, gendered bodies are *discursively constructed*. They are the way they are, at least to a substantial extent, because of what is attributed to bodies or practices and how those bodies are classified. These assignments—calling someone, some properties, or some acts as female or male—are normative.

What can be observed on the normative level in terms of the regulations involved with and the poses demanded of the practice?

9.8 Female Bodybuilding Contests: Demands for Femininity

Bodybuilding contests are social rituals structured and normatively regulated in a quite formal and rigid way. Regarding regulations on the level of evaluation—on which expert and reward power are applied—there are, apart from gender neutral criteria, several gender specific ones. These are used by members of a jury, which is usually monopolized by men, to arrive at their judgments.

"Demands for femininity" are explicitly formulated, and in order for a participant to be successful in a contest, she must meet those demands. This turnabout evidently brings the bodily capital into a situation where it is presented and analyzed from a perspective that looks at the body's sexual appeal. "Femininity" in this context, "is synonymous with sexuality, or what used to be known as sex appeal" (Heywood 1998, p. 98). Heywood is not criticizing sex or its marketability, but the fact that the sexualizing of female bodybuilders makes sex (rather than the aims of the practice) the center of attention, which signifies a clear difference from male bodybuilding contests (Liang 2011).

These femininity requirements include "hairstyle, make up, (nail decoration, bikinis with sequins and rhinestones, jewelry) expression, posture (and in most of the cases a breast implant as well)" (St Martin and Gavey 1996, p. 55). These demands operate as a "balancing act"—like a traditional corset—between the competing expectations of femininity and muscularity. They are a mechanism, introduced to control what are seen as the exaggerations of more extremely shaped, muscled female bodies. Though there are some differences in these regulations between bodybuilding federations, they are all attempts to control and mold the preferred female image the federations would like to promote. For example, high heels and bikinis (their measurements, form, material, kind of rhinestones, etc.), jewels, tanning agents, and the glossy material used to dye the body are all elements that are measured and controlled (Fig. 9.3).

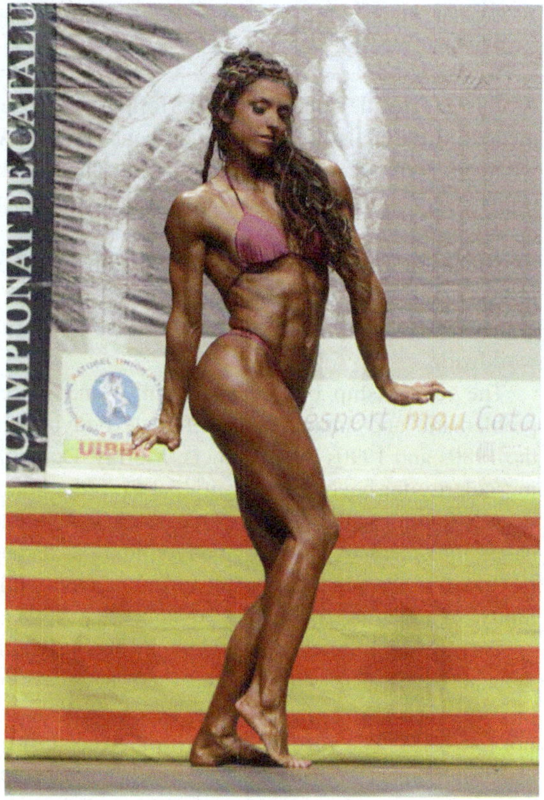

Fig. 9.3 Catalonia Championship 2014. Isabel Fontbona performing in female bodybuilding category (+52 kilos). Cornellà de Llobregat (Barcelona), May 24th 2014. *Photo*: Manuel Balaguer

But perhaps the most important aspects of these regulations are the standard poses required of participants, the so-called "attitude" of the practice. WNBB, for example, has established five poses for women and eight poses for men. In terms of the way the actors are supposed to look and behave when they present on stage, the regulations point out a number of conventions. They are most specific for women, to ensure the reproduction of gender dimorphism:

> In Bodybuilding for women and men: no jewelry except alliance. Note for women: Female athletes should be well proportioned without exaggeration hypertrophy or dry, have mass, volume, and muscle shape. They must imperatively have biceps significantly larger than the forearms, shoulders well rounded, a back well "V" and massive, thighs curved quadriceps, ischios and calves clearly developed and well projecting. Epilation or shaving of the entire body is mandatory (women and men).

> General attitude: Tell-standing, "dignified" head, smiling. (World Natural Body Building (WNBB) n.d.)

As this excerpt demonstrates, a performative gender dimension has been implemented in the male-dominated judgment process. A similar phenomenon occurs when one looks at the evaluation criteria. The criteria attempt to control the "grotesque" and "extreme" female bodybuilders by directing more attention to aspects of "beauty".

The institution that acts as an international gatekeeper for the rules and regulations of men's and women's bodybuilding competitions is the International Federation of Bodybuilding (IFBB), the most powerful association within the field worldwide.

The relationship between femininity and muscularity has been the central problematic and antagonism of female bodybuilding. Its main practitioners in the 1980s and 1990s (e.g., Francis, Murray, Dunlap, and Everson) pushed female muscularity far beyond its beauty-pageant-like boundaries. However, this did not occur without growing moral panic, to the extent that gender-bending bodies were called the "death of women's bodybuilding as a spectator sport" (Williams 2000, p. 105). The bodybuilding gatekeepers reacted with "the formalization of 'femininity' as a judging criterion" (Boyle 2005, p. 134).

Wayne DeMilla, vice president of the IFBB, described the heteronomous, demand-based factors that forced these new developments: "We saw that as the physiques became more extreme, we couldn't market it" (quoted in Boyle 2005, pp. 134–135). The IFFB decision to actively institute "femininity" as a judging criterion seems to mirror a broader shift within cultural and sports politics. It is a response to the gains made by the cultural left, and it is part of a broader cultural effort by the (populist) right to protect normative sex, gender, and heterosexual identity against a perceived attack.

With regard to the open and hidden construction of a gender binary and dimorphism, Bell suggests there should be a change to the expert power structure:

> Either women's competitions should be judged by female judges only (at least during a transitional period), or all judging panels should consist of half men and half women for both men's and women's events (excluding the head judge, to allow an odd number). Additionally, either 50% of all head judges should be women, or women's competitions should have female head judges. (Bell 2008, p. 45)

This could facilitate judgments that depart from "feminine beauty" and instead open the possibility for judging based on the standard of "muscular beauty."

The largest impediments to the kinds of changes Bell proposes are the male-dominated commercial interests controlling female bodybuilding. Practicing these changes in an alternate venue where the profit motive is not a primary force, which is the case, for example, in community and college teams, is thus another potential solution.

9.9 Conclusions

As this essay suggests, bodybuilding is part not of one social field only, but as an effect of its social differentiation over time embedded in several fields besides sport.

With regard to the potential of bodybuilding, it seems possible to first suggest that it is a terrain in which gender borders might be blurred and new interpretations might be made possible. In fact, bodybuilders are beings who often make people feel uncomfortable and irritated. Bodybuilding allows bodies to be built as "sites for a radical de- and (re)construction of gender to the limits" (Lambert 2017, p. 132). On stage in the field of sport, however, female bodybuilders are heavily controlled by regulations that are enforced by male-dominated institutions reproducing traditional gender hierarchies.

Regarding the political question of reproduction vs. resistance or subversion, the fields of sports, fashion and beauty, entertainment, eroticism and pornography do not really show possibilities for subversive practices, so long as they are based on emphasizing male defined femininity and sex appeal (whether in fact or already by definition). The attack on masculinity represented by muscles also has to account limits of this strategy for the fact that the ideal types of mighty men in the societal field of power are not emphasizing big muscles. Rather, they are agents with high economic, social, cultural, or symbolic capital, agents who do not perform the kind of work which produces muscles. Mainly working-class (and some middle-class) men show these physical signs. Already Bourdieu (1984, p. 210) visually contrasted the photo of a muscular bodybuilder from the working class with that of a slim aristocrat playing tennis, Valéry Giscard d'Estaing, president of France from 1974 to 1981. Class difference and distance was demonstrated on the level of incorporated habitus.

Instead of talking about bodybuilding as a sport only, we have considered its submissive and complicit use in eroticism and pornography, but also its potential strengths as a tool to destabilize and deconstruct widespread gender stereotypes which might be realized in the arts, not least by using the symbolic capital of these fields and the media attention they get. Figures like Francesca Steele and

Cassils have been examples. From a Butlerian perspective, they have been able to develop interventions against structures of oppressive power based on heteronomous demands, requested ways of being and acting in the world.

The process of metamorphosis, which a body experiences through bodybuilding preparation, seems like a natural spearhead for attacks against the gender binary and the stereotyping of bodies that are classified according to that binary's values. This is particularly true when that kind of preparation can occur in certain artistic subfields, whether in performative or contemporary visual art. Both the practice and the signs transmitted by it seem to be more efficient in cultural fields than in the highly regulated terrain of sports competition, which demands demonstrating femininity in a very objectivizing way. Of course, the arts also face the problem of only reaching certain publics and they are domains of symbolic, rather than real, politics. Bourdieu has rightly noted, that "definite social and economic conditions must be assembled for the critique of categories to become social critique and that we must struggle collectively to realize these conditions" (Bourdieu 1996–7, p. 201).

Ideas for this essay were developed for conference presentations at Lancaster University (England) in July 2018, at Leuphana University of Lüneburg (Germany) in November 2018, and at the Euroacademia conference in Dublin (Ireland) in June 2019. We thank Cornelia Kastelan (Stanford University, Berlin) for participating in the Lancaster University Bourdieu Study Group conference of the British Sociological Association, and Michael Jauchen (Warner, New Hampshire, USA) for edits of the last version of our text for this book.

References

Aoki, Doug. 1996. Sex and muscle: the female bodybuilder meets Lacan. Body & Society 2(4): 59–74.

Aranyosi, István. 2018. Body, skill, and look: is bodybuilding a sport? Phenomenology and the Cognitive Sciences 17: 401–410.

Becker, Howard S. 2008. Art Worlds. Berkeley: University of California Press.

Bell, Melina Constantine. 2008. Strength in muscle and beauty in integrity: Building a body for her. Journal of the Philosophy of Sport 35: 43–62.

Beutler, Jack. 2017. Muscles and Fishnets: Exploring the Similar Autoerotic Elements within Body Image, ESSAI: Vol. 15, Article 11. https://dc.cod.edu/essai/vol15/iss1/11.

Bolin, Anne. 2012. Buff bodies and the beast: Emphasized femininity, labor, and power relation among fitness, figure, and women bodybuilding competitors 1985–2010. In Critical readings in bodybuilding, eds. Adam Locks and Niall Richardson, 29–57. New York: Routledge.

Bordo, Susan. 1990. Material girl: The effacements of postmodern culture. Michigan Quarterly Review 29: 653–677.

Bourdieu, Pierre. 1979. La Distinction. Paris: Seuil.

Bourdieu, Pierre. 1984. Distinction. A social critique of the judgement of taste. Cambridge, MA: Harvard University Press.

Bourdieu, Pierre. 1993. The field of cultural production. Essays on art and literature, ed. Randall Johnson. Cambridge, UK: Polity.

Bourdieu, Pierre. 1996–7. Masculine domination revisited. Berkeley Journal of Sociology 41: 189–203.

Bourdieu, Pierre. 1998. Praktische Vernunft. Frankfurt/Main: Suhrkamp.

Bourdieu, Pierre. 2001. Masculine Domination. Trans. R. Nice. Stanford, CA: Stanford University Press.

Bourdieu, Pierre, and Loïc J. D. Wacquant. 1992. An Invitation to Reflexive Sociology. Cambridge: Polity Press.

Boyle, Lex. 2005. Flexing the tensions of female muscularity: How female bodybuilders negotiate normative femininity in competitive bodybuilding. Women's Studies Quarterly 33: 134–149.

Butler, Judith. 1999. Gender Trouble. London: Routledge.

Connell, Raewyn. 1987. Gender and Power: Society, the Person and Sexual Politics. Stanford: Stanford University Press.

Crumb, Robert. 1996 Art & Beauty Magazine. Northampton, MA: Kitchen Sink Press.

Dobbins, Bill. 1994. The Women: Photographs of the Top Female Bodybuilders. New York: Artisan.

Dobbins, Bill. 1999. Kim Chizevsky: The best female bodybuilder of all time? Bill Dobbins Photo. The Creative World of "The Body Photographer" blog. http://www.billdobbins.com/PUBLIC/pages/coolfree/Kimbest.html. Accessed: 11 September 2019.

Featherstone, Mike. 1999. Body modification: An introduction. Body & Society 5: 1–13.

Femuscleblog 2017. Has female bodybuilding died? Femuscleblog Sports, Fitness, and Health Science blog. https://femuscleblog.wordpress.com/2017/08/26/has-female-bodybuilding-died/. Accessed: 10 July 2020.

Foucault, Michel. 1988. Technologies of the self. Eds. Luther H. Martin, Huck Gutman, and Patrick H. Hutton. Technologies of the Self: A Seminar with Michel Foucault. Amherst, MA: University of Massachusetts Press, 16–49.

Fowler, Bridget. 2003. Reading Pierre Bourdieu's masculine domination: Notes towards an intersectional analysis of gender, culture and class, Cultural Studies, 17(3–4): 468–494, doi: https://doi.org/10.1080/0950238032000083908.

Franck, Georg. 1999. The economy of attention. Telepolis. Trans. S. Plaza. https://www.heise.de/tp/features/The-Economy-of-Attention-3444929.html. Accessed: 10 July 2020.

Frueh, Johanna, Laurie Fierstein, and Judith Stein, eds. 2000. Picturing the Modern Amazon. New York: Rizzoli Books.

Gardiner, Judith Kegan. 2012. Female masculinity and phallic women—Unruly concepts. Feminist Studies 38, no. 3, 584–611.

Garratt, D. 2015. Queer and uncanny: An ethnographic critique of female natural bodybuilding, Qualitative Inquiry 21(9): 776–786, doi: https://doi.org/10.1177/1077800415574910.

Getsy, David J., and Julia Steinmetz. 2015. Cassils: Artist Book. United States: Publisher Mu.

Google Books N Gram Viewer. n.d. female bodybuilding 1970–2019. https://books.google.com/ngrams/graph?content=female+bodybuilding&year_start=1970&year_end=2019&corpus=26&smoothing=3. Accessed: 10 July 2020.

Hakim, Catherine. 2011. Erotic Capital: The Power of Attraction in the Boardroom and the Bedroom. New York: Basic Books.

Heywood, Leslie. 1998. Bodymakers: A Cultural Anatomy of Women's Body Building. New Brunswick, NJ and London: Rutgers University Press.

Heywood, Leslie, ed. 2006. The Women's Movement Today. A Handbook of Third Wave Feminism. Volume 2. Westport, CT and London: Greenwood Press.

Ian, Marcia. 1991. Popular culture: From abject to object: Women's bodybuilding. Postmodern Culture 1(3):1–11. doi: https://doi.org/10.1353/pmc.1991.0014.

Ian, Marcia. 2001. The primitive subject of female bodybuilding: Transgression and other postmodern myths. Differences: A Journal of Feminist Cultural Studies 12(3): 69–100.

IFBB 2001. Amateur rules for bodybuilding. 2001 edition. http://www.ifbb.com/amarules/indes.html. Accessed: 6 October 2004.

IFBB 2009. IFBB rules men and women bodybuilding, men classic bodybuilding, women fitness, men fitness, women body fitness. 2009 edition. http://www.ifbb.com/pdf/IFBBrulebook.pdf. Accessed: 15 December 2009.

Instagram #femalebodybuilders. n.d. #femalebodybuilders. https://www.instagram.com/explore/tags/femalebodybuilders/. Accessed: 10 July 2020.

Jagodzinski, Jan. 2003. Women's bodies of performative excess: Miming, feigning, refusing, and rejecting the phallus. Journal for the Psychoanalysis of Culture and Society, Volume 8(1): 23–41. doi: https://doi.org/10.1353/psy.2003.0013.

Kimmel, Michael S. 2005. The Gender of Desire. Albany, NY: State University of New York Press.

Klein, Alan. 1985. Pumping Iron, Society, 22(6), 68–75.

Klein, Alan. 1994. The Cultural Anatomy of Competitive Women's Bodybuilding. ed. Nicole Sault. Many Mirrors: Body Image and Social Relations. New Brunswick: Rutgers University Press, 76–104.

Klein, Alan. 2001. Pumping irony: crisis and contradiction in bodybuilding. eds. Andrew Yiannakis and Merrill J. Melnick. Contemporary issues in sociology of sport. Leeds: Human Kinetics, 413–426.

Klein, Alan. 2007. Comic book masculinity. Sport in Society, Volume 10(6), 1073–1119. https://doi.org/10.1080/17430430701550512.

Kunitz, Daniel. 2017. How art has depicted the ideal male body throughout history. Artsy. https://www.artsy.net/article/artsy-editorial-art-depicted-ideal-male-body-history. Accessed: 10 July 2020.

Kunsthalle Zürich 2015–2016. Exhibition: Building modern bodies. The Art of Bodybuilding Curated by Jörg Scheller. http://kunsthallezurich.ch/en/building-modern-bodies-art-bodybuilding. Accessed: 10 July 2020.

Laqueur, Thomas. 1990. Making Sex: Body and Gender from the Greeks to Freud. Cambridge: Harvard University Press.

Lambert, Catherine. 2017. Queering identity: becoming queer in the work of Cassils. In Reconstructing Identity. A Transdisciplinary Approach. eds. Nicholas Monk, Mia Lindgren, Sarah McDonald and Sarah Pasield-Neofitou. Melbourne: Palgrave Macmillan.

Liang, Emily. 2011. The Media's Sexualization of Female Athletes: A Bad Call for the Modern Game. Inquiries Journal/Student Pulse, 3(10). http://www.inquiriesjournal.com/articles/587/the-medias-sexualization-of-female-athletes-a-bad-call-for-the-modern-game. Accessed: 10 July 2020.

Liokaftos, Dimitrios. 2019. Natural bodybuilding: An account of its emergence and development as competition sport. International Review for the Sociology of Sport 54: 753–770.

Locks, Adam, and Niall Richardson, eds. 2012. Critical Readings in Bodybuilding. New York: Routledge.

Lowe, Maria R. 1998. Women of Steel. Female Bodybuilders and the Struggle for Self-Definition. New York and London: New York University Press.

Mapplethorpe, Robert, and Bruce Chatwin. 1983. Lady, Lisa Lyon. New York: St. Martin's Press.

Marcuse, Herbert. 1974. Marxism and feminism. Women's Studies, 2: 279–288.

McLish, Rachel, and Bill Reynolds. 1984. Flex Appeal by Rachel. New York: Grand Central Publishing.

McLish, Rachel, and Joyce L. Vedral. 1987. Perfect Parts: A World Champions Guide to Spot Slimming Shaping and Strengthening Your Body. New York: Grand Central Publishing.

McTavish, Lianne. 2015. Feminist Figure Girl. Look Hot While You Fight the Patriarchy. Albany: Sunny Press.

Mikkola, Mari. n.d. Feminist Perspectives on Sex and Gender. In The Stanford Encyclopedia of Philosophy, ed. Edward N. Zalta. https://plato.stanford.edu/entries/feminism-gender/. Accessed: 10 July 2020.

Morton-Brown, Marla. 2004. Artificial ef-femination: Female bodybuilding and gender disruption. Philosophy in the Contemporary World 11: 27–34.

Parsi, Novid. 1997. Don't worry Sam, you are not alone: Bodybuilding is so queer. In Building Bodies, ed. Pamela Moore. New Brunswick, NJ: Rutgers University Press: 103–134.

Pumping Iron [Film]. 1977. Directed by Butler, George, and Robert Fiore. New York: White Mountain Films.

Pumping Iron II: The Women [Film]. 1985. Directed by Butler, George, and Charles Gaines. New York: White Mountain Films.

Raven, Bertram H. 2008. The bases of power and the power/interaction model of interpersonal influence. Analyses of Social and Public Policy 8(1): 1–22.

Richards, Christina, Walter Pierre Bouman, and Meg-John Barker, eds. 2017. Genderqueer and Non-Binary Genders. London: Palgrave Macmillan.

Richardson, Niall. 2004. The queer activity of extreme male bodybuilding: gender dissidence, autoeroticism and hysteria. Social Semiotics 14: 49–65.

Richardson, Niall. 2008. Flex-rated! Female bodybuilding: feminist resistance or erotic spectacle? Journal of Gender Studies 17(4), 289-301. https://doi.org/10.1080/09589230802419930

Russell, Bertrand. 1938. Power: A New Social Analysis, New York: Norton.

Scheller, Jörg. 2017. Pretty buff women—Gender trouble in female bodybuilding. In Faktor X—the chromosome of art. Third Artistes' Biennale. Haus der Kunst München. eds. Albert Coers and Alex de Vries, 42–48. Berlin: Revolver Publishing.

Schwarzenegger, Arnold, and Bill Dobbins. 1998. The New Encyclopedia of Modern Bodybuilding. New York: Simon & Schuster.

Schippert, Claudia. 2007. Can muscles be queer? Reconsidering the transgressive hyper-built body. Journal of Gender Studies 16: 2007.

Scott, Amy. 2011. Pumping up the pomp: An exploration of femininity and female bodybuilding. Vis-à-vis: Explorations in Anthropology 11: 70–88.

Shilling, Chris, and Tanya Bunsell. 2009. The female bodybuilder as a gender outlaw. Qualitative Research in Sport and Exercise 1: 141–159.

St Martin, Leena, and Nicola Gavey. 1996. Women's bodybuilding: Feminist resistance and/or femininity's recuperation? Body & Society 2: 45–57.

Steele, Francesa. 2009. Bodybuilding project. Francesca Steele blog. http://francesca-steele.blogspot.com/2009/. Accessed: 10 July 2020.

Steele, Francesca. 2019. Breaking the Mirror. Performance Research, 23(8): 51–58.

Steinbock, Eliza. 2014. Photographic Flashes: On Imagining Trans Violence in Heather Cassils' Durational Art. Photography & Culture, Volume 7(3): 253-268. doi: https://doi.org/10.2752/175145214X14153800234775.

Stiegler, Bernard. 2012. Within the limits of capitalism, economizing means taking care. Telemorphosis: Theory in the Era of Climate Change, 1. ed. Tom Cohen. Ann Arbor: University of Michigan. Open Humanities Press: 104–120.

Szymanski, Dawn M., Moffitt, Lauren B. and Erika R. Carr. 2011. Sexual Objectification of Women: Advances to Theory and Research. The Counseling Psychologist 39(1): 6–38.

The National Review of Live Art database n.d. Francesca Steele. Routine. https://nrla30.com/the-artists/francesca-steele/?fbclid=IwAR2SwZeXAvATq5fpeNcnJREGAaesiizd3ZV0AK1OuEVQxZ30RdP9GTShPtM. Accessed: 10 July 2020.

Thirer, Joel, and Donald L. Greer. 1978. Competitive bodybuilding: Sport, art, or exhibitionism? Journal of Sport Behaviour 1: 186–194.

Todd, Jan. 2000. Bring on the amazons. An evolutionary history. In Picturing the Modern Amazon. Frueh, Johanna, Laurie Fierstein, and Judith Stein, eds. New York: Rizzoli Books: 48–61.

Too Big for the World. Irene Anderson. 2019. http://www.toobigfortheworld.com/?fbclid=IwAR3gkZKYTwgpoWBXrlatwXo4O69UqJm_ArQumtHxChT-xWGXGswwj8o-lrI. Accessed: 10 July 2020.

van Maanen, Hans. 2009. How to Study Art Worlds: On the Societal Functioning of Aesthetic Values. Amsterdam: Amsterdam University Press.

van Ouwerkerk, Sarah. 1999. Sarah van Ouwerkerk Photography. Bodybuilders II. https://valeriebelin.com/works/bodybuilders-ii. Accessed: 10 July 2020.

van Ouwerkerk, Sarah. n.d. Sarah van Ouwerkerk Photography. http://vanouwerkerkpictures.com/. Accessed: 10 July 2020.

Wacquant, Loïc J. D. 1995. Review article: Why men desire muscles. Body & Society: 163–179.

Wickstrom, Maurya. 2014. Desire and Karios: Cassils Terisias. The Drama Review (TDR) 58: 46–55.

Williams, Carla. 2000: Hardcore: The radical self-portraiture of black female bodybuilders. In: Picturing the Modern Amazon, eds. Johanna Frueh, Laurie Fierstein, and Judith Stein, 104–110. New York: Rizzoli Books.

World Natural Body Building (WNBB). n.d. Sport Regulation. https://www.wnbb.eu/crbst_5.html. Accessed: 10 July 2020.

Youtube. 2012 [1976]. Body Building as Art—Whitney Museum 1976. https://www.youtube.com/watch?v=ok-FhIOPw1g&t=4s. Accessed: 10 July 2020.

Youtube. 2016. Lisa Lyon / Robert Mapplethorpe. https://www.youtube.com/watch?v=l0Qfog9rGwA. Accessed: 10 July 2020.

On the History of the Blues in the Field of Music Production in the United States Beginning in the 20th Century: A Field Theoretical Approach

10

Christoph Behnke

Abstract

In this article certain aspects of the history of blues music are analysed using a field theoretical approach. The analysis shows that blues music – in particular in the 'classic blues' of the 1920s – had a far-reaching impact on various subfields of cultural production in the United States, whether in the form of 'race records' in the popular music segment or of the Harlem Renaissance, which was located in the literary-artistic field. A fundamental element of all aspects of cultural production in the United States in the 1920s was the structural effect of the 'colour line', itself a consequence of racist segregation in American society. Alongside the musical colour line with its mechanisms of exclusion and accompanying cultural appropriation, one of the field effects examined is also the generation of 'racial pride' and the invention of a music that was to have an enormous influence in the course of the 20th century. It is thus recommended that a broader perspective be taken of Bourdieu's heteronomous pole in the field of cultural production.

Keywords

Blues music · Field theoretical perspective · Musical colour line · Black aesthetics · "race"

C. Behnke (✉)
Leuphana University of Lüneburg, Lüneburg, Germany
e-mail: behnke@uni.leuphana.de

L. Gaupp et al. (eds.), *Arts and Power*, Kunst und Gesellschaft,
https://doi.org/10.1007/978-3-658-37429-7_10

Although the history of the blues has been an object of research for decades and is documented in a vast literature from different disciplinary perspectives, there are only few studies that take a field theoretical approach.[1] On the state of research in music sociology, Regev writes the following about Bourdieu's field theory, "It seems that the sociological study of popular music has hardly started to excavate the potential richness and relevance of this work for research" (2015, p. 37). In the following discussion, I will not attempt to construct an autonomous subfield of blues music but instead show how in a particular historical constellation the blues gravitated toward a number of different subfields of cultural production.

Bourdieu draws a now well-known distinction between the relatively autonomous field of restricted cultural production and the field of large-scale cultural production, the two in opposition to each other but, through the *illusio* of "legitimate" culture, belonging to a common classification system in which the relatively autonomous pole of the field is given the power of defining the legitimacy of culture (Bourdieu 1993, p. 125). His treatment of the popular arts is particularly controversial as he assumes that these arts occupy a fundamentally inferior and illegitimate position (Hesmondhalgh 2006; Fowler 1997, p. 157; Shusterman 2000, p. 175). This assumption has been criticised, in different variants, by proposing that a subfield of restricted popular production be established within the popular arts, one subject to the same forces operating on Bourdieu's autonomous and restricted field of the high arts. Such a restricted field in the popular arts would have to have a "popular aesthetic", a term that Bourdieu in his classic work *Distinction* put in inverted commas with the argument that in popular arts the function of form is prioritised: recipients are "involved" and react aversively to formal experiments, that is, they will not tolerate any aesthetic for its own sake (1984, p. 32). One of the authors who attempts to reconstruct a popular restricted field together with a popular aesthetic, Paul Lopes, investigates the rise of a new form of jazz in the 1940s as bebop succeeded swing (2000). Even though Lopes does not discuss the blues, his reconstruction of the popular music field in the United States contains important insights that can help us to better understand the blues.

An appropriate starting point for a field theoretical perspective on blues music would be the historical period in which the blues first emerged, namely in the

[1] Field theory is understood here as the research tradition initiated by Pierre Bourdieu. For a more current discussion of field theory, see for example Hilgers and Mangez (2015).

1920s in the segregated commercial field of "race music" and in the high culture subfield of the Harlem Renaissance, in particular in its music and literature. Part of the history of this field structure is the musical "colour line" from the Jim Crow era, which established a symbolic order that was fundamental for the development of blues music. Beyond its visibility in these specific field contexts, the music that at the beginning of the 20th century gradually became known as the blues was a vernacular music typically played by amateurs in red-light districts or clubs or on the streets. This musical practice, which was located outside professional music production, became visible in the constituting disciplines of anthropology and sociology, which scientized this practice as "folklore", making it a legitimate object of research (Odum 1911; Miller 2010).

In this article I discuss whether the commercial and heteronomous pole of professional blues music—linked as it is with names like Ma Rainey, Bessie Smith, Ethel Waters, Ida Cox, Victoria Spivey, to name but a few—can be analysed with the Bourdieuian dichotomy of high (consecrated) and low (illegitimate) and how this musical practice relates to the high culture adaptation of the blues in the Harlem Renaissance. I will also consider whether proposed extensions and revisions of Bourdieuian field theory are applicable to the field of popular arts.

Radano writes that during the slave trade, since all capacities related to the body of the enslaved were property of the slaveholder, the commercial value of enslaved African people increased if they could make music (2010, p. 365). This did not change until after the Civil War, when "black music" became part of the market for popular music. In this period, some individual African Americans were able to penetrate the field of cultural production, not least to escape an existence as a sharecropper. Once the Emancipation Proclamation entered into force, white supremacy, which had been preserved by the "peculiar institution" of slavery, was now called into question. This impending loss of power led, in particular for the white population of the South, to a dramatic increase in "racecraft" (Fields and Fields 2012), which was to become a foundation for the Jim Crow era and ensured a continuation of the antebellum racial order (Fredrickson 1971, 2002; Fields and Fields 2012).

In this time of violent racial discrimination, the entertainment industry with its many opportunities emerged, and availed itself of this same "racial identity". All subfields of cultural production in this industry were formed by racially motivated segregation from their beginnings. The forces acting in a particular field divided the actors and their audiences by the colour of their skin, with the result that, to this day, as Radano writes, "American listeners remain committed to the belief that racial differences could be heard in music, and this belief motivated performers to produce styles that affirmed colour line thinking" (2010, p. 368).

One of the beginnings of the American culture industry was the "blackface minstrel shows". In these shows, which had taken place since the 1840s, white actors and musicians parodied African Americans by blackening their skin with burnt cork and, not least, by making use of musical material from the black vernacular tradition. These stereotyped racist performances in the medium of entertainment were parodies that allowed white audiences to internalise their supposed superiority. A prime example is "jumping Jim Crow". This portrayal of African Americans as "dancing fools" was appropriated by T.D. Rice in the 1830s from his supposed observations while traveling in the South. The starting point of what was to become known as blackface minstrelsy "was little more than cultural robbery, a form of what Marx called expropriation", which, as Lott adds in his study *Love and Theft*, "troubled guilty whites all the more because they were so attracted to the culture they plundered" (1993, p. 9).

After the abolition of slavery in the 1860s, African American actors also performed in black minstrelsy shows since, until the end of the century, it was only possible for African Americans to enter professional entertainment behind the minstrel mask: "at that moment, minstrelsy offered black entertainers the most lucrative opportunity available to them; they may have felt that they had no better alternative than to imitate the white performers imitating them" (Garofalo and Waksman 2014, p. 20).[2] One of the most controversial appropriations by African American musicians for the increasingly profitable culture industry was the "coon song", a genre presenting racist stereotypes of black people. In "All Coons Look Alike to Me" the African American composer Ernest Hogan made a pact that "was like contracting with the devil" (Gilbert 2012, p. 434). By catering to an affectively charged form of racial domination, as Lott points out, blackface minstrelsy was also based on an experience of "racial crossing": it combined the racist stereotype with fear, panic, as well as desire (Lott 1993).

In Bourdieu's terminology the situation of African American musicians at the end of the 19th century was characterized by racist delegitimization, since their accumulation of symbolic capital was, more than anything else, connected with the socially prevalent term of "race". Racial affiliation decided who had access

[2] An example for the encoded imitations of these imitations is given by Amiri Baraka. The "cakewalk", which became a famous dance in the minstrel shows, was an imitation by African Americans of the sometimes pretentious way their white "masters" would walk. Baraka comments: "I find the idea of white minstrels in blackface satirizing a dance satirizing themselves a remarkable kind of irony—which, I suppose, is the whole point of the minstrels" (Baraka 1963, p. 86).

to all other types of capital. Which is why the experience of a musical colour line preceded the formation of any field-specific symbolic capital. The symbolic violence of this structure gradually changed as racialized "black music" became successful in certain subfields of music production, for example in ragtime music, which dominated the entertainment market at the turn of the twentieth century. Pioneering innovations were also associated with the name of James Reese Europe and the Clef Club, he incorporated in 1910. His "fusion of classical music with black-derived music" gave an impetus to the professionalization of African American musicians (Gilbert 2012). In 1912 the Clef Club Symphony Orchestra was the first African American ensemble to perform in Carnegie Hall, playing with great success a mixture of classical music, ragtime dance, spirituals, and popular ballads (Gilbert 2012, p. 438). Europe accomplished this project with the help of white New York patrons of the arts, who had encouraged him to present black vernacular music. In a strategically clever move, he met the expectations a white audience had of a racialized "black music" while at the same time advancing the professionalization of African American musicians in order to become competitive in a subfield of high culture dominated by whites.

Lopes emphases that music production in the United States around 1900 was above all concerned with forming a professional ethos (1999). This happened not least because of the rapidly growing subfield of popular music production, which attracted a broad audience with a business model based on sheet music and the spatial monopolizing of commercial music in New York (Tin Pan Alley) along with the large-scale establishment of vaudeville shows, concert halls, and musical revues. A peculiarity of the American development is seen by Lopes in the American (that is, European and African American) vernacular traditions being "cultivated" by professional musicians and Tin Pan Alley composers. The professional ethos of "good music" was thus transferred to popular music as classically trained musicians also played popular music (Lopes 2004, p. 13). African American musicians entering the professional musical field were only tolerated when they played music that was connoted with their race—that is, racialized "black music"—or when they accepted a sign of their inferiority, for example not being allowed to use "music stands or written music" in performances before a white audience (Lopes 1999, p. 27).

It was in this context that blues music emerged. Inspired in part by the traditions of vernacular African American music but largely a re-invention of popular music, it led to a "blues craze" in the commercial music field in ca. 1910. One of the legendary blues stories is of the African American W.C. Handy, a professionally trained musician who called himself the "father of the blues".[3] While performing with his dance orchestra in Cleveland, Mississippi, he was asked during the break to let a local trio on stage, and they played "one of those over-and-over strains that seem to have no very clear beginning and certainly no ending at all" (Handy 1941/1991), a kind of music that was for Handy not only illegitimate as high art music but also did not fulfil the professional standards of the popular field. When the trio finished, "the audience showered coins on the stage", more money than the nine musicians in Handy's band had earned for the whole night. He decided: "My idea of what constitutes music was changed by the sight of that silver money cascading around the splay feet of a Mississippi string band" (Handy 1941/1991, pp. 76–78). The "beauty of primitive music" was, as Miller noted, its profit potential (2010, p. 148). The compositions that made Handy famous, "Memphis Blues" and "St. Louis Blues" (1912), were based, as he said himself, on melodic "snatches" he had heard while travelling in the South. The blues craze, heard mainly by black audiences on the streets of the red light districts or on the newly built vaudeville stages, was exploited by white entrepreneurs near the commercial pole and made into a legitimate part of the heteronomous pole of the field. However, the colour line in the production apparatus meant that many of the first blues recordings were made by white musicians. Elijah Wald calls attention to the difficulty of deciding whether singers like Marie Cahill or Marion Harris, both called "queens of the blues" by their record label, saw themselves as part of the minstrel tradition or whether they saw the blues as something apart from that racialized music (2004, p. 47). The Italian-American anarchist Antonio Maggio had already published in 1908, and so before W.C. Handy, a piece for the piano called "I Got the Blues". In the first phase of the adaptation of blues elements to the commercial pole of the field, the dominant business model involved the sale of sheet music, after 1917 it was the record. However, the vaudeville stages in the South were even more important. Butler "String Beans" May, the "commander-in-chief of the real blues", was one of its most famous performers. There are no recordings of his music, only admiring comments from other musicians (Abbott and Seroff 2017, pp. 67–68).

[3] For criticism of Handy, see for example the comment by George: "One of worst offenders was black middle-class composer W.C. Handy, who is called the father of the blues only because he was one of the first to commit it to sheet music" (George 1988, p. 20). See also the comment by Hamilton on the origin myth of the blues: "No one knows who sang the first blues, or where they sang it, or when" (2001, p. 22).

In its "cultivated" form, the blues had always been attractive to a certain white audience. A recurring motif in statements by whites about "black music" was the desire to enter into a state freed from the constraints of white "civilisation": "black music was a path to white ecstasy", writes Miller and gives the white singer Marion Harris as an example (2010, p. 153). From a field theoretical perspective, she is also a good example as she benefited from her whiteness to gain access to the production apparatus while using the blues craze to position herself at the commercial and professional pole of the field and deepen the schism between high and low by devaluing high art: "Honey don't play me no opera".[4]

From the perspective of the white owners of the production apparatus, a structural ambivalence in the field consisted in African American musicians occupying the lowest status of the hierarchy between high and low. The characteristics of being commercial and illegitimate only appeared in *their* music. This high culture position towards music was articulated at the time by the magazine Metronome in this way:

> To show its demoralizing tendency on the public discrimination, it is simply necessary to call attention to the fact that when in music rhythm is dissociated from melody and harmony, the tendency is decadent and leads directly back to the primitive, the negro, and the savage. (quoted in Lopes 2004, p. 25)

This rhetoric was part of a racist ideology in which the black population was seen to be in a "process of degenerating, or 'reverting to savagery'" (Fredrickson 2002, p. 90). However, this ideology was undermined by the adaptation of the blues in the practice of the commercially motivated entertainment industry while its "cultivation" by the musical standards of Tin Pan Alley would undo its contamination as illegitimate.

Though W.E.B. Du Bois did not explicitly reject the blues, he did condemn the "debased melodies" of the African American music played in the world of Tin Pan Alley while valuing the tradition of the "sorrow songs", the spirituals (1903/1986). His main concern was to ward off the accusation of "primitivism" directed at African American music, even though as a critique of (white)

[4] "Honey, don't play me no opera/ Play me some blue melody/ I don't care nothing about Carmen/ When I hear those harmonies" Marion Harris, "Paradise Blues" (Victor 18,152, 1916) as quoted in Miller (2010, p. 153).

civilisation it had a positive connotation.[5] The background to this attitude towards African American cultural heritage would seem to be a process of the black middle-class repositioning itself, which led to a segmentation of blues music. The upwardly mobile black middle-class preferred jubilee singing as it could, Booker T. Washington believed, serve to motivate the hard-working masses while they felt an aversion to the blues as it was reminiscent of the situation they were trying to escape (George 1988, p. 18). To this extent, the blues were caught up in the symbolic order of hierarchically structured race relations and negatively expressed class differences in the African American population.

In its field structure, the adaptation of the blues in the period from 1910 to 1920 was a contentious affair. The colour line ensured that, when it was in a "cultivated" form and commercially profitable, the blues became a part of the entertainment field. However, it was first in 1920 that the white-dominated recording industry engaged an African American singer, Mamie Smith, whose recording of "Crazy Blues" was a surprising commercial success for the OKeh Records label (Abbott and Seroff 2017, p. 264). The blues singers who were recorded in the market for "race records"[6] at the beginning of the 1920s had little to do with the country blues musicians who were performing outside the field of popular music in the South. As a rule, they did not write their own songs but let professional songwriters and musicians, such as Perry Bradford, provide them with material. They were entertainers and came from the world of theatre or comedy. African American composers were professional musicians familiar with the structure and conventions of popular music. Blues queens like Ma Rainey and Bessie Smith, who before their career as blues singers had performed as "coon shouters", generally performed in a small jazz combo with five musicians (Abbott and Seroff 2017, p. 161). They occupied the commercial pole of the African American entertainment subfield, although some of their audience were white. Whites-only concerts, so-called "midnight frolics", were staged so that Southern whites could

[5] For this position, Fredrickson introduced the term "romantic racialism" (see Fredrickson 1971, pp. 97–99). See also Tom Lutz (1991) and, from the perspective of the discovery of "African" art at the turn of the twentieth century, Achille Mbembe (2017, p. 43).

[6] Albert Murray (1976) has pointed out that the term "race" in the 1920s did not have a negative connotation for African Americans: "The fact remains—oblivious as certain critics may be to it—that in the black press of the 1920s the most prominent Negro leaders and spokesmen referred to themselves as race spokesmen and race leaders … the indignation over the race terminology as applied to records has been misguided at best" quoted in Nelson George (1988, p. 20).

hear the blues without subverting the racial divide. Ethel Waters commented: "So we found ourselves applauded by the ofays in the theater and insulted by them on the streets" (Abbott and Seroff 2017, p. 263).

Two different positionings can be found for musicians performing in this field. Bessie Smith, a superstar of African American popular culture, became successful in 1923 with her recording of "Down Hearted Blues" (Columbia A 3844, 1923), which was followed by 160 recordings with record sales in the millions. After being turned down by the African American label Black Swan "because of her grassroots sound", she signed a contract with Columbia (Davis 1998, p. 152).[7] Black Swan had been founded by the African American Harry Pace (on each album was printed the slogan "The Only Genuine Colored Record— Others Are Only Passing for Colored") and had W.E.B. Du Bois on their board of directors and William Grant Still as musical director (Suisman 2009, p. 221). However, Black Swan did engage Ethel Waters, another blues queen, because she was said to be compatible with the white singers of popular music (Cherry 2006). This crossover logic was used by Black Swan to position blues music in the market for mainstream productions and achieve a degree of legitimacy. Like Bessie Smith, Ethel Waters became a famous singer, but with a different style. She had renounced "blues shouting", a style that in the racialized discourse of the folklore paradigm was characterised as "primitive" (Miller 2010).[8]

At the time, the influential New York music critic Carl van Vechten said that Ethel Waters "refines her comedy, refines her pathos, refines even her obscenities" (quoted in Davis 1998, p. 148). Her music, just as the version of "Crazy Blues" sung by Mamie Smith, was considered an example of a New York style called "jazz sentimental" (Abbott and Seroff 2017, p. 261). The fact that even the Black Swan label was concerned about signing a contract with Bessie Smith was mainly due to the fear of some African Americans who occupied the restricted pole of high culture being designated as "primitives". "Successful blacks disliked Bessie", and treated the blues like an "indelicate stepchild" wrote Angela Davis (1998, p. 154). Ethel Waters—like other African American musicians in

[7] Suisman writes about Bessie Smith: "When she stopped abruptly in the middle of her first test recording and said, 'Hold on, let me spit,' she thrust herself outside the 'respectable' model of musicianship that Black Swan endorsed, and the company declined to add her to its roster" (Suisman 2009, p. 221).

[8] The term "folklore paradigm" is used by Karl Hagstroem Miller (2010) to draw attention to the scientization of racialized differences that he feels have made an important contribution to legitimizing the musical colour line (2010).

the professional world, such as W.C. Handy or Jelly Roll Morton—avoided any kind of "folk expression" in her adaptation of the blues and created a new, nationally acceptable "American" music in her production field, which was "a commercial narrative in keeping with white concepts of black culture and creativity" (Abbott and Seroff 2017, p. 264). The music of Bessie Smith, on the other hand, was based on a mixture of "art and life". She took up themes from the experiences and feelings of the African American working class in a critical time when "double-consciousness" (Du Bois) and social impoverishment were at the forefront (see, for example, Charles S. Johnson's classic 1922 study of the race riots in Chicago). No less important were the disruptive anomic effects of the Great Migration on African American family structures. Bessie Smith's music was both heavily criticised and praised, with comments ranging from insinuations that it was commercial, semi-pornographic, or unpolitical (Charters 1963, p. 152; Murray 1976, pp. 67–68; Hamilton 2000, p. 2) to interpretations that described her treatment of sexuality as revolutionary and the criticism in her texts of a male-dominated world as "ahead of their time" (Davis 1998; Danaher 2005, p. 1465).

In a field theoretical perspective, both of these different blues adaptations, by Ethel Waters and by Bessie Smith, are fundamentally structured by segregation, which confined them to performing in the subfield "race records". In this subfield, as in white music, commercial success was paramount. To achieve this success, Ethel Waters chose a strategy based on a kind of acculturation of "black music" in the white-dominated music industry; while Bessie Smith, in the space available to her, took up the everyday life of the African American working class. Her lyrics are about the struggle against male domination, the sexual emancipation of the African American woman, and social inequality (Davis 1998; Hale 1998, p. 38). She used musical means of expression from urban red light districts that could be "read" in this milieu: raw, stomping, shouting, and vulgar while at the same time sophisticated because there was often a "black" reading of the music inaccessible to whites.

Both blues queens were successful at the commercial pole of the field. Their success was accompanied by a theoretically paradoxical effect, since their music strengthened the formation of symbolic capital for "black music", a currency that in the course of the 20th century was to become dominant in popular music. "Black music", which was supposedly inferior and was thought to be contained and under control in the ghetto of "race records", used its specific field conditions to form a "black aesthetics", which slowly but surely took its place in the white-dominated entertainment industry (Baraka 1991; Gayle 1972). However, the African American musicians had to expect that along with their music becoming a business model for white entrepreneurs, their musical innovations would be appropriated by white musicians and diminish their return on symbolic capital. Baraka remarks sarcastically that "African American music did not become a completely American expression until the white man could play it" (1963, p. 155). Elsewhere he writes of "the great music robbery".[9]

In the case of Bessie Smith, the paradoxical field effect was that her economic success was not confined to the smaller segment of African American entertainment: it was also part of the mainstream cultural industry. Moreover, she became a voice in the social avant-garde by questioning gender relations in her own community, making her an early protagonist of black feminism (Davis 1998). In a narrow space of possibilities that required racialized music to be economically successful, she was nevertheless able to achieve a degree of autonomous quality in her work. This interpretation is not meant to suggest that Bessie Smith was consciously a social revolutionary or, to use Du Bois' phrase, part of the "talented tenth"; her position under these specific conditions in the field of popular segregated music allowed her to go beyond the production of music as a commodity and use it as a vehicle for self-development, thus serving as a role model for black women emancipating themselves from white perceptions of them as "black nannies". In a framework of intersectionality, these field conditions combine the themes of race (segregated race records), class (racial uplift versus working-class blues), and gender (blues queens as an expression of black feminism) (Crenshaw 1991).

In the 1920s, a further blues adaptation took place as part of the Harlem Renaissance, an intellectual and artistic African American movement that occupied the field of relatively autonomous and restricted cultural production in the United States (Baker 1987; Hutchinson 1995; Plessner 1977; Sherrard-Johnson 2015). As in the commercial world of entertainment, this field was fundamentally structured by racist segregation, with the difference however that African American positioning in this subfield received its legitimacy not in commercial success but in the production of symbolic capital. The struggle over whether and how this "currency" should be recognised—that is, of the value of the African American cultural heritage—has yielded a profusion of assessments and reassessments of the Harlem Renaissance, even that it was failure (Hutchinson 1995, p. 16). From an African American perspective, the restricted field of cultural production was marked by a structural ambivalence: the actors found themselves in a racialized

---

[9] In this context, the career of white jazz musician Paul Whiteman comes to mind (Lemke 1998, pp. 59–61).

position that was "activated" when they entered the field. They knew that the logic of the field would only allow a transformation if a new aesthetic like "The New Negro" of Alain Locke (1925) showed an awareness of the history of the field before questioning the dominant positions of white aesthetic modernism, a logic that explains the discovery of the "miscegenated style" (Hale 1998, p. 38). As Gayle writes, "the Renaissance writers were forced to choose between two worlds—one of opulence and splendor, already in existence, the other of poverty and hope, waiting to be born" (1972, p. 84). He proposes, in a discussion typical of the 1960s, attributing oppositional poles to "nationalists" and "assimilationists", the first being those African Americans who articulated a self-awareness "that black people all over the world share an origin and a heritage, that the welfare of black people everywhere are inexorably linked, and that the cultural products of blacks everywhere should express their particular fundamental beliefs" (Floyd 1995, p. 100). The assimilationists, on the other hand, saw highbrow culture as essential to any artistic productivity they could hope would be successful.

The blues now became acceptable as music from an African American tradition, but it needed an "effect of refraction" (Bourdieu 1995, p. 220) for this raw material to be "readable" in the relatively autonomous field of restricted production. Both in entertainment circles and in the field of high art, folk expressions of the blues were rejected as "old Negro music". However, in their urban versions, jazz and blues formed the aesthetic environment of African Americans, not least in the "race capital" of Harlem (Locke 1925). In literature, Sterling A. Brown and Langston Hughes in particular adapted elements of the blues, not in writing about the blues but as a method to create a "black aesthetic" (Tracy 1988; Hollenbach 2015; Sherrard-Johnson 2015; Thomas 1997; Ellison 1945/1999). Given the ambivalent positioning of this field, these experiments were controversial in the Harlem Renaissance. The "African American Symphony" composed by William Grant Still in the 1920s is an example of an adaptation of the blues in the restricted field of art music (Spencer 1997; Abbott and Seroff 2008, p. 434; Floyd 1995, p. 109). In the first movement, *moderato assai*, even casual listeners immediately hear the twelve-bar blues progression. William Grant Still's position in the field is remarkable because, unlike many white composers, he also worked in the field of entertainment. While engaged as an arranger for W.C. Handy and Fletcher Henderson, he was privately tutored by Edgar Varèse, one of the leading avant-garde composers of the time.

The colour line also ran through high cultural contexts, triggering discussions about whether the rules of European "art music" should, or must, be followed in order for African American music to become acceptable or whether, giving African American musicians a special positioning, elements of black vernacular music could be transformed into highbrow music. Everything depended on the consecration of this music, on its transformation into art music. Although a more detailed musicological analysis would likely show that all types of African American music were in fact based on transcultural transformation, in this field there was always the danger that their music would be racialized. This conflict was, as discussed above, homologous in the field of popular music. The differing assessments of the music of Bessie Smith and Ethel Water is an exemplary case, with the positioning of Bessie Smith, who drew strongly on working class traditions, being especially problematic in the Harlem Renaissance.

It would be inaccurate to label blues music in the entertainment field, the "classic blues", as an exclusively heteronomous music production. The question is whether the characteristics described here (which should be subject to a much deeper analysis) can be categorized as a "popular restricted field", in Lopes's sense of the term. A difference is that the bebop musicians he studied had an anti-commercial impulse, as many earned money as big band musicians in dance halls while giving jazz concerts in the tradition of European art music (Lopes 2000, p. 172). "Bebop, unlike much of the swing music that preceded it, was concert rather than dance music, intended to appeal to a sophisticated listener" (Laver 2009, p. 4). This type of jazz music was not legitimated because of its black heritage: it was not racialized, as Laver writes, it was Eurocentric (2009).

By contrast, the race records of the 1920s, especially in the case of Bessie Smith, were about production restricted by the musical colour line; they went beyond entertainment and, by combining "art and life", had elements of the social avant-garde. In general, the label race records was positively perceived by the African American population as increasing the visibility of their race, a self-perception that can be seen as an expression of "racial pride" or "racial uplift". This changes the conditions in which symbolic capital is evaluated in terms of "race". In certain social subfields, there is now a possibility that "whiteness" is no longer the predominant social power. The question arises whether the term "relative autonomy" from Bourdieuian field theory applies in the case of classic blues music. Although the heteronomous pole of this field is transcended by the production of symbolic capital, there is no need for a "popular aesthetic" to create this effect. Even if there are different aesthetic strategies to meet the expectations of record labels, these do not function as specific instances of consecration but instead serve to achieve economic success. The function of the social avant-garde is also an effect external to the music, an effect that goes far beyond the function of entertainment, but does not entail its musical production, which in Lopes's framework functions as a "popular aesthetic". In this respect, the analysis presented here provides a broader perspective on the heteronomous pole of the cultural production field in the United States in the 1920s. The paradoxical effect described above stimulated segregated African American musicians in the race records subfield to create highly innovative music, which fed back into white-dominated areas of the music production field while at the same time it entered the African American community as a social avant-garde movement and, by heightening racial pride, led to an increase in symbolic capital.

As seen here in the example of the adaptation of the blues in the context of the Harlem Renaissance, the relatively autonomous pole of music production was, similar to the heteronomous pole, structurally determined by racial segregation. This racism is seen in the fact that, contrary to other impressions at the time, "the overwhelming majority of publishers showed absolutely no interest in publishing the work of black authors" (Hutchinson 1995, p. 342). The adaptation of the blues, a form of music considered "popular", led not only to a dedifferentiation of high and low, but was also an expression of the ambivalence of the cultural colour line: the "blacker" the literary and musical avant-garde positioned itself, the farther it distanced itself from an aesthetic format seen as legitimate throughout the history of the field. If the black avant-garde had accepted the requirements of the field, then they would have faced racist exclusion. Nevertheless, both of the subfields discussed here are interrelated in many ways. The entertainment production apparatus used by African Americans was largely in white hands as was the patronage underlying the Harlem Renaissance. Under these conditions, it seemed highly unlikely that a form of music would emerge that over the course of the 20th century would be enormously influential.

An advantage of using field theory as a frame of reference is to be able to focus on the history of the blues in the United States while drawing attention to what the (contested) rules of the professional field of music made out of the blues. The different variations of blues music, and especially their use in the literature and music of the Harlem Renaissance, are owed to positionings that can be explained by their gravitation to different musical segments. A consequence of this is that the question as to the origin of the "authentic blues", as the folklore paradigm would have it, recedes into the background, a shift that at any rate has taken place in recent research into the blues. However, this problematic origin myth reasserts itself in the field with the guitar recordings of country blues beginning in 1926, when male artists such as Papa Charlie Jackson, Lonnie Johnson, and Blind Lemon Jefferson first achieved success, and the rediscovery of this subfield by white blues musicians in the 1960s (Garabedian 2014; Adelt 2010).

Acknowledgements For their helpful criticism I would like to thank Alenka Barber-Kersov, Lisa Gaupp, Annette Grigoleit and Volker Kirchberg. Special thanks go to Paul Lauer, who translated the text from german to english.

References

Abbott, Lynn and Doug Seroff. 2008. They Cert'ly Sound Good to Me: Sheet Music, Southern Vaudeville, and the Commercial Ascendancy of the Blues in Ramblin' on My Mind: New Perspectives on the blues, ed. David Evans, 49–104. University of Illinois Press.

Abbott, Lynn and Doug Seroff. 2017. The Original Blues: The Emergence of the Blues in African American Vaudeville. Jackson: University Press of Mississippi.

Adelt, Ulrich. 2010. Blues Music in the Sixties: A Story in Black and White. New Brunswick, New Jersey and London: Rutgers University Press.

Baker, Houston A. 1987. Modernism and the Harlem Renaissance. The University of Chicago Press.

Baraka, Amiri (Jones, LeRoi). 1963. Blues People. New York: William Morrow.

Baraka, Amiri. 1991. The "Blues Aesthetic" and the "Black Aesthetic": Aesthetics as the Continuing Political History of a Culture. Black Music Research Journal, 11(2), Autumn 1991, 101–109.

Bourdieu, Pierre. 1984. Distinction: A Social Critique of the Judgement of Taste. Cambridge: Harvard University Press.

Bourdieu, Pierre. 1993. The Field of Cultural Production. Columbia University Press.

Bourdieu, Pierre. 1995. The Rules of Art: Genesis and Structure of the Literary Field. Stanford: Stanford University Press.

Charters, Samuel. 1963. The Poetry of the Blues. New York: Avon.

Cherry, Randall. 2006. Ethel Waters. Long, Lean, Lanky Mama in Nobody Knows Where the Blues Come From: Lyrics and History, ed. Robert Springer, 264–282. Jackson: University Press of Mississippi.

Crenshaw, Kimberlé. 1991. Mapping the Margins: Intersectionality, identity politics, and violence against women of color. Stanford Law Review, 43(6), 1241–1299.

Danaher, William F. 2005. Gender Power: The Influence of Blues Queens, 1921 to 1929. American Behavioral Scientist. 48(11), 1453–1467.

Davis, Angela Y. 1998. Blues Legacies and Black Feminism: Gertrude 'Ma' Rainey, Bessie Smith, and Billie Holiday. New York: Vintage.

Du Bois, W.E.B. 1986 [1903]. The Souls of Black Folk. New Haven and London: Yale University Press.

Ellison, Ralph. 1999 [1945]. Richard Wright's Blues. The Antioch Review, 57(3), Jazz (Summer 1999), 263–276.

Fields, Barbara J. and Karen E. Fields. 2012. Racecraft: The Soul of Inequality in American Life. Verso: London, New York.

Floyd, Samuel. 1995. The Power of Black Music. Oxford: Oxford University Press.

Fowler, Bridget. 1997. Pierre Bourdieu and Cultural Theory: Critical Investigations. London: Sage Publications.

Fredrickson, George M. 1971. The Black Image in the White Mind: The Debate on African American Character and Destiny, 1817–1914. Hanover: Wesleyan University Press.
Fredrickson, George M. 2002. Racism: A Short History. Princeton: Princeton University Press.
Garabedian, Steven P. 2014. 'The Blues Image in the White Mind: Blues Historiography and White Romantic Racialism'. Popular Music and Society, 37(4), 476–494.
Garofalo, Reebee and Steve Waksman. 2014. Rockin' Out: Popular Music in the U.S.A. Somerville, Massachusetts: Pearson.
Gayle, Addison, Jr., ed. 1972: The Black Aesthetic. New York: Anchor.
George, Nelson. 1988. The Death of Rhythm and Blues. New York: Pantheon Books.
Gilbert, David. 2012. Clef Club Inc. James Reese Europe and New York's Musical Place. Journal of Popular Music Studies, 24(4), 430–456.
Hale, Grace Elisabeth. 1998. Making Whiteness: The Culture of Segregation in the South, 1890–1940. New York: Vintage Books.
Hamilton, Marybeth. 2000. Sexuality, Authenticity and the Making of the Blues Tradition. Past and Present No. 169, 132–160.
Hamilton, Marybeth. 2001. 'The Blues, the Folk, and African American History'. Transactions of the Royal Historical Society 11, 17–35.
Handy, W. C. 1991 [1941]. Father of the Blues: An Autobiography. New York: Da Capo Press.
Hesmondhalgh, David. 2006. Bourdieu, the media and cultural production. Media, Culture & Society 28, 211–231.
Hilgers, Mathieu and Eric Mangez, ed. 2015. Bourdieu's Theory of Social Fields: Concepts and Applications. London: Routledge.
Hollenbach, Lisa. 2015. Phonography, Race Records, and the Blues Poetry of Langston Hughes in A Companion to the Harlem Renaissance, ed. Cherene Sherrard-Johnson, 301–316.
Hutchinson, George. 1995. The Harlem Renaissance in Black and White. Cambridge: Belknap/Harvard UP.
Johnson, Charles S. 1922. The Negro in Chicago: A Study of Race Relations and a Race Riot, ed. Chicago: University of Chicago Press.
Laver, Mark. 2009. The Greatest Jazz Concert Ever: Critical Discourse, European Aesthetics, and the Legitimization of Jazz. Critical Studies in Improvisation/Études critiques en improvisation, 5(1), 1–12.
Lemke, Sieglinde. 1998. Primitivist Modernism: Black Culture and the Origins of Transatlantic Modernism. Oxford: Oxford University Press.
Locke, Alain. 1925. Enter the New Negro in Survey Graphic, March 1925, 1–6.
Lopes, Paul. 1999. Diffusion and Syncretism: The Modern Jazz Tradition. Annals, AAPSS, 566, November 1999, 25–36.
Lopes, Paul. 2000. Pierre Bourdieu's Fields of Cultural Production: A Case Study of Modern Jazz. In Pierre Bourdieu: Fieldwork in Culture. Nicholas Brown & Imre Szeman, eds., 165–185. Boston: Rowman & Littlefield Publishers.
Lopes, Paul. 2004. The Rise of a Jazz Artworld. Cambridge UK: Cambridge University Press.
Lott, Eric. 1993. Love and Theft: Blackface Minstrelsy and the American Working Class. New York: Oxford University Press.

Lutz, Tom. 1991. Curing the Blues: W. E. B. Du Bois, Fashionable Diseases, and Degraded Music. Black Music Research Journal, 11(2), Autumn 1991, 137–156.
Mbembe, Achille. 2017. Critique of Black Reason. Duke University Press.
Miller, Karl Hagstrom. 2010. Segregating Sound: Inventing Folk and Pop Music in the Age of Jim Crow. Durham and London: Duke University Press.
Murray, Albert. 1976. Stomping the Blues. New York: McGraw-Hill.
Odum, Howard W. 1911. Folk-Song and Folk-Poetry as Found in the Secular Songs of the Southern Negroes. The Journal of American Folklore 24893) 255–294.
Plessner, Monika. 1977. Ich bin der dunklere Bruder. Die Literatur der schwarzen Amerikaner. Frankfurt: Fischer.
Radano, Ronald. 2010. On Ownership and Value. Black Music Research Journal, 30(2), Fall 2010, 363–370.
Regev, Motti. 2015. Notes on Sociological Theory and Popular Music Studies in The SAGE Handbook of Popular Music, Andy Bennett and Steve Waksman, eds., 33–45. London: Sage Publications.
Sherrard-Johnson, Cherene, ed. 2015. A Companion to the Harlem Renaissance. Oxford: Wiley Blackwell.
Shusterman, Richard. 2000. Pragmatist Aesthetics: Living Beauty, Rethinking Art. Boston: Rowman & Littlefield.
Spencer, Jon Michael. 1997. The New Negroes and Their Music: The Success of the Harlem Renaissance. Knoxville: The University of Tennessee Press.
Suisman, David. 2009. Selling Sounds: The Commercial Revolution in American Music. Cambridge: Harvard University Press.
Thomas, Lorenzo. 1997. Authenticity and Elevation: Sterling Brown's Theory of Blues. African American Review, 31(3), 409–416.
Tracy, Steven C. 1988. Langston Hughes and the Blues. Urbana: University of Illinois Press.
Wald, Elijah. 2004. Escaping the Delta: Robert Johnson and the Invention of the Blues. New York: HarperCollins.

"It's Almost like Greeting"—Artistic Development Projects and Value Making of Traditional Art Practices in Ugandan Civil Society

11

Anna-Lisa Klages

Abstract

This paper investigates the linkages between artistic practices and civil society dynamics in Uganda. While it briefly touches upon modern practices, it focuses on "traditional" art forms. Preliminary findings indicate that traditional art forms often negotiate social and cultural dimensions. In addition, and although they may be considered of little value with regard to economic growth, artistic and project leaders consider them significant for the economic development of Uganda. The literature further indicates a scarcity of studies investigating forms of traditional art and their linkages to *empowerment* and *development*, often funded by international actors. Findings suggest that the discourse on local traditional visual culture production is dominated by the notions of economic development, which may be reinforced through the way these art products are promoted and used today in Uganda's civil society.

Keywords

Traditional art · Cultural hegemony · Civil society · Development · Empowerment · Uganda

A.-L. Klages (✉)
FHWS Würzburg, Würzburg, Germany
e-mail: anna-lisa.klages@fhws.de

L. Gaupp et al. (eds.), *Arts and Power*, Kunst und Gesellschaft,
https://doi.org/10.1007/978-3-658-37429-7_11

11.1 Introduction

According to Andrew Odoki[1], a Ugandan-Tanzanian artist, scholar and researcher, "traditional" art practices in Uganda are given a similar value as a greeting whenever two people meet. No one has to be trained or learn to greet, one does not value a greeting through payment. It is not given much value, but taken for granted, says Odoki. At the same time, however, it is an everyday ritual of significant value for members of any society, and a carrier of norms and values. Through greeting others, individuals exchange and communicate and pay respect to their counterparts. Greetings may vary according to the status of the other and/or the relationship between the individuals involved.

Local forms of visual culture that make use of "traditional" art forms may not necessarily be perceived as "art" among the local population. Wolunkau-Wanambwa (2019) and colleagues were unable to identify an equivalent to the English word "art" in any indigenous African language. This finding suggests very different perceptions of the value and meaning making of creative expression (Pinther and Weigand 2018). Today, some scholars, the national Ugandan government and practitioners regard forms of local visual culture as a catalyst for poverty eradication and economic growth. This paper acknowledges this notion, but seeks to look beyond the obvious. Taking a broad definition of civil society, as discussed below, allows for a discussion of the linkages between visual culture and its relevance for meaning making processes in Uganda's civil society, thereby taking power aspects specifically into account.

11.2 The Question of Power

Lamont (1989, quoted in Lamont and Thévenot 2000) delimited four areas of investigation of culture in the exercise of power (Engelstad 2003):

> Cultural industries as an arena for power struggles; Knowledge and aesthetic competence serving as resources in the exercise of power; Cultural power exerted indirectly by modes of definition; and finally, Power by means of exclusion—exclusion of people or ideas. (p. 1)

[1] Name changed.

In a Gramscian sense, power manifests itself in form of (cultural) hegemony, where a dominant group seeks to maintain power or rule through ideological, social, cultural or economic influence (Fonseca 2016). The question of power here is thus a question of definition of cultural and societal norms, knowledge production and of inclusion and exclusion, negotiated, among others, through the various forms of visual culture.

11.3 Methodology

This paper is part of a dissertational research project that investigates the linkages between visual culture and civil society development in the context of (international) development efforts in Uganda. The expert interviews, the empirical material that informs this paper, were conducted during an explorative study in Uganda in August 2018. Experts here are understood as consultants, and, especially in the context of an explorative study, deliver relevant content and enable access to a scientific discipline and an area of knowledge (Helfferich 2019). Given the postcolonial framework of the research project at hand it was particularly important to embed and ground the inquiry in local expertise and the situatedness of the phenomenon, as to avoid the construction of a purely "Western" narrative.

11.4 Sample

In the context of the explorative study, nine interviews were conducted in August 2018, varying between 38–57 min in length. Interviewees were comprised of three groups. The first group included practicing and lecturing artists from two major universities in Kampala, while the second group were socially engaged artists involved in artistic projects with a focus on (social) development. Interviewees in the third group were involved with development projects of an international East African development agency. All interviews were transcribed and analyzed using open coding strategies according to Strauss and Glaser (1967).

The theoretical foundations of the research and this paper lay in postcolonial theory (Mbembe 2001), and is particularly informed by power theory according to Foucault and Gramsci's concept of cultural hegemony, and visual culture theory. Visual culture theory seeks to overcome categories that differentiate forms of artistic expression between "masterpieces" or "high art", and "low arts", such as handicrafts and many indigenous forms of creative expression (Dikovitskaya

2005). Here, all visual image practices are considered as art, however, where applicable, terms used in the discourse of the situation of inquiry will be used (Clarke 2007; Clarke et al. 2018).

11.5 Visual Culture and Civil Society in Uganda

Before elaborating on Uganda's contemporary art scene, it is worthwhile to take a look at the establishment of the Margaret Trowell School of Industrial and Fine Arts (MTSIFA) at Makerere University. The school was founded in 1937 by Trowell herself; an art educator who moved from the United Kingdom to Uganda in 1929, at the time a Protectorate under British rule. The school became the first art department in any East African university (Siegenthaler 2018). Her legacy is controversial and has been discussed at length (e. g. Wolukau-Wanambwa 2014; Kyeyune 2004). In spite of the ambivalent reasons for including artistic crafts into the art school's curriculum, Trowell's motivation, among others, came from a deep concern that the local art forms and other manifestations of the cultures of the region may be "submerged under western materialism for several generations, at the most it may even go altogether" (Trowell 1936, quoted in Wolukau-Wanambwa 2014, p. 101).

At the same time, however, Trowell found cultural craftmanship to be a sound path to teach the local population to become "civilized citizens" of the protectorate. Through teaching craftmanship the local population would gain a sense of work ethics according to British standards at the time. In addition, it would show them that the wealth and prosperity of the "white men" was not born of merit, but the consequence of dedicated labor (ibid.).

Formal education according to the "Western" educational system had been brought to East Africa by missionaries, and was later manifested through the colonial system (Kayamba and Kwesiga 2016). Prior to colonization, one interviewee explained, people were taught either at home or at the monarchs' palaces. Information and knowledge were passed on orally and through formal and informal vocational trainings (Kayambar and Kwesiga 2016). With the arrival of the first missionaries, however, formal education was introduced, and children began to study in schools. Local art forms did not find their way into the school curricula, as children were sent to school to "learn the skill of the European" (Koeune 1937, quoted in Kyeyune 2004), not what they could learn at home. The findings from the interviews suggest this notion to be dominant until this day, in spite of several significant changes, as will be discussed below.

11.6 Art Education Today

While today other academic institutions have established art departments[2], Uganda's contemporary art scene has been criticized to still be dominated by a Makerere-centric art school (Garrido Castellano 2018). Many artists, curators, scholars and researchers were trained at Makerere University, and/or are affiliated to it (Nakazibwe 2010; Garrido Castellano 2018). It may well be partly due to this domination that curator Simon Njami opted for a new and alternative concept with the 2018 Kampala Art Biennale (KAB). Instead of displaying the works of Ugandan and international artists alike, he chose a studio format, inviting seven acclaimed (mostly African) artists[3] to Kampala to teach and train young artists, apprentices, in a ten-day studio workshop (KAB18 2018).

Whereas critics of Njami's curatorial choices argue his decisions regarding the concept of the biennale to be neo-colonialist (ibid.), Njami himself finds the studio concept to always having been essential in art history, remarking that by taking in the most talented apprentices as assistants, they would eventually become the Masters of the next generation. In addition, he states that

> Modern practices, notably in Europe, have turned the artist into a solitarian genius who creates masterpieces in the silence of his studio ... Africa is still a space where the community plays a critical role. It is, through this third edition of the Kampala Art Biennale, our aim to revitalize ancient practices that are more than needed in our contemporary world. Practices that would bring back notions like transmission and togetherness. (KAB18 2018)

While international contemporary art festivals such as KAB and KLA ART[4] are used to negotiate the role of contemporary African art in urban Uganda and its

[2] Today there are eight universities, and a few academies, that offer art programs: Kyambogo University; Nkumba University; The East African International University; St. Lawrence University; Kampala University; Buganda Royal University; Uganda Christian University, Mukono University and Ndejje University.

[3] The "Masters" were: Bili Bidjocka (Cameroun), Godfried Donkor (Ghana/UK), Abdoulaye Konate (Mali), Myriam Mihindou (Gabon), Radenko Milak (Bosnia Herzegovina), Aida Muluneh (Ethiopia), Pascale Marthine Tayou (Cameroun).

[4] KLA ART is a biennial art festival that was inaugurated in 2014, after a dispute among the organizers of the KAB. For further information, *see also* Garrido Castellano (2018). The boda-moment. Positioning socially engaged art in contemporary Uganda.

significance to the civil society discourse situated in urban Africa in the 21st century, the role and meaning of "traditional" art forms are often negotiated in ethnological museums of the North (Sarr and Savoy 2018) and, in the case of Uganda, additionally, through development projects. Often overseen, global players, including governments, embassies, international NGOs, and multinational companies play an "influential role as integral parts of the postmodern and postcolonial setting for much of the African … art world" (Siegenthaler 2018, p. 2). This observation is not only true in an urban setting, but also in rural areas. Here, art activities are often coordinated and implemented by or in cooperation with (international) NGOs, typically funded by governments of the North, private NGOs or so-called well-wishers, and art is used as a tool that serves a social or societal cause (Stupples 2011). This practice creates a donor-dependency which leads to a power imbalance, and raises the question whether artistic project funding from the North in a development context are not, ultimately, an "attempt to help the non-Western world to get culture, and if so, whose culture on whose terms?" (Stupples 2011, p. 5).

11.7 Civil Society

Community, transmission and togetherness are among the terms civil society literature refers to frequently (Edwards 2011). And while definitions and approaches to the civil society concept(s) have common parameters, the debate around the notion continues to be controversial. This may be partly due to the fact that the understandings of civil society have changed over time, continue to change, and are politically constructed in response to the culture and society they relate to. It is an elastic concept that is seen by "many as a part of society (the world of voluntary associations), by some as a kind of society (marked out by certain social norms), and by others as a space for citizen action and engagement (described as the public square of sphere)" (Edwards 2011, p. 7). Since civil society is a concept constituted by political, cultural and historical circumstances, many post-colonial African scholars urge a "critical distance", the mere adaptation, they argue, may lead towards a misapplication of "Western political constructs to African circumstances" (Obadare 2011, p. 185).

In a Gramscian sense, civil society, interconnected with his theory of hegemony, cannot be separated from the state (Buttigieg 1995; Gramsci 1999). The organic relationships between civil society and political society "enable certain strata of society … to gain dominance within the state and … to maintain it, perpetuation the subalternity of other strata" (Buttigieg 1995, p. 4). Civil society here can be described as the sphere of hegemony in which relations of power and influence are negotiated.

For the purposes of this paper civil society is understood as an ecosystem, which includes a "variety of types of association, engagements in the public sphere, and relationships between civil and political action" (Obadare 2011, p. 191). Obadare contends that a civil society definition that moves beyond the reduction to certain kinds of voluntary associations, often urban-based NGOs, will enable a more diversified analysis that may ultimately inspire "subaltern alternatives and actions" (ibid.).

11.8 Visual Culture

11.8.1 Socially Engaged Artistic Expression

Art festivals such as KAB, and even more so KLA ART[5], aim to connect political and societal debates with the artistic discourse. In its 2014 edition, KLA ART integrated a boda boda project into its festival design. Boda bodas are motorcycle taxis and the most common form of (semi) public transportation in urban Uganda. They symbolize the "invisibility of the informal sector within the configuration of urban imaginaries" (Garrido Castellano 2018, p. 4). It was partly due to their practical relevance to everyday urban life in Kampala that a total of 20 boda boda motorcycles were converted into "innovative, experimental and dynamic artworks, each representing their interpretation of the unmapped" (Namakula 2014, p. 3).

One topic addressed by the boda boda project by DAPU (Disability Art Project Uganda) was the invisibility of people with mobility impairments. According to DAPU founder and director, Fred Batale, people with physical disabilities are on the margins in Uganda. DAPU seeks to raise awareness. In addition, it aims at training people with disabilities in art and design techniques so they can sustain

[5] Both, KLA ART (Kampala Art Festival) and KAB (Kampala Art Biennale) are biennales that were initiated 2012 and 2014, respectively. KLA ART in particular seeks to find new and alternative approaches to approach the general urban public, e. g. through establishing studio spaces in urban centers, or through the boda boda project. For more details, *see* Peters-Klaphake (2015).

themselves. Here, individuals are equipped with artistic skills for them to generate an income through their practices (DAPU 2019).

Another example of a socially oriented art initiative is the soap sculpture project of Lilian Nabulime. In her dissertational art project she aimed to raise awareness and break societal taboos concerning HIV/AIDS in Uganda's rural communities (Nabulime and McEwan 2011; Nabulime and McEwan 2014). Nabulime created small, inexpensive sculptures in the form of female and male genitals, and inserted a number of everyday objects, such as nails, beans, and seeds. These objects, Nabulime explains, "convey messages that are familiar to Ugandans … beans communicate ideas about germination, life and fertility … Lychee skin and seeds symbolize the multiplication of the virus, infection, and rashes. Nails illustrate the worst stages of the disease: injury, infection, danger, and full-blown AIDS; they also act as metaphors" (Nabulime and McEwan 2014, p. 276).

While these sculptures are not the sole artistic production in which she addresses HIV/AIDS (Nabulime and McEwan 2011), however, unlike her other work, the soap sculptures primarily served a social cause as it meant to break taboos and stimulate conversation about the virus among community members (see Fig. 11.1). One of the major challenges, Nabulime states, was that the target population she worked with is usually not addressed by artists and their artworks. Finding forms, objects, and materials that were familiar to the communities she sought to reach out to, was thus determined crucial by the artist (ibid).

This notion, to find culturally appropriate art forms familiar and aesthetically appealing to the target populations of development and/or awareness projects, also drove the development of the Design, Health and Community Project, a cooperation between Northumbria University (UK), the Durban University of Technology (SA), and Makerere University (Uganda) (Guille 2012). Using a South African model developed by Kate Wells in 1997, which "interweaves the health education, life skills and economic development of rural women through

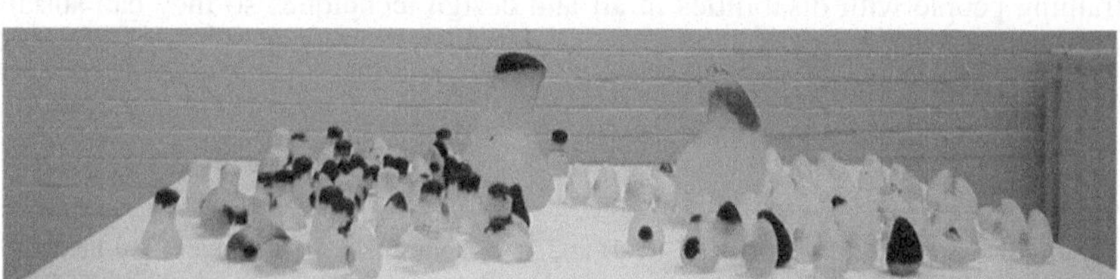

Fig. 11.1 Lilian Nabulime, Soap Family, 2011, mixed materials. Image curtesy of the artist. With permission from the artists

the regeneration of traditional bead crafts" (White 2009, p. 159)[6], rural women in Uganda were invited to share their experiences with HIV/AIDS through traditional art forms and materials, namely wickerwork, raffia and bark-clot. Strengthening the community and the empowerment of women affected through HIV/AIDS was among the goals of the project. Yet, emphasis was also put on the salability of the products, as explained by an involved interviewee:

> So, what we did here, ... because we have some links with some women groups, and some of them had skills of making basketry. So, we said we can use basketry and then the bark-cloth, because we also have the tradition of making bark-cloth. So, we also started putting messages, linking up with this women groups, and surprisingly some of them had similar challenges like ladies in Kwazulu-land[7]. ... For us, the issue was not only the messages, but the quality we're paying attention to the quality, we're paying attention to the actual products, because we're interested in sharing and then producing products which could sell.[8] (Stephen Mukiibi[9] pers. comm. 2018)

He then continues to elaborate on how market trends influence the artistic products created in the context of this one-year project. Mukiibi does not perceive this problematic or limiting the creativity or freedom of design of the project participants. Quite the contrary appears to be true; the access to an international market is understood as added value to the project, and therefore emphasis was put on the quality and design of the products. Consequently, it is only logical that perceived market demands have a significant impact on the designs of the artworks. He states:

> Because if you are producing here ..., most of the basketry here is colorful. And you are targeting maybe a market in [name of location] and you plan for them. And this season they're interested in natural products, things which don't have color. So

[6] In 1996, Wells became involved with rural craftspeople in KwaZulu-Natal, and quickly discovered that the women's beadwork was a way of communicating intimate and emotional matters (hlonipha). Bead-work and the making of dolls with beads is understood as a symbol of Zulu culture. Among others, it was "meant to transmit love messages secretly" (Wells 2016, p. 131) and concludingly was identified as most suitable for the Siyazama project (Wells 2016).

[7] Kwazulu-Natal Land, SA, is where the Siyazama-project, developed by Kate Wells, was first implemented. For further details *see also* Wells 2012.

[8] All quotes in this article have been edited for clarity.

[9] Name changed.

> instead of putting color then you make them natural, without color. [Or] you may find the season in that area is green or blue. So instead of making colorful products then you make blue products. So that is how we were making an intervention, apart from putting the messages. (ibid.)

Unlike the South African project, which is long-term in its nature, the Design, Health and Community Project, which was funded by the England-Africa Partnership[10], operated for one year (White 2009). Today, a number of art products that were created in the context of the project are owned by the British Museum, which, to the project leadership in Uganda, is an indicator of the success and the importance of the project.

11.8.2 Visual Culture Between Neglection and Appreciation

Artistic practices perceived as "traditional" in Uganda include not only wickerwork and bark cloth products, but also jewelry making from (rolled paper) beads and pottery, in addition to blacksmithing and wood carvings, though the latter is rather peripheral to Uganda's cultures (Kayamba and Kwesiga 2016; Kyeyune 2004; Nakazibwe 2005; Wachsmann and Trowell 1953).

It is noteworthy that in rural areas some artistic practices, such as pottery in south-western Uganda, have remained almost unaltered for generations (Kayamba and Kwesiga 2016, 2017). In addition to serving practical functions in peoples' daily activities, they perpetuate the visual culture, social dimensions, and symbolic meanings associated with the production and use of pottery (Kayamba and Kwesige 2016, 2017). In their study, Kayamba and Kwesiga investigate traditional pottery with regard to its production, use and impact on the environment in the Ankole region in western Uganda. They find that ceramics play an essential role both in the families and the community as a whole, and argue that ceramics are not only used to meet the needs as functional household items. The forms and decorations of the pots are "symbolic and serve to transmit culture; it encodes, meditates and reinforces the pattern of social relationship" (Kayamba and Kwesige 2016, p. 49). Pottery making among the Ankole, as described by

[10] The England-Africa Partnership is a program of the British government's Department for Innovation, Universities and Skills.

Kayamba and Kwesiga, is generally understood as a female role or activity. Since the knowledge and skills are passed on from mother to daughter, pottery making is not included in formal education.

According to the authors, and in spite of its apparent agency in family and community contexts, the positioning of pottery making is ambivalent. On the one hand, the process and the products themselves are important to the dynamics and the functioning of the community; certain practices, such as the production of smoking pipes, are limited to certain members of the societies only. They carry important symbolic messages during ceremonies and rituals, and the involved communities negotiate a social order through the process of making and using pottery.

Pinther and Weigand fin "material" in Africa to be an "oral tradition" (Pinther and Weigand 2018, p. 8). They understand that "certain materials are reserved tor privileged groups or people, and skeumorphic principles can be found, for instance, in wickerwork, various weaving techniques and architecture" (Pinther and Weigand 2018, p. 8).

On the other hand, Kayamba and Kwesiga find a correlation between the introduction of Christianity and "Western" education, and an altered perception of traditional practices, such as pottery for food. The use of pottery in homes other than for cooking became "despised practices and [the local population] associated them with paganism" (Kayamba and Kwesiga 2017, p. 42). Furthermore, they propose that it was "Western" education, too, that claimed to bring a "higher form of civilization to Uganda", which ultimately resulted in the lowering of "the traditional standards of order and beauty in many of the material possessions of African life, especially crafts" (Kayamba and Kwesiga 2017, p. 43) (Fig. 11.2).

In spite of the low status of the local (rural) visual culture within the communities, the tradition of pottery making has remained intact, and passing the skill on to the next generation considered pivotal (Fig. 11.2). Many, potters favor traditional methods over newer technologies. Some potters interviewed by Kayamba and Kwesiga believe that some of the pots produced could not be imported. They are stated to be unique in their nature, and only a person familiar with the culture would be able to create an item of the required shapes and forms. In addition, the researchers conclude, food may be perceived to have a better taste if cooked in a locally produced pot, which resonates with the findings of the empirical study at hand. This apparent ambivalence between perceived value and lived value of traditional art forms, which appears to be true for other art forms as well, will be further discussed in the next section.

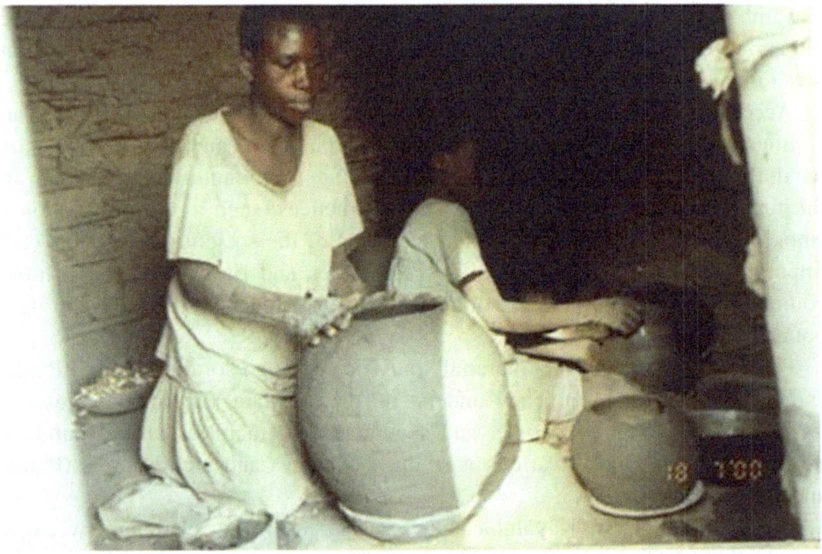

Fig. 11.2 A young potter learning from her mother (Kayamba and Kwesiga 2017, p. 44), Image curtesy of the authors. With permission from the authors

11.8.3 Framing of Visual Culture

In Uganda's past and contemporary visual culture paper beads play a significant role, and traditionally, the craftspeople that created them were respected in their communities even though they did not work in professional studios or manufactories (Kasozi 2019). Today, beadwork is particularly popular among projects and initiatives that aim to empower marginalized women[11] economically (ibid.). Local initiatives and "Western" charities alike identify paper beads as suitable for empowering activities, classified as "appropriately feminine behavior" (Kasozi

[11]The National Association of Women Organizations Uganda defined "marginalized" women to be women who "did not have an opportunity to enrol, stay and complete formal education because her society does not value girlchild education or discriminates girls and women with disabilities and those living with HIV/AIDS, [women] without access to land and finances … [and] mother[s] and her children" NAWOU (2017).

2019, p. 41). Kasozi finds that through paper beading, women are able to negotiate their positions within the family, thereby enabling alternatives to traditional power-relations within families and communities. Simultaneously, and while power-positions at micro level may be re-negotiated through the artistic activity, thereby altering boundaries of the private and the public, new forms of dependencies on individuals (artists) and associations that promote and sell the products (internationally) are created. Artistic activity here is framed within the discourse of gender roles, development and poverty eradication (Kasozi 2019; Tumusiime 2012), and products, unless further processed, sold as cultural crafts or souvenirs on websites, craft markets and others, hence, creating a power imbalance between the artistic creators of the products and those who promote and sell them and whose perception of the product is, ultimately adopted.

11.9 It's Almost like Greeting—Value Making of Traditional Art Practices

Art and culture in Uganda are often funded through international organizations, corporations, expats and other individuals, who have a significant impact on the artistic landscape in the country (Siegenthaler 2018).

The findings of the literature study and the expert interviews conducted suggest that, in addition to the relevant actors mentioned above, the "Western" education system, too, significantly influences the perception of art and visual culture in Uganda.

During the interviews, a number of individuals reflected on the advantages of learning the skills of making cultural crafts. They are convinced that, if people were to appreciate and learn cultural crafts, they would become creative and innovative in other areas of life, too. One interviewee explained:

> They're looking at new things. Literature, they're looking at technology. But they're not looking at technology of crafts, you know. And they don't see the connection, because it is then that if they knew that there was a connection between making a better craft and making a good technology, in order to make a good craft, then they would have embedded it. ... But they don't. They don't think like that. Very, limited, few people think like that. (Andrew Odoki, pers. comm., August 28, 2018)

Artistic activity here is closely connected to education, learning skills and learning to be innovative. If people were to closely study "the crafts" and the technological processes involved in the production thereof, so the argument, people could translate this creative knowledge into other areas, and catalyze development in an economic sense. In other words; if more people would learn "the crafts", the country could and would develop other industries as well. In his statement, Odoki then proceeds towards indirectly criticizing the citizens of Uganda for not appreciating their own visual cultural heritage. And while he understands the correlation between colonization and the perception of traditional art forms being of low value playing a pivotal role, he also, potentially unwillingly, re-emphasizes Margret Trowell's position with regard to teaching civilization to peasants through teaching cultural crafts:

> In Trowell's view, exposure to European goods and technology under colonialism was not spurring East Africans into innovation and self-improvement. … Teaching "good" craftsmanship would, she argued, meet an "obvious", and "psychological need": firstly it would teach the colonised to appreciate the value of labour—"the worth of doing"—within the capitalist colonial economy. Secondly, the effort of learning to make things themselves would help to instill in Africans the idea that "the white man's" social, economic and political advantages were borne of merit. (Wolukau-Wanambwa 2014, pp. 105–106)

11.10 Poverty Eradication Through Craft-Making

Cultural craft-making has been identified as a tool to foster (economic) development and alleviate poverty by the interviewed experts, project developers and the Ugandan government. In its 2040 Vision for "socioeconomic transformation" and development, the Uganda Vision 2040, the government envisions the society transformed from a "peasant to a modern and prosperous country within 30 years" (Government of Uganda 2007). The government here regards the arts and crafts industry as a pillar to enhance tourism activity, stating:

> With improvement in research and development, efforts will be geared towards diversifying tourism products to meet the needs for the tourism sector. This will include the development of the souvenir art and craft industry, tangible and intangible heritage. (Government of Uganda 2007, p. 35)

The role of visual culture in the sense of "traditional" art forms or "cultural crafts" is thus multifold; first, it is a tool that can be used for economic development and growth. The findings from the interviews indicate that, indeed, "cultural craft-making" has been identified by international and national development actors for the (economic) empowerment of marginalized populations, in particular women.

Second, it is understood as carrying the potential to teach creative and thinking skills, and thus as an important soft skill for the innovation required in the dynamics of the 21st century. Findings further indicate that the interviewed experts do not think that the majority of the Ugandan people acknowledge this, thereby referring to citizens and policy makers alike. Neither of them two seem to appreciate the potential of the creative industries in general, and "cultural crafts-making" in particular.

11.11 Lived and Felt Value of Visual Culture

The statement "it's almost like greeting" captures this sensation that "traditional" art practices are not highly valued. Their existence, just like greeting, is a given. The codes of greeting, similar to the skill of making "cultural crafts", are embedded in informal education. Children learn to greet through interaction with their immediate environment and subconsciously absorb the cultural and societal norms embedded within such rituals they ultimately take for granted:

> It's almost like greeting. You don't pay for someone greeting you, so if you draw me a face, that's good. That is your talent. It's like magic. So, that's what I think could be the biggest challenge. That we have not come to terms with how we use our own local skills to make the development. When the few policy makers think that development must come from outside (Andrew Odoki, pers. comm., August 28, 2018).

Yet, in spite of the criticism towards the perception of the local visual culture, Odoki indirectly positions "traditional" art practices at the center of societal functioning, as greeting is one of the many daily activities through which social functioning and social order are negotiated. After all, they are important during wedding ceremonials (Chikukwa et al. 2016), as a "marker of social hierarchy among the Baganda"[12] (Nakazibwe 2005, p. 49), in funerals and other "cultural occasions" (ibid.), here and elsewhere.

[12] The Baganda are an ethnic group from central-southern Uganda, which includes the urban centers Kampala and Entebbe.

11.12 Conclusions

"Traditional" art forms in Uganda may, at first sight, not appear as significant in contemporary society and in their aesthetic value. They are, however, significant and powerful actants in cultural and social and practices. To be more precise, the available literature and the findings from the explorative study at hand indicate that so-called "traditional" art forms are pivotal actants in Uganda's civil society, as many societal structures, (power and speaking) positions, (gender) roles and notions of aesthetics are, among others, negotiated through the art forms.

Although some scholas consider demarcations between artforms to be artificial and "Western", the findings here indicate that, partly because of the different roles and functions of visual culture, there is indeed a need to sensitively investigate the various artistic practices individually to be able to reconstruct linkages and interconnectedness.

In addition, the analyses of the interviews indicate a hierarchy within artistic practices. Whereas local art forms, such as wickerwork, may serve as an inspiration for contemporary forms of visual artistic expression (such as painting, drawing, and sculpting), they themselves are seldom considered to be art, but rather "cultural crafts". Their association to local cultures, as household items, souvenirs and an activity that can be learned at home place them at the lower end of the scale. Artisans here are almost exclusively women, as the giving nature of wickerwork is traditionally associated with female roles. This positioning, however, is ironic as those very artefacts are crucial in many ceremonies and rituals essential for (collective) identity construction, including divination rituals, weddings, and hospitality. Furthermore, the findings suggest a strong cultural hegemony[13] in which "traditional" art forms are pushed to the margins, initially consciously and subconsciously placed there by missionaries and colonizers. Today, the government and, though potentially unwillingly, artists who engage in empowerment programs such as the Design, Health, and Community Project, may reinforce the hegemonic structures, and with it, the continuation of the marginalization of local art practices and forms.

The current findings do not allow for any conclusions regarding how artistic practices are perceived by the local populations with regard to their aesthetic functioning or on the role of NGOs on the perception of the local visual culture. Both will have to be investigated in order to better understand the dynamics,

[13] *See also* Protte, this volume.

agency, meaning making and power structures involved, particularly in the context of empowerment and development discourses in the broader framework of civil society development.

References

Buttigieg, Joseph A. 1995. *Gramsci on Civil Society*. boundary 2(22): 1–32.

Chikukwa, Raphael, Katharina Hoffmann, Ingmar Lähnemann, Katrin Peters-Klaphake, and Lydia Potts. 2016. *Kabbo ka Muwala – The Girl's Basket: Migration and Mobility in Contemporary Southern and Eastern Africa*. Berlin: Revolver Publishing.

Clarke, Adele. 2007. *Situational Analysis – Grounded Theory after the Postmodern Turn*. Thousand Oaks: SAGE.

Clarke, Adele, Carrie Friese, and Rachel Washburn. 2018. *Situational Analysis – Grounded Theory after the Interpretative Turn*. Thousand Oaks: SAGE.

DAPU. 2019. *Disability Art Project Uganda: About us*. http://www.freetocharities.org.uk/dapu/aboutus.html. Accessed: 31 October 2019.

Dikovitskaya, Margaret, ed. 2005. *Visual Culture: The Study of the Visual After the Cultural Turn*. Massachusetts Institute of Technology Press.

Edwards, Michael. 2011. Introduction: Civil society and the geometry of human relations. In *The Oxford handbook of civil society*, Michael Edwards ed., 3–14, 1st edn. New York: Oxford University Press.

Engelstad, Fredrik. 2003. *Comparative studies of culture and power*. Amsterdam: Elsevier JAI.

Fonseca, Marco. 2016. *Gramsci's Critique of Civil Society*. Routledge.

Garrido Castellano, Carlos. 2018. The Boda Moment: Positioning Socially Engaged Art in Contemporary Uganda. FIELD. *Journal of Socially Engaged Art Criticism* 10. http://field-journal.com/issue-10/the-boda-moment-positioning-socially-engaged-art-in-contemporary-uganda. Accessed: 31 August 2020.

Government of Uganda. 2007. *Uganda Vision 2040 – A Transformed Ugandan Society from a Peasant to a Modern and Prosperous Country within 30 years*. https://www.greengrowthknowledge.org/sites/default/files/downloads/policy-database/UGANDA%29%20Vision%202040.pdf. Accessed: 28 July 2021.

Gramsci, Antonio. 1999. *Selections from Prison Notebooks. Essential Classics in Politics*. London: Elec Book Company.

Guille, Jackie. 2012. Design, health and community: transferring the Siyazama model to Uganda. In *Siyazama: Art, AIDS and education in South Africa*, Kate Wells ed., 73–79. Scottsville, South Africa: University of KwaZulu-Natal Press.

Helfferich, Cornelia. 2019. Leitfaden- und Experteninterviews. In *Handbuch Methoden der empirischen Sozialforschung*, 2nd edn., Nina Baur and Jörg Blasius eds., 669–686. Wiesbaden: Springer Fachmedien Wiesbaden.

KAB18. 2018. Curatorial Concept: *"The Studio"*. http://kampalabiennale.org/the-studio-concept/. Accessed: 28 July 2021.

Kasozi, Dorah. 2019. *Paper Beads on the Move: Mobilizing Trajectories and Subjectivities to Shape Contemporary Art in Uganda*. https://www.mitpressjournals.org/doi/abs/10.1162/afar_a_00459. Accessed: 2 November 2019.

Kayamba, Willam K., and Philip Kwesiga. 2016. The role of pottery production in development: A case study of the Ankole region in Western Uganda. *Net Journal of Social Sciences* 4(4): 81–90.

Kayamba, Willam K., and Philip Kwesiga. 2017. Gender and traditional pottery practice in Ankole region, western Uganda. *Net Journal of Social Sciences* 5(3): 42–54.

Kyeyune, George. 2004. *A century of modern art in Uganda*. Dissertation, University of London, London.

Koeune, Esther. 1937. Our African schools. In Uganda Teachers Journal, 4. Quoted from: Kyeyune, George. 2004. Art in Uganda in the twentieth century. University of London (unpublished dissertation)

Lamont, Michele, and Laurent Thévenot. 2000. *Rethinking Comparative Cultural Sociology: Repertoires of Evaluation in France and the United States*. Cambridge: Cambridge University Press.

Mbembe, Achille. 2001. *On the Postcolony*. Berkeley: University of California Press.

Nabulime, Lilian, and Cheryl McEwan. 2011. Art as social practice: transforming lives using sculpture in HIV/AIDS awareness and prevention in Uganda. *cultural geographies* 18(3): 275–296. doi: https://doi.org/10.1177/1474474010377548.

Nabulime, Lilian M., and Cheryl McEwan. 2014. Using art and everyday objects to fight HIV/AIDS in Uganda. *Development in Practice* 24(2): 272–285. doi: https://doi.org/10.1080/09614524.2014.884539.

Nakazibwe, Venny. 2005. Bark-cloth of the Baganda people of southern Uganda. A record of continuity and change from the late eighteenth century to the early twenty-first century. Dissertation.

Nakazibwe, Venny. 2010. *An assessment of the status of arts' education in Uganda*. http://education.go.ug/files/downloads/REPORT%20ON%20ARTS%20EDUACTION.pdf. Accessed: 30 October 2019.

Namakula, Elizabeth. 2014. Mapping Kampala with KLA ART 014. *Start Journal* 14. https://startjournal.org/2014/11/mapping-kampala-with-kla-art-014/. Accessed: 31 August 2020.

NAWOU. 2017. *NAWOU Handmade*. http://nawouganda.ug/our-programmes/economic-empowerment/the-marketing-department/. Accessed: 11 May 2019.

Obadare, Ebenezer. 2011. Civil society in sub-Saharan Africa. In *The Oxford handbook of civil society*, Michael Edwards ed., 183–194, 1st edn. New York: Oxford University Press.

Pinther, Kerstin, and Alexandra Weigand eds. 2018. Flow of *Forms / Forms of Flow: Design Histories between Africa and Europe*. Bielefeld: transcript Verlag.

Sarr, Felwine, and Benedicte Savoy. 2018. *The restitution of African cultural heritage. Towards a new relational Ethics*. http://restitutionreport2018.com/sarr_savoy_en.pdf. Accessed: 28 July 2021.

Siegenthaler, Fiona. 2018. Co-optation as an imaginative act: art-related initiatives and social space in Kampala. *Social Dynamics* 44(3): 528–546. doi: https://doi.org/10.1080/02533952.2018.1507544.

Stupples, Polly T. 2011. *Breaking the frame: art in international development*. Dissertation, Massey University, Turitea, New Zealand.

Strauss, Anselm, and Barney Glaser. 1967. *The Discovery of Grounded Theory – strategies for qualitative research*. Chicago: Aldine.

Trowell, Margaret. 1936. Suggestions for the treatment of handwork in the training of teachers for workin Africa. *In Oversea Education* 7(2): 78–84.

Tumusiime, Amanda Evassy. 2012. *Art and gender. Imag(in)ing the new woman in contemporary Ugandan art*. Dissertation. University of South Africa, Pretoria. http://hdl.handle.net/10500/9036. Accessed: 31 August 2020.

Wachsmann, K. P., and Margaret Trowell. 1953. *Tribal crafts of Uganda*. Oxford: Oxford University Press.

Wells, Kate, Marsha Macdowell, C. Kurt Dewhurst, Marit Dewhurst eds. 2012. *Siyazama: Art, AIDS and education in South Africa*. University of KwaZulu-Natal Press.

Wells, Kate. 2016. Siyazama in South Africa: Zulu beadwork, HIV/AIDS, and the consequences of culture. In *Oxford Textbook of Creative Arts, Health, and Wellbeing: International Perspectives on Practice, Policy and Research*, Stephen Clift and Paul M. Camic eds. 129–133. Oxford University Press

White, Mike. 2009. *Arts Development in Community Health: A Social Tonic*. Radcliffe Pub.

Wolukau-Wanambwa, Emma. 2014. Margaret Trowell's School of Art. A Case Study in Colonial Subject Formation. In *Wahrnehmung, Erfahrung, Experiment, Wissen: Objektivität und Subjektivität in den Künsten und den Wissenschaften*, Susanne Stemmler ed. 1st edn. Zürich: Diaphanes.

Wolukau-Wanambwa, Emma. 2019. This word "art". In *Dissident Lines: lis rhodes*, ed. Irene Aristizábal and Lis Rhodes. Zagreb: kerschoffset.

Part III

The Arts and the Dominance of Politics and Economic Order

Art Politics' Tendencies of the Documenta Exhibition Series

12

Lutz Hieber

Abstract

Exhibition visitors expect to see art on display. But what they also see is a representation of the exhibition organisers' conception of art. The documenta series of exhibitions, launched in 1955, are regarded by the German art world as key events on the international contemporary art scene. Yet they show only a limited section of the art spectrum. A comparison between documenta exhibitions of the first four decades with exhibitions mounted by two leading New York museums, showing art from their own holdings of the same period, brings some fundamental differences to light. It emerges that the documentas are symptomatic of the specific character of the German art world.

Keywords

Documenta (Kassel) · Museum of Modern Art New York (MoMA) · Whitney Museum New York · Museumsufer Frankfurt (Main) · Psychedelic poster · ACT UP · National habitus

L. Hieber (✉)
Leibniz Universität Hannover, Hannover, Germany
e-mail: hieber@ish.uni-hannover.de

L. Gaupp et al. (eds.), *Arts and Power*, Kunst und Gesellschaft,
https://doi.org/10.1007/978-3-658-37429-7_12

The *documenta* series of art exhibitions is the recurring pinnacle of the visual art world[1] in Germany, and of international importance. Founded in 1955 and held in Kassel initially at four-year, now at five-year intervals, it has—according to the host city's website—evolved into "the world's leading regular exhibition of contemporary art. As a medium of information on the latest tendencies in the international art scene, over the course of its history it has attended, documented and commented on all stages of the changing conception of art" (Geschichte der documenta, n. d.). Each forthcoming edition is scrutinised and discussed in art journals and newspapers. The exhibitions have been mostly positively received by German-speaking commentators. Prior to the opening of the last documenta in 2017, for instance, Catrin Lorch wrote for the nationwide newspaper *Süddeutsche Zeitung* that the exhibition heralded Kassel's revival as a "centre of activity … and not Paris, London or even Berlin" (Lorch 2017).

Below, I will consider curatorial policy in the first four decades of documenta. During this time, the exhibition's popularity has steadily grown, drawing larger attendances with each successive edition (documenta Retrospective, n. d.). The number of visitors increased from 130,000 in 1955 to 615,640 in 1992.

The first four documenta exhibitions (in 1955, 1959, 1964 and 1968) were supervised by Arnold Bode in the position of artistic director. In 1972, this task was assumed by Harald Szeemann, in 1977 by Manfred Schneckenburger, in 1982 by Rudi Fuchs, in 1987 by Manfred Schneckenburger again and in 1992 by Jan Hoet. They were all assisted by working committees or advisory boards.

In retrospect, these editions of documenta can be viewed—regardless of the contemporary art workers' gripes—in terms of the curators' work. Insofar as they sought to document the state of contemporary art, they are representative of post-war tendencies in curatorial policy in (West) Germany. Bearing this in mind, I would like to consider the first documenta in 1955, the 4th documenta in 1968 and documenta IX in 1992.

These three editions are emblematic of the documenta series. The first edition aimed to herald a new era in the arts in post-war West Germany. The editions in 1968 and 1991 were landmarks because they took place at times of artistic revolution, in the sense of revolution as analogous to Thomas S. Kuhn's concept of scientific revolution. I would like to consider this concept briefly. Kuhn shows that the "normal" progress of science is occasionally shaken by "revolutions".

[1] The concept "art world" refers to the productive apparatus and also the ideas about art that prevail at a given time and that determine the reception of works.

He describes the latter as processes in which "a reconstruction of the field from new fundamentals" occurs (Kuhn 2012, p. 85). Every revolution is preceded by a crisis that challenges existing conventions. Kuhn sees a "parallel between political and scientific development …: Political revolutions aim to change political institutions in ways that those institutions themselves prohibit" (Kuhn 2012, p. 93). Similar transformations occur in the field of the visual arts. Here, too, revolutions bring entirely new artistic practices to the surface and shift the foundations of the concept of art. Subsequent to the period of innovation and change, however, the new-found practices can develop along a calmer path of a reformulated "normality". As the eruptive changes of the late 1960s and the period around 1990 led to the formation of new foundations, the documentas that were held at those times mark outstanding events in the series.

12.1 Documenta 1955

Arnold Bode initiated the first documenta exhibition in 1955 as a cultural addition to the West German horticultural show held in Kassel that year. The artistic concept was largely devised by Werner Haftmann, a renowned art historian and member of the advisory board.[2] After the rupture caused by Nazism, he saw the exhibition as an "initial attempt to broadly reconnect with the world" (Haftmann 1955a, p. 23). In this spirit, much space was given to the presentation of works created before 1933. In his opening speech, Haftmann addressed the "question of the social function of modern art", remarking that it was "entirely redundant to attach humanitarian, social or religious demands to it" (Haftmann 1955b).

Yet this first documenta, with which Germany appeared "as a notable host to the international art community … for the first time since 1945" (Fitzke 2015, p. 166), made some glaring exclusions that caused the exhibition to miss its declared target on three counts. A wave of emigration had taken place during the Nazi era in the 1930s, draining Germany of leading minds, including many from the art world. The documenta organisers failed to invite any of them to contribute

[2] In 2019 it emerged that Werner Haftmann had been a member of the Nazi Party. Moreover, while conducting research for her doctoral thesis, Kassel-based cultural studies expert Mirl Redmann found that landscape gardener Hermann Mattern, who had organized the West German Garden Show in 1955, had previously been responsible for the Reich Garden Show in 1939, and joined the Nazi Party in 1940 (Hinrichsen and Trebing 2020).

to preparing the exhibition. Secondly, although the exhibition made up for lost time by acknowledging many works of pre-1933 modern art, *Jewish artists* such as Otto Freundlich, whose work *Der neue Mensch* had been used, unfavourably photographed from below, as the cover illustration for the Nazis' 1937 *Degenerate Art* exhibition guide (Fig. 12.1), was not represented. Other missing artists were "Jankel Adler, Felix Nussbaum, Gerd Wollheim, Ludwig Meidner" (Fitzke 2015, p. 167). Thirdly, the exhibition was based on a narrow artistic premise that focused on the beaux-arts and the paradigm of autonomous art. True, sixty architecture photographs formed a "small group attached to the main exhibition" (Hirzel 1955). Nonetheless, the exhibition was essentially limited to the beaux-arts. The Bauhaus school was shrunk to paintings, sculpture and architecture photos. The Berlin Dadaists including John Heartfield (who created photo-montages for book jackets and the newspaper *Arbeiter Illustrierte Zeitung*), George Grosz and Rudolf Schlichter (not only painters but also illustrators of literature) were not featured. "Apart from Otto Dix—just two portraits by whom were displayed—none of the artists who engaged in social commentary or criticism were exhibited at the documenta" (Fitzke 2015, p. 167).

The phase of ignoring Jewish artists ended in the latter decades of the last century; their work has now been reappraised and exhibited. But the first documenta's conception of art was grounded in the 19th century and neglected such developments. Below, I would like to comment on the exhibition's exclusive focus on the beaux-arts in the light of two different factors: firstly, the avant-garde of early twentieth century past, and secondly, the situation across the Atlantic.

12.2 Avant-Garde Approaches

The avant-garde of the early twentieth century was not merely concerned with aesthetic demands but also with acute psychological and social need. The mass slaughter of the First World War had destroyed the old order and along with it the old canons of education.

The idea of the beaux-arts that emerged in the 19th century had at first been persuasive. Art was to contribute to a humanitarian education by allowing people whose everyday lives were confined to pragmatic considerations and actions to develop their other inclinations and talents to their full potential. Art was to be independent to guarantee that it was not subverted by other purposes and demands. Artistic independence "fed on the idea of humanity" (Adorno 1970, p. 9). But the rampant nationalist agitation of the First World War showed that the art world's idealism could easily be harnessed for purposes of jingoistic

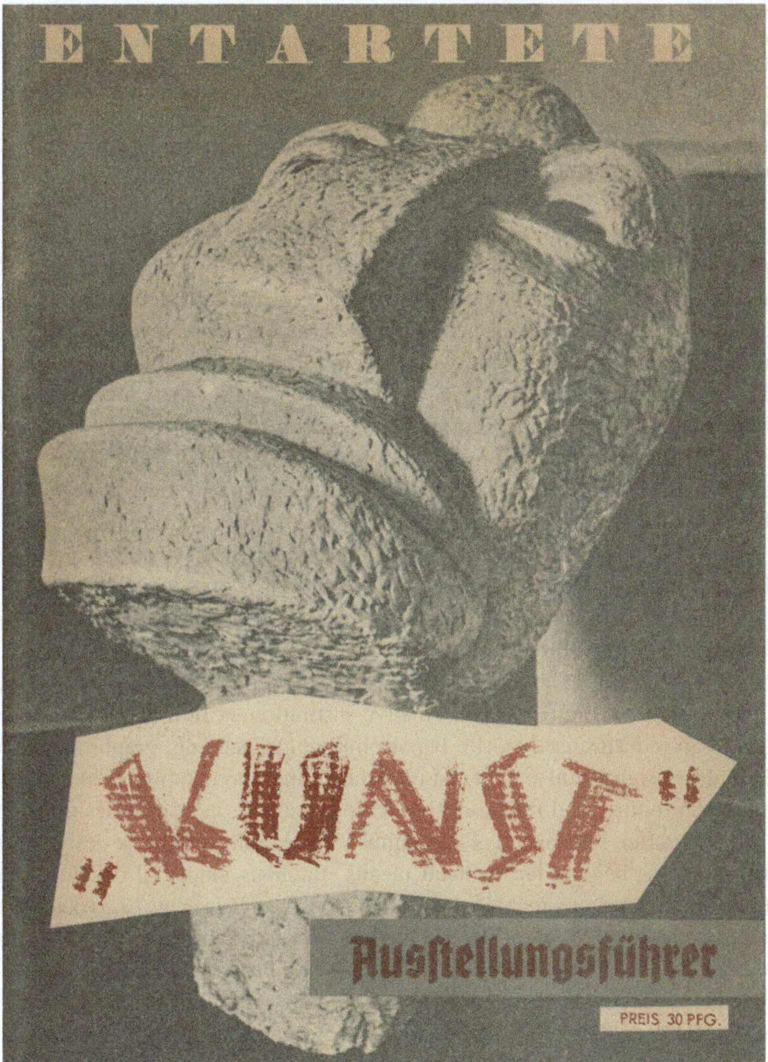

Fig. 12.1 Cover illustration for the Nazis' 1937 Degenerate Art exhibition guide

patriotism and brutal warmongering. Dada and Bauhaus are two examples of the conclusions that were drawn.

The "German Dada movement", wrote George Grosz,

> was rooted in the realisation, to which I and some associates came at the same time, that it was complete madness to believe that the intellect or any kind of spiritual leaders ruled the world. Goethe under fire, Nietzsche in the haversack, Jesus in the trenches ... Dadaism was not a "made" movement, it was an organic product that evolved in reaction to the tendency of so-called sacred art to keep its head in the clouds, reflecting on cubes and Gothicism, while generals painted with blood. (Grosz and Herzfelde 1925, p. 22–23)

Dada in Berlin, especially, gained a sharp political edge. Dissatisfied with art-world conventions, the Dadaists shifted their focus on to popular culture. "Condemning the press, condemning advertising, condemning sensations" wrote the Dada almanac of 1920, "is for people whose armchair is more important to them than the sounds of the street and who make a virtue of being duped by every nit-picker," (Huelsenbeck 1920, p. 37). In the first years after the war, what we now call the arts industry was in its infancy. For entertainment, there was literature, in the form of illustrated magazines and books, and cinemas showing silent movies. But it was precisely this that the Berlin Dadaists' gravitated towards in the immediate post-war period. One example of their involvement in popular culture was the novel *100%* by Upton Sinclair, for which John Heartfield designed the cover and George Grosz provided lithographic illustrations (Fig. 12.2).

The Bauhaus school led the second onslaught on the hegemonic art paradigm in Germany after the First World War. Under constant attack from the political right-wing, but with international connections, it gradually evolved into a multi-faceted educational, research and creative institution, with a hands-on approach to life in an industrialised world. It not only taught painting, graphics, sculpture and theatre design but also engaged in architecture, photography, commercial art, jewellery, furniture, and textile design, carpet-making, industrial design (including tea and coffee sets, ashtrays and lamps), typography and book design. A film department was beyond its financial means. Bauhaus industrial design products that gained iconic status include the tubular steel armchair (Fig. 12.3) designed by Marcel Breuer in 1925. It was later manufactured and sold by the Berlin company Standard-Möbel (Wingler 1975, p. 429) and is still produced today by Knoll.

Insofar as they were engaged in building a new culture, Dada and the Bauhaus school moved in the same direction. In 1927, Lazlo Moholy-Nagy formulated the Bauhaus idea (Fig. 12.4) in response to conventional attitudes: "The previous generation's idea of life was: Man must live his everyday life and in his leisure time he may engage with the phenomena of 'artistic' design and their 'inspired' works". He and the Bauhaus school, in contrast, asserted that "what we need" is not an artworld alongside which "life flows separately, but the self-expanding synthesis of all instances of life into an all-encompassing complete work (life) that

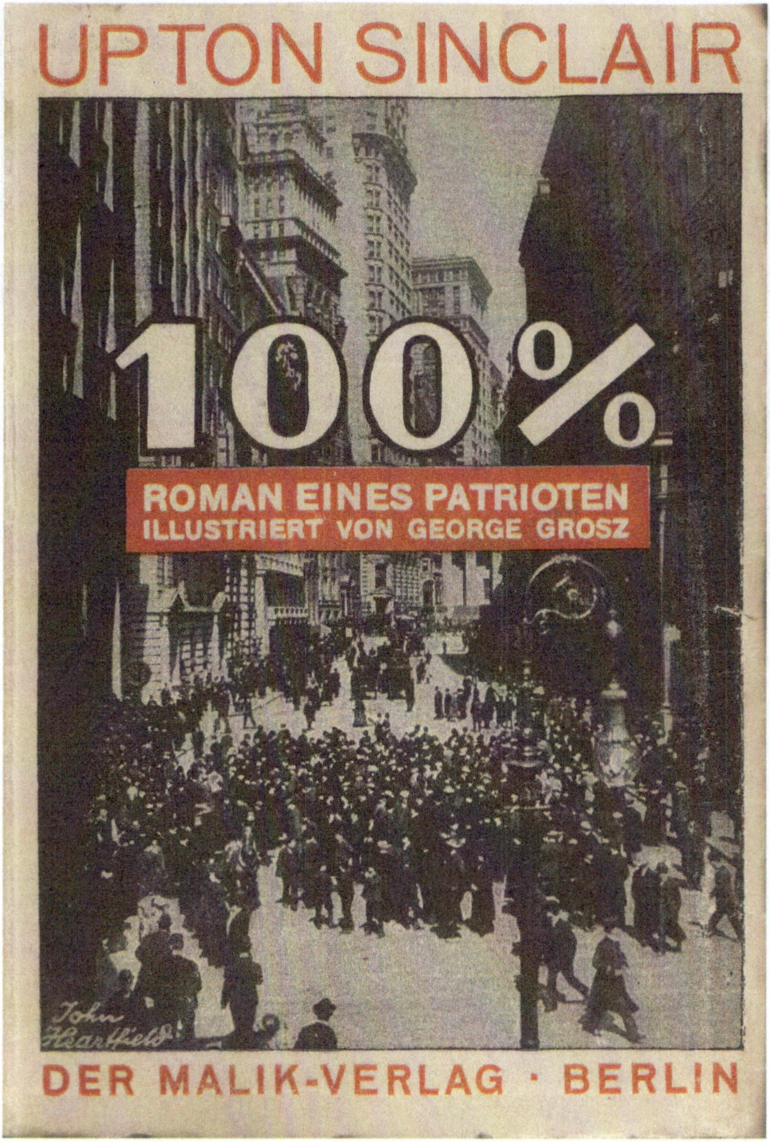

Fig. 12.2 Novel 100% by Upton Sinclair with cover designed by John Heartfield and lithographic illustrations by George Grosz

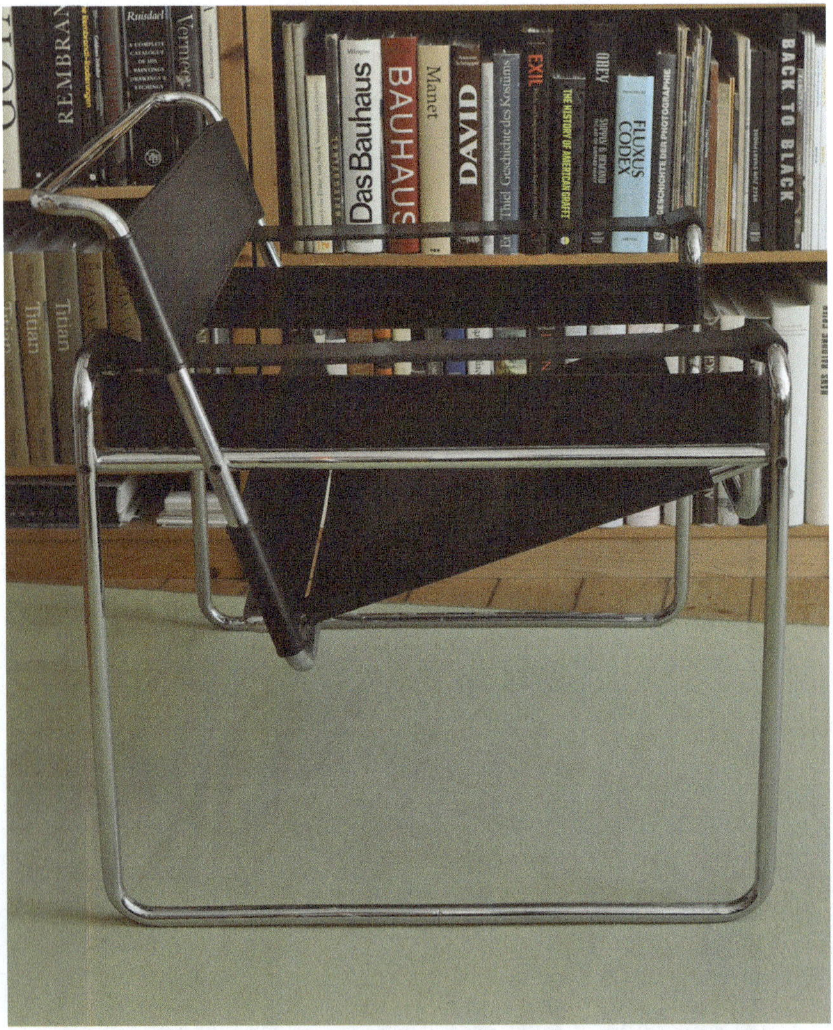

Fig. 12.3 Tubular steel armchair designed by Marcel Breuer 1925

suspends all segregation" (Moholy-Nagy 1927, pp. 14–15). Extending its conception of art beyond the beaux-arts to include consumer and industrial design, the Bauhaus school joined in the choir of voices that made up the historical avant-

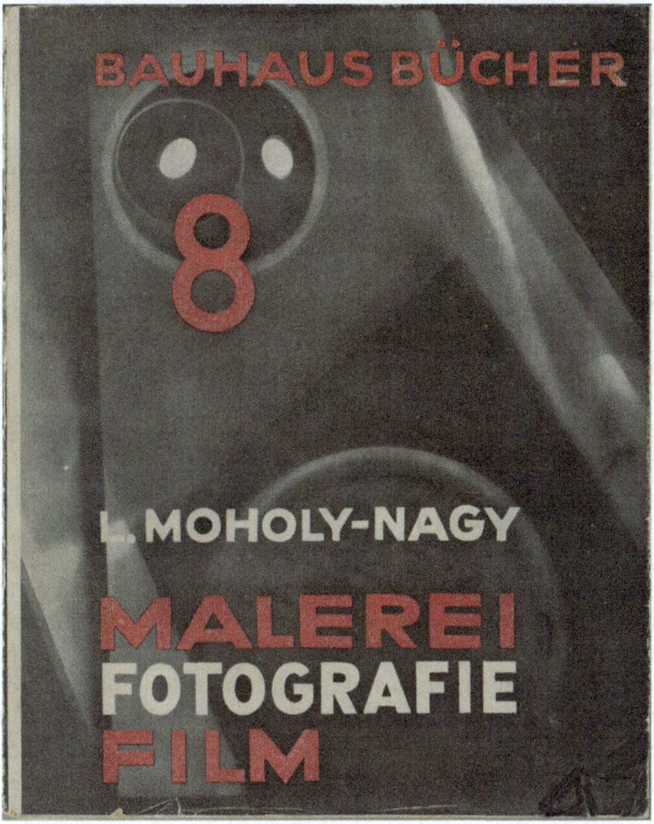

Fig. 12.4 Bauhaus Bücher 8. München 1927: Albert Langen. Bauhaus publication by Lazlo Moholy-Nagy 1927

garde, which drew the logical conclusions from the demise of Europe's old order in the wake of the First World War and attempted a complete renewal of the arts.

In 1933, political forces put an abrupt end to these efforts. The Nazi regime crushed the avant-garde. Most of Germany's avant-garde actors emigrated to the United States, where they found new places to work. Erwin Panofsky (art historian at Hamburg University who emigrated in 1933) later recalled the chairman of the Institute of Fine Arts at New York University saying: "Hitler is my best friend … he shakes the tree and I gather the apples" (Panofsky 1978, p. 389).

Under the auspices of its founding director Alfred H. Barr Jr., the Museum of Modern Art (MoMA) in New York challenged the conventional concept of the art museum. Opened in 1929, it was conceived of not only as an institution for preserving and presenting art but also as a site of research. In line with the ideas of the internationally cooperating avant-garde, the MoMA extended its ambit beyond the fine arts to include the entire spectrum of creative work. Barr's policy for the MoMA was to collect and present not only works of fine art but also "prints, photography, typography, arts of commerce and industry, architecture, stage designing, furniture, decorative arts" as well as to set up a "department of motion pictures" (Breidecker 1996, p. 212). While the art world of the 19th century had still neatly distinguished between the *applied arts* and *fine art*, in the 1930s Barr managed to create an institution that encompassed all genres of the visual arts under one roof. Thanks to the generous support of sponsors, he was able to realize the idea of an *avant-gardist* museum.

12.3 Documenta 4 (1968)

The experts responsible for the first documenta in 1955 had ignored the prewar avant-gardist tendencies and adhered to conventions from the preceding century. And in the late 1960s, the documenta organisers continued to uphold the conventional beaux-arts paradigm. Meanwhile, in the major cities of the United States, artists had disengaged from the art world's lofty detachment from life praxis. Influenced by the European emigrants' ideas for alternative artistic practices, they held happenings and designed posters and record covers to explore fields of art beyond the gallery. A climate prevailed which generated vital innovations in all aesthetic areas. But the 4th documenta in 1968 was not interested. Once again, it was confined to autonomous works: paintings on canvas, sculptures, assemblages, and printed graphics. A page of figures from the exhibition catalogue can convey a general idea (Fig. 12.5): Top left is the painting *Große Sonntagslandschaft mit 3 Figuren* by Horst Antes, alongside the screen print *As is when* by Eduardo Paolozzi; below is a work from the *Anthropometry* series by Yves Klein and below that *M-Maybe* by Roy Lichtenstein and Claes Oldenburg's *Soft Washstand*.

Despite the resistance of some of the organisers, the 4th documenta had embraced Pop Art. This decision prompted Werner Haftmann, a crucial voice up to that point, to leave the supervisory board (Rosebrock 2015, p. 222). But even so, the exhibition included only the small section of autonomous art from the tremendously expanded range of artistic practices of the 1960s. A comparison with the MoMA's approach to presenting art of this era reveals two starkly differing strategies.

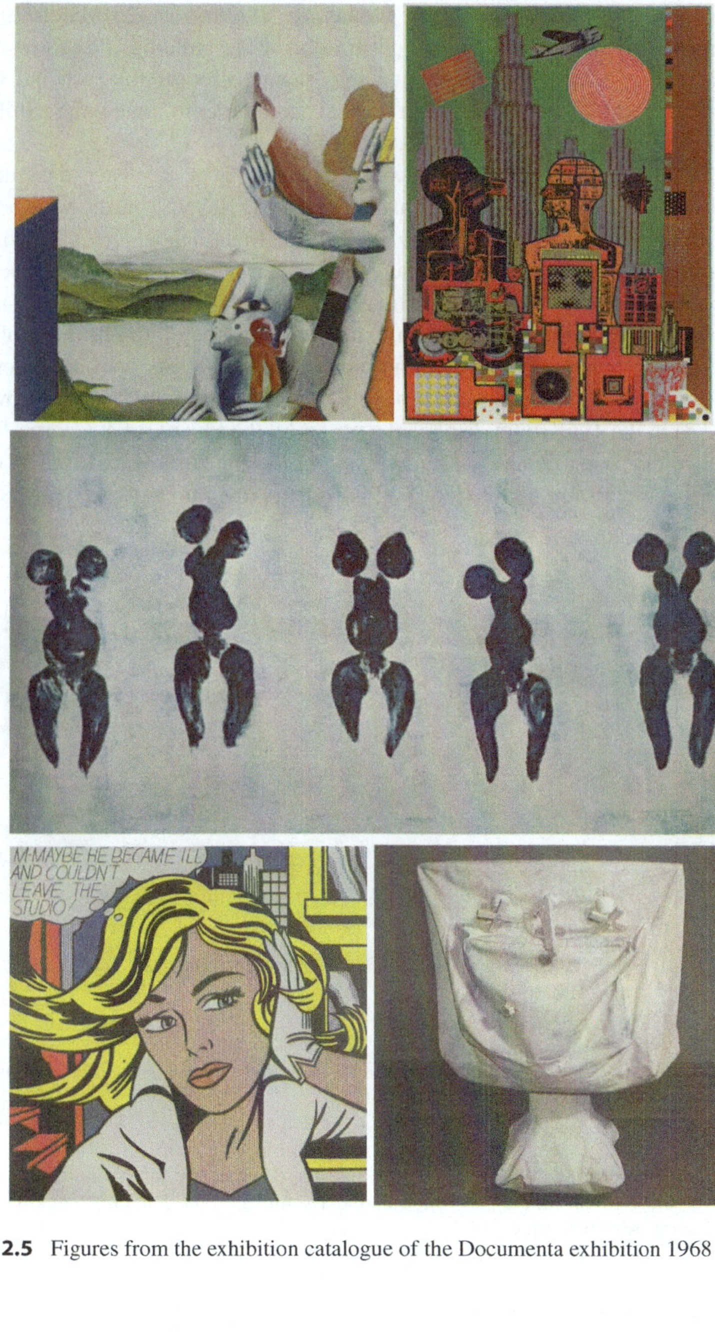

Fig. 12.5 Figures from the exhibition catalogue of the Documenta exhibition 1968

The MoMA exhibition *From the Collection: 1960–1969* (26 March 2016 to 12 March 2017) covered the entire fourth floor of the building. The show of the museum's own holdings was arranged into separate rooms for each individual year and the works that originated in them. I would like to review the exhibition with reference to some illustrative photographs.

In the middle of the *1961* room was a Jaguar E-Type Roadster, designed by Sir William Lyons, Malcolm Sayer and William M. Heynes; on the wall behind, the assemblage *First Landing Jump* by Robert Rauschenberg (Fig. 12.6). The subsequent years followed in a serene vein. In the *1966* room, however, the visitor encountered Joseph Beuys' *Eurasia Sibirian Symphony 1963* (Fig. 12.7). And the last rooms took the visitor on an exhilarating journey through the proliferation of artistic practices in the second half of the 1960s. As well as gallery art and consumer design, there were products of political activity. A high showcase (Fig. 12.8) in the *1968* room displayed a poster by the Fluxus artist George Macunias above the *Swiss Officer's Knife Champion* designed by Karl Elsener (manifactured by Victorinox Switzerland), alongside a *Nizo Movie Camera* by Robert

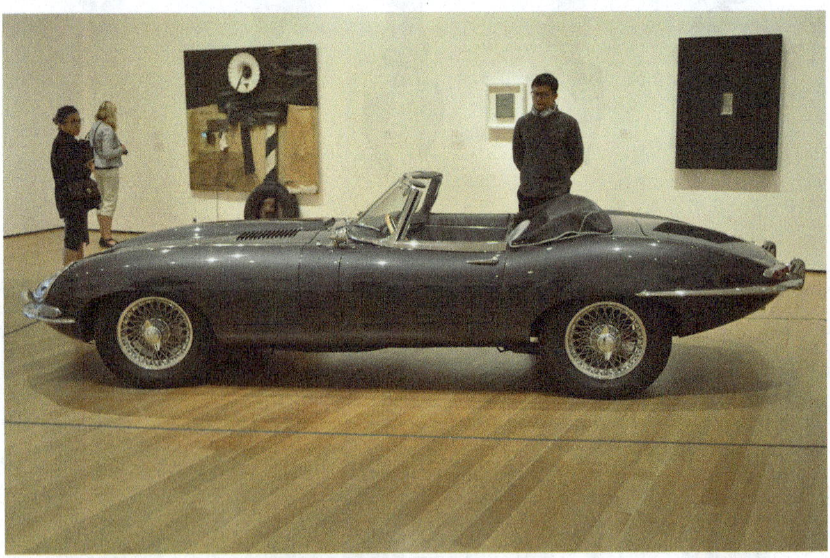

Fig. 12.6 *1961* Room with Jaguar E-Type Roadster, designed by Sir William Lyons, Malcom Sayer and William M. Heynes (middle) and assemblage *First Landing Jump* by Robert Rauschenberg (wall behind)

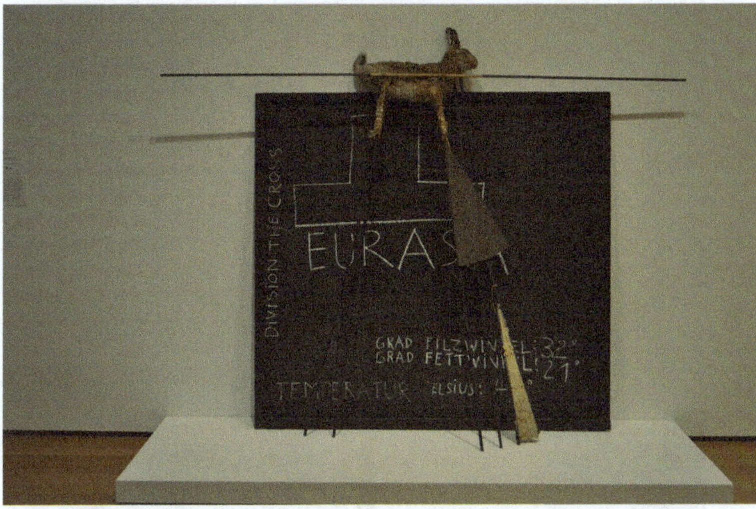

Fig. 12.7 Joseph Beuys' *Eurasia Sibirian Symphony 1963*

Oberheim (manufactured by Braun AG, Frankfurt), and underneath the multiple *Portrait of Antonin Artaud* by Lucio Fontana (who also created ground-breaking works, transcending illusionist representation, by slashing, drilling into and perforating canvasses); at the bottom was *Weigt Scale (model HW1)* by Dieter Rams (manufactured by Braun AG, Frankfurt) next to the the *Chronotime Clock* by Pio Maczù (manufactured by Italora, Milan) and the *Valentine Portable Typewriter* by Ettore Sottsass and Perry King (manufactured by Olivetti, Ivrea).

The Maciunas poster (Fig. 12.9) was mistakenly assigned to the year 1968; in fact, it originates from the year 1966. It is based on the design of the U.S. flag but has skulls and crossbones instead of stars and text making up the stripes. This work of artistic activism, which was one of the earliest in protest at the Vietnam war and white-man power politics, starts with the words "U.S.A. surpasses all the genocide records!" and ends with an offer to supply information on sources and references.[3]

[3] The poster can be dated from this double-sided printed text.

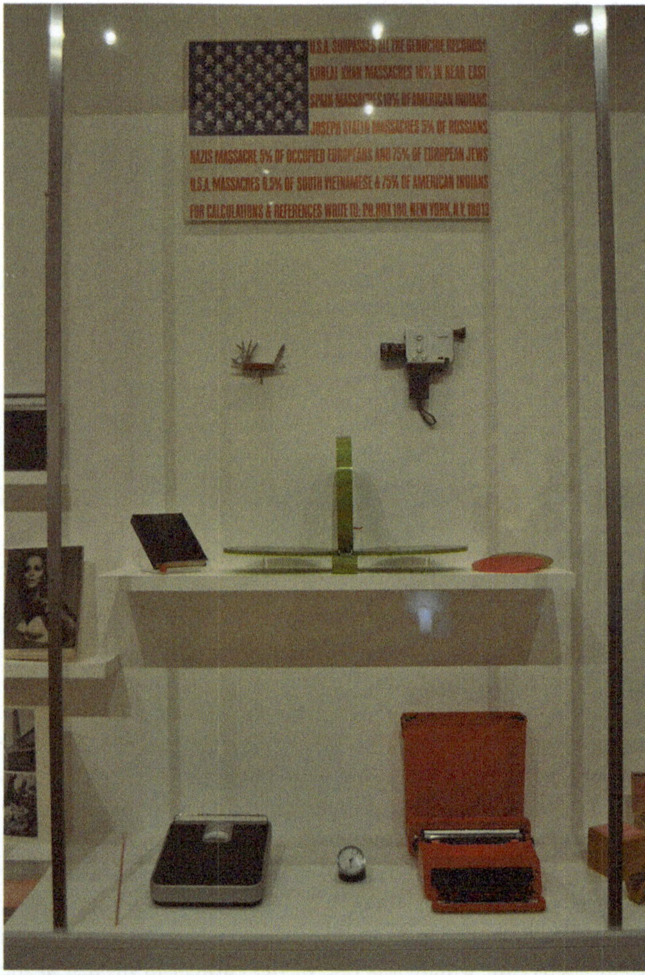

Fig. 12.8 High showcase in 1968 room displaying a poster by Fluxus artist George Macunias above *Swiss Officer Knife Champion* designed by Karl Esener, alongside a *Nizo Movie Camera* by Robert Oberheim, underneath the *Portrait of Antonin Artaud* by Lucio Fontana and *Weigt Scale* by Dieter Rams next to *Chronotime* Clock by Pio Maczù and *Valentine table Typewriter* by Ettore Sottsass and Perry King (bottom)

The *1967* room was the main attraction of the museum's show of its holdings from the sixties. Psychedelic posters were displayed on one wall while a light show of a rock concert in Bill Graham's Fillmore East by Joshua White ran in

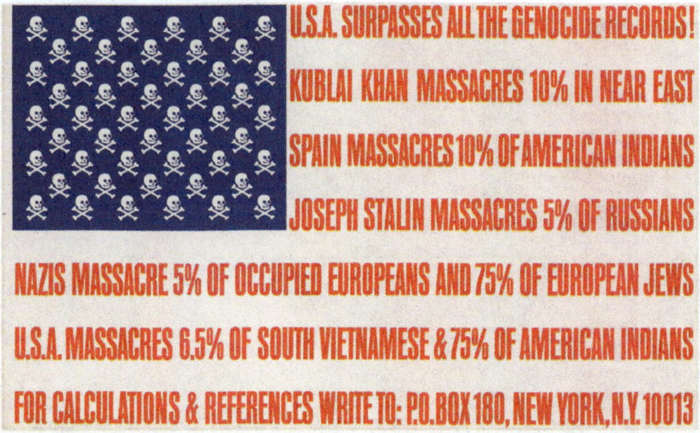

Fig. 12.9 Poster by George Maciunas 1966

the background; next to this were record covers and a dress from the period. A selection of eleven posters by Victor Moscoso and one by Wes Wilson convey an impression of the day's publicity for rock concerts in San Francisco (Fig. 12.10). Moscoso had studied with Josef Albers (who had taught at the Bauhaus school and emigrated to the United States in 1933) and combined this knowledge with a hippie counterculture aesthetic to express a new esprit in psychedelic posters. As we showed in our 2017 exhibition titled *San Francisco 1967* in Folkwang Museum Essen, this movement had a strong political dimension (Hieber 2017).

The above outline of the MoMA exhibition on the sixties, which dealt with the same period of art history as the 4th documenta, highlights the differences between the two shows. Although the 1968 documenta devoted some space to innovative developments in the field of gallery art, it basically adhered to the policy of its predecessors by focusing on autonomous art.

12.4 Documenta IX (1992)

Since the 4th documenta had failed to include any happenings or Fluxus, psychedelic or political posters or even consumer design, the inclusion of advertising images by Charles Wilps in documenta 5 in 1972 seemed like a rather awkward attempt to correct the previous four exhibitions' narrow policy. Unsurprisingly, the subsequent documentas resumed their conventional course.

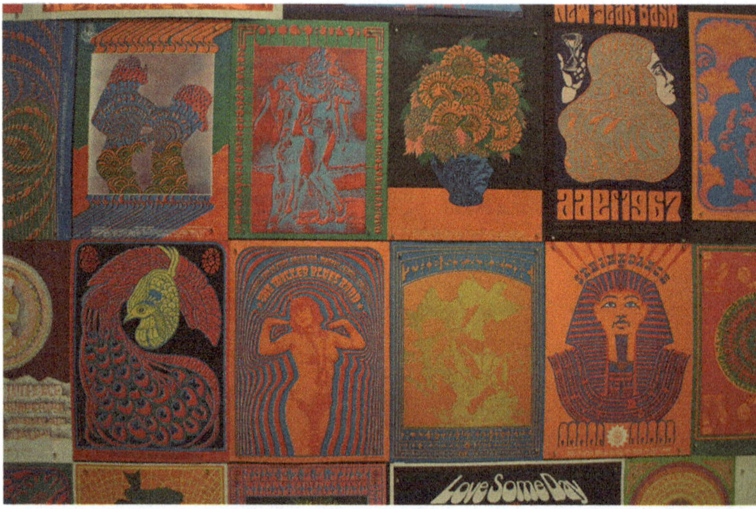

Fig. 12.10 Selection of eleven posters by Victor Moscoso and one by Wes Wilson in *1967* room

Documenta IX of 1992 also remained within the bounds of the beaux-arts. This implied excluding cutting-edge practices of *activist art* that responded to problems which had become virulent during the preceding decade. These innovative practices had evolved on New York's art scene and produced milestone works on a par with those of the historical avantgarde and the sixties.

As this artistic activism has been largely neglected in the German-language countries, I will look to another leading art museum in New York, the Whitney Museum, for an instance of its presentation. Here, the genre was given apt attention—alongside "traditional" artworks—in the exhibition *An Incomplete History of Protest: Selections from the Whitney's Collection*,[4] which ran from August 2017 to August 2018. Both the "traditional" artworks on political subjects and the works of "artistic activism" arose from a sense of social catastrophe and deep concern. They addressed the AIDS crisis. Among the "traditional" works in the Whitney exhibition was *AIDS* (1988), an installation of backlit coloured foils in

[4] In am grateful to Andreas Urban, Hanover, for providing the photographic documentation of the exhibition, which was not accompanied by a catalogue.

a display case by the Toronto collective General Idea (AA Bronson, Felix Partz, Jorge Zontal). It quoted Robert Indiana's famous *LOVE* image of the 1960s with the letters AIDS placed in a similar arrangement against a multi-coloured background. Alongside works such as this, works of artistic activism from the Whitney's own holdings were displayed.

They were created at a time when neither the world of politics nor the pharmaceuticals industry could muster the conviction to take steps against the AIDS crisis, to which the public, moreover, seemed largely indifferent. In this pressing situation, creative minds launched a political movement to deal with it. Named ACT UP (AIDS Coalition To Unleash Power), artists played a key role in the movement. They used any aesthetic strategies at their disposal to their political ends. Without being able to go into depth here, I would like to comment on this development by considering three works from the Whitney exhibition.

In 1987, Donald Moffett (Fig. 12.11) placed the Republican US president Ronald Reagan next to a target. In this way, he implied that the US government's neglect of promoting medical research was ultimately responsible for the death of many people. A short time later Moffett became co-founder of the artists' group *Gran Fury*, which used artistic practices and direct actions to attempt to end the AIDS crisis (Gober 2011, p. 12). His poster, incidentally, was immediately taken into the MoMA's permanent exhibit.

In 1989 the activists came upon the idea of using publicity in subway trains (Fig. 12.12). They produced prints in the same formats as the product advertising

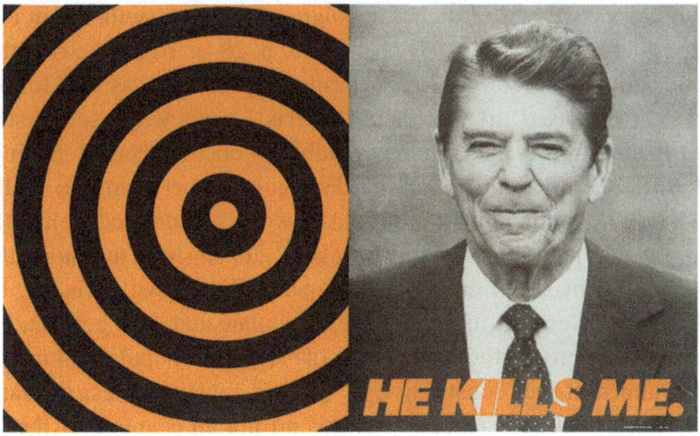

Fig. 12.11 Work by Donald Moffett with Republican US president Ronald Reagan next to a target 1987

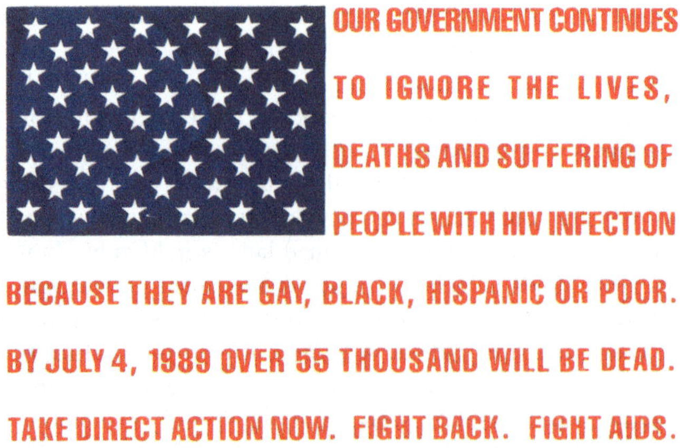

Fig. 12.12 Publicity published by activists in subway trains 1989

inside the carriages and exchanged them late at night when there were no passengers. One of their prints, made to fit the advertising bands at the eye level of standing passengers, reads: "Our government continues to ignore the lives, deaths and suffering of people with HIV infection because they are gay, black, Hispanic or poor. By July 4, 1989 over 55 thousand will be dead. Take direct action now. Fight back. Fight AIDS".

Aware that politics is fought out in the media, the ACT UP artists sought television and press coverage of their campaigns. And to carry them out effectively, they needed financial resources (Hieber 2015, p. 284). To help cover the cost of printing and actions, a poster by Keith Haring (Fig. 12.13) was sold for $15.

The New York-based art historian Douglas Crimp witnessed first-hand the events that this section of the Whitney exhibition documented. Referring to what he saw during this turbulent period, he distinguished between two aesthetic tendencies: One, which "includes traditional artworks" addressed political subjects and is expressed in paintings, poems, novels etc., or in short, autonomous art. The other tendency "consists of cultural participation in activist politics". Integrated into a collective movement, these practices articulate and even generate the movement's politics. "Appropriating techniques of 'high art', popular culture, and mass advertising, aimed at and constitutive of specific constituencies, relevant only to local and transitory circumstances … is this art not an example of the 'sublation of art into the praxis of life'?" (Crimp 1993, p. 22). To Crimp, these artistic practices seemed "to continue the unfinished avant-garde project" (Crimp 1993, p. 19).

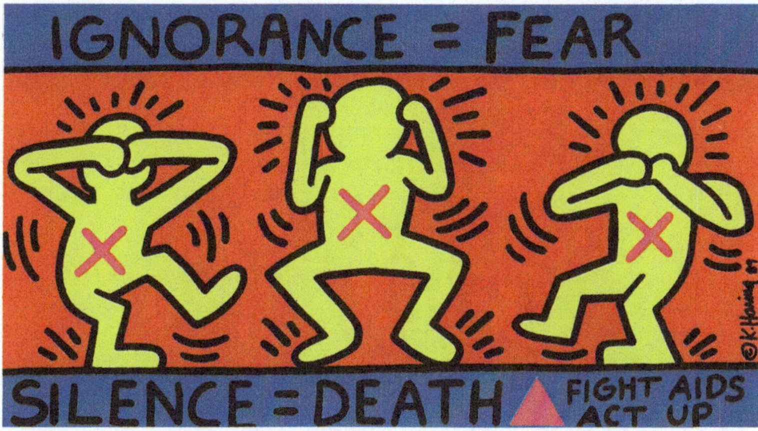

Fig. 12.13 Poster by Keith Haring 1989

Documenta IX ignored artistic activism. In the art magazine *Texte zur Kunst*, Stefan Germer commented on the documenta approach of these years with the words: "Rather than exploring the relationship between activism and art production, they invited Zoe Leonard to exhibit" (Germer 1992, p. 61). To grasp the significance of this, I will now recall the artist Zoe Leonard.

Leonard created personal works in her studio. At the same time, she was involved in both the *fierce pussy*[5] group, which fought for lesbian acceptance in society, and the *Gang*[6] group, dedicated to issues of body politics. She not only created autonomous art, then, but also participated in works that pursued political ("heteronomous") objectives. I would like to present three examples of the work of these groups before returning to consider documenta IX.

A *fierce pussy* poster (Fig. 12.14) showing a photo of a school-class addressed "queer liberation" and gender stereotypes; Zoe Leonard is one of the pupils, seated at the centre of the first row. For a poster (Fig. 12.15) created during the first Gulf

[5] The *fierce pussy* group comprised Pamela Brandt, Nancy Brody, Jean Carlomusto, Diana Dietrich, Joy Episalla, Donna Evans, Lisa Hubbard, Zoe Leonard, Carrie Yamaoka, Suzanne Wight.

[6] The *Gang* group comprised Peter Bowen, Holly Hughes, Zoe Leonard, Wellington Love, Loring McAlpin, Martin McElhiney, Michael Perelman, Maria Perez, Adam Rolston, Daniel Wolfe, Suzanne Wright.

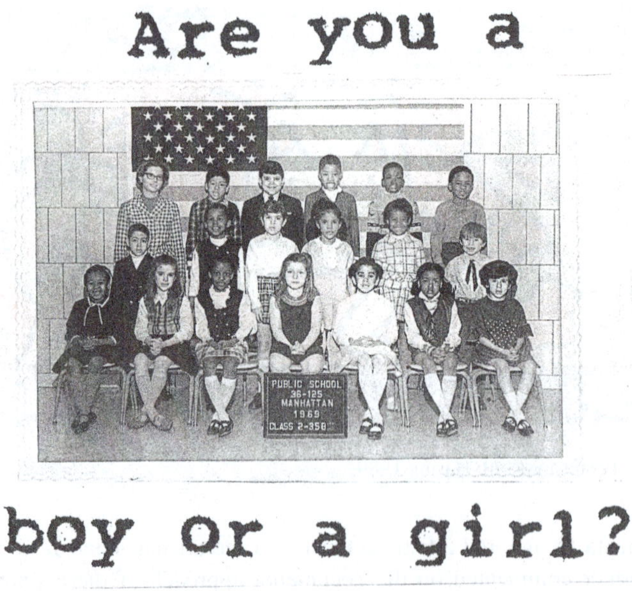

Fig. 12.14 Fierce pussy: Are you a boy or a girl? 1992

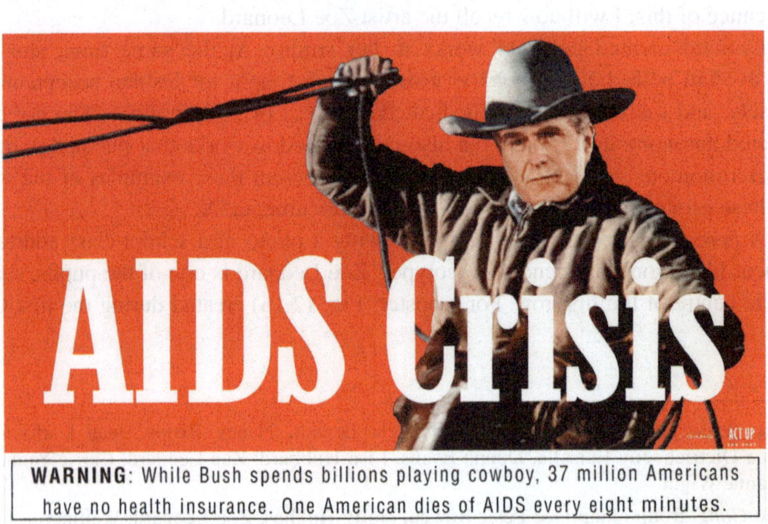

Fig. 12.15 Poster by the *Gang* group appropriating the Marlboro cigarette brand 1991

War, the *Gang* group appropriated the Marlboro cigarette brand's well-known advertising image, transporting ideas of freedom and the untamed life of the cowboy. But *Gang* placed George H. Bush in the role of the cowboy. The text-box at the base of the image visually conforms to the warning requirement but contains the words: "While Bush spends billions playing cowboy, 37 million Americans have no health insurance. One American dies of AIDS every eight minutes." Another *Gang* poster (Fig. 12.16) joined in the abortion debate. The idiomatic expression "read my lips" can be interpreted figuratively or literally; the latter reading is suggested by the poster's combination of a photograph of a vagina with the continuation of the message with the words "before they're sealed". The poster not only fuelled protests at planned legislation to constrain family counselling centres but also sparked debates with conservative feminists who found it too "pornographic". Leonard used the master for this poster in her installation at documenta IX.

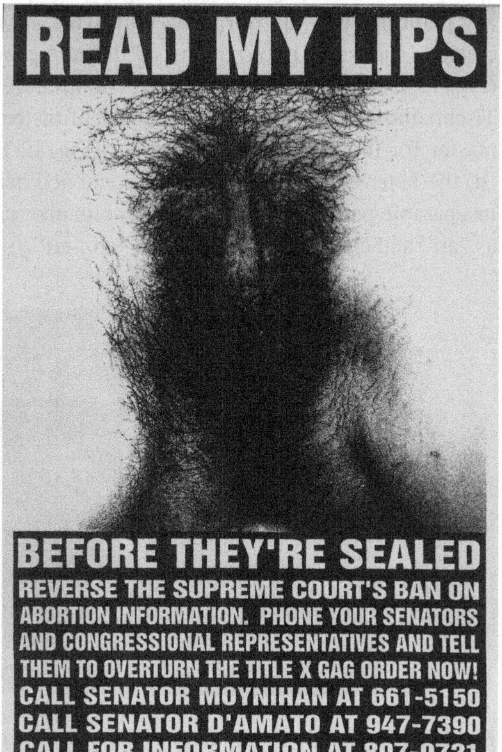

Fig. 12.16 "Read my lips" poster by *Gang* group 1992

In the run-up to the documenta, Leonard had visited the Neue Galerie in Kassel. In a later interview, she recalled: "The paintings were what made me want to work in those rooms" (Leonard and Blume 1997, p. 15). Initially, Leonard planned a complex response to these paintings. But things turned out differently.

> A week before I was supposed to go to Kassel to install, I was monkeying around here in my studio, and I looked down and I saw the print for the Gang poster on the floor. I looked at that, and I thought, "Oh my God. That is everything in it." It's the most aggressive image and the most passive image. It's the thing that people are most afraid to look at, most afraid to talk about, most eager to look at, most eager to talk about, and it is the very thing that's both missing and overly present in all those paintings. It focused everything. (Leonard and Blume 1997, p. 16)

The artist was offered the use of all the ground-floor rooms of the Neue Galerie in Kassel for a site-specific installation. Normally, these rooms house an extensive collection of regional paintings from the last centuries. Zoe Leonard had any paintings that did *not* show portrayals of women removed. In the gaps thus created, she hung photographs of different vaginas. A detail of the exhibition (Fig. 12.17) conveys a rough impression: Close to a portrait by Johann Heinrich Tischbein the Elder from the second half of the 18th century (Fig. 12.17a) is the master for the "read my lips" poster (Fig. 12.17b). To the New York-based artist, the heteronomous works of political activism and the autonomous works are inseparable partners. The documenta organizers, however, accepted only the latter as "art" and excluded the former as "non-art".

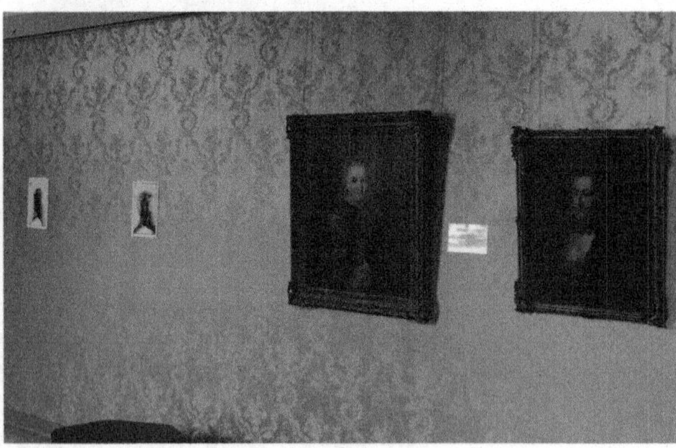

Fig. 12.17 **a** Detail of exhibition in the ground-floor rooms of the Neue Galerie in Kassel 1992. **b** Portrait by Johann Heinrich Tischbein the Elder. **c** Master for the "read my lips" poster 1992

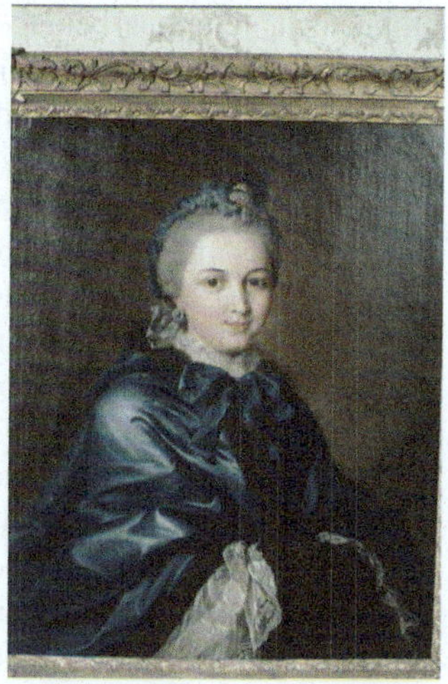

Fig. 12.17 (continued)

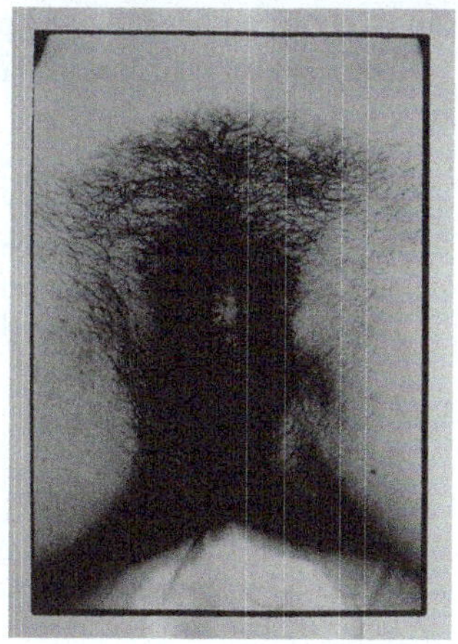

Fig. 12.17 (continued)

12.5 Different Art Worlds

Pierre Bourdieu found that "art today is a site of obscurantism" and inferred that "if Durkheim or Weber came back, they would not pursue the sociology of religion anymore but the sociology of art, because many of the logics evidenced in prophetism, religious belief etc., which can be observed in the religious field, now prevail in the artistic field" (Bourdieu 2015, p. 183). Long-held convictions, adhered to unchallenged, play a key role in this context. As Gerhard Panzer has shown in his art-sociological analysis, the first documenta was the work of a regional network of associates which stretched back to the Weimar Republic (Panzer 2015, p. 80). This group eschewed additional support from the outside to organise the exhibition. Not only the artistic director of the first documenta but also his successors— almost right up to the present day—abided by the beaux-arts paradigm, which only recognised autonomously created works as art. Documenta has consistently followed the prevalent cultural tendency in the German-speaking countries, as manifested, for instance, in the *Museumsufer* chain of museums in Frankfurt (established in the 1980s). Walking along the south bank of the River Main, starting at the Städel museum of art, one passes the museum of communication, the architecture museum, the film museum, the museum of world cultures, and lastly, after about eight minutes, the museum of applied arts. This arrangement of distinct museums exemplifies the hegemonic, narrow conception of art which is bound to the 19th century and neatly separates all genres, as if autonomous art needed to be protected from infection.

The art historian Erwin Panofsky, on the other hand, criticised the beaux-arts paradigm:

> If commercial art be defined as all art not primarily produced in order to gratify the creative urge of its maker but primarily intended to meet the requirements of a patron or a buying public, it must be said that non-commercial art is the exception rather than the rule, and a fairly recent and not always felicitous exception to that,

and he went on to say that

> while it is true that commercial art is always in danger of ending up as a prostitute, it is equally true that non-commercial art is always in danger of ending up as an old maid.[7] Non-commercial art has given us Seurat's "Grande Jatte" and Shakespeare's Sonnets, but also much that is esoteric to the point of incommunicability. Conversely, commercial art has given us much that is vulgar or snobbish (two aspects of the same thing) to the point of loathsomeness, but also Dürer's prints and Shakespeare's plays. (Panofsky 1947, p. 18)

[7] The meaning of the term "old maid" has changed since Panofsky wrote this text. He meant it in the sense of someone or something so cautious or non-dynamic as to be unattractive.

The historical avant-garde took the approach of not only cultivating the field of autonomous art but also of tending those of politics and commerce. It tried "to organise a *new* life praxis from a basis in art" (Bürger 1984, p. 49). The Dadaists sprang from this intention, as did the Bauhaus school's efforts to extend the conception of art beyond the fine arts.

Before returning to the documenta, I would like to emphasise once again that both avant-garde approaches evolved further in the US art world. Consider, for example, the Lower East Side in Manhattan in the year 1983, where Joseph Nechvatal produced the exhibition *Photomontages of the Nazi Period John Heartfield* in ABC No Rio (Fig. 12.18) for the artists' group *Colab* (short for *Collaborative Projects*).[8] That can stand here for the political dimension. For an illustration of the continuation of a broader conception of art, such as the Bauhaus school represented, we need only recall the MoMA exhibition on the sixties (Fig. 12.19).

The organisers of the first documenta persisted with *pre*-avantgarde conventions. By not making any effort to include the *entire spectrum* of positions destroyed by Nazi dictatorship, they cemented the traditional conception of art.

My sketch of the differences between leading institutions on both sides of the Atlantic, between the documentas here in Europe and the major New York museums in America, is by no means intended as an illustration of New York's role model function. Rather, it is intended as a contribution to explorations of German history.

Norbert Elias speaks of a "national habitus" to explain the characteristics that are stamped upon the individual members of a society by the political system in which they live. "Sociologists face a task here which distantly recalls the task which Freud tackled. He attempted to show the connection between the outcome of the conflict-ridden channelling of drives in a person's development and his or her habitus. But there are also analogous connections between a people's long-term fortunes and experiences and their social habitus at any subsequent time" (Elias 1996, p. 19).

---

[8] Nechvatal later wrote that he used the strategy of appropriating art: "I produced the Colab sponsored show (with performances) simply called John Heartfield at ABC No Rio, held from November 1st to November 18th in 1983. Xeroxes and photomechanical blowups of John Heartfield's (1891–1968) art, torn from a book, were the maquettes this time. Reproductions of his anti-Nazi/anti-Fascist photomontages were wheat-pasted on the walls of ABC No Rio, walls that had been painted a sinister black from top to bottom. I filled the space with audio collages by Bradley Eros and hit the streets with Mitch Corber, putting posters all over downtown: advertisements for the show along with powerful John Heartfield images" (Nechvatal 2016).

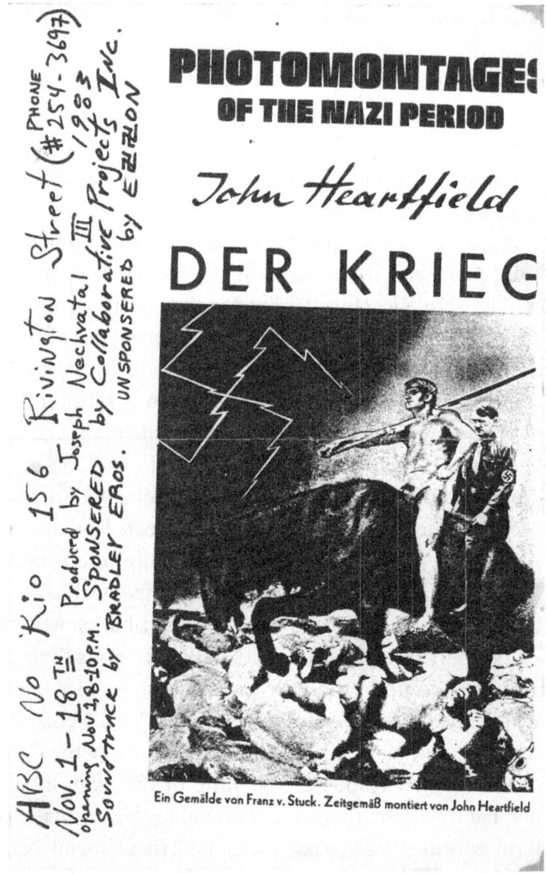

Fig. 12.18 *Photomontages of the Nazi Period John Heartfield* in ABC NO RIO by Joseph Nevchvatal 1983

It was certainly fateful for the formation of a national habitus in Germany that many critical intellectuals were compelled to leave Germany by the Nazi regime and war. The wave of emigrations had a negative impact on Germany as it left only the officeholders and conservative fellow-travelers supporting the dictatorship to exert influence—and so set the course for the future. On the other hand, many emigrants took professorships or other functions in the U.S. that enabled them to contribute to enriching the arts in their new home, such as Herbert Bayer and Erwin Panofsky, to mention just two. Herbert Bayer, once head of the Bauhaus workshop for printing and commercial art, designed the MoMA's Bauhaus

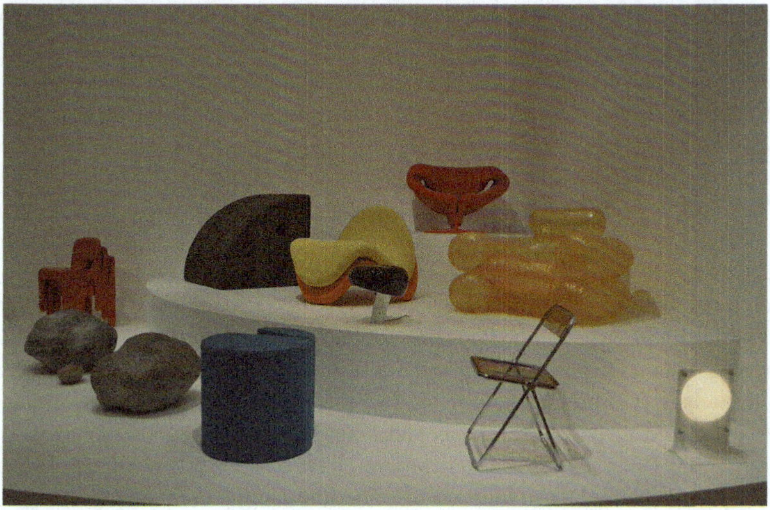

Fig. 12.19 MoMa exhibition on the sixties 1967

retrospective in 1938. And Erwin Panofsky contributed to setting up the MoMA's film department in 1936 as art theory specialist.

West German culture initially drew on the potential remaining in the country. The New York museums, meanwhile, showed how the students of avant-gardists continued to develop their activist art practices and extend concepts of art. In the process, they served as reminders that excluding such actors results in an impoverishment of the art world. To put it bluntly, by disregarding all avant-gardist positions, the first documenta perpetuated *one aspect* of the Nazis' repression of the avant-garde—namely, the return to the conventional beaux-arts paradigm. As subsequent editions adhered to this basic premise, the documenta prolonged the cultural bloodletting begun by the Nazis.

Exhibition visitors expect to see art on display. But what they also see is a representation of the exhibition organisers' *conception of art*.

Insofar as exhibitions convey a conception of art through the works they present and the way in which they present them, they play a role in the public's aesthetic education. That is why the exclusion of the historical avantgarde from the first documenta in 1955 is not only a question of art history but also of what is deemed art and what is not. Subsequent documentas have sustained and cemented this approach to evaluation. For this reason, these popular exhibitions, which have attracted ever larger attendances over the years, set the parameters for what art can be and what it cannot. But drawing boundaries between "independent art" and "applied art" comes at a cost. Restricting movement between the different fields of activity has consequences. Excluding practices that seek real-life relevance, especially, robs art of its power.

To close, I would like to mention a personal insight gained through decades of engaging with the New York art world. Initially, I was surprised that New York artists were not unequivocally positive about the MoMA but often critical of it. Over time, however, I and my associate Gisela Theising became increasingly familiarised with artistic practices that we had not seen on the conventional German arts scene (Theising 2008). And we began to understand the New York artists' criticism. It is precipitated by the fact that while artistic practices in the city have continued to develop in leaps and bounds since the 1960s (Avgikos 1995; Crimp 2003; Guerrilla Girls 1995; Moore 2011; Stather 2014), the MoMA consistently trails behind these innovations. The following picture emerges of curatorial policy either side of the Atlantic: In Germany, the documenta has adhered to the beaux-arts paradigm, perpetuating its roots in a 19th century conception of art. In the United States, meanwhile, the MoMA became established as a thoroughly 20th century institution under its founding director Alfred H. Barr because it ignored the formerly customary boundaries between the genres. But the US art world continued to evolve after the 1960s. From the point of view of contemporary artists in New York, the MoMA more-or-less stagnated in the mid-20th century Barr era. For a public shaped by German culture, however, it is a far more positive prospect. It can provide the unfamiliar experience of visiting a museum that has at least arrived in the 20th century.

List of Figures

Fig. 12.1 Ausstellungsführer »Entartete Kunst«. 1937
Fig. 12.2 John Heartfield: Upton Sinclair, 100%: Roman eines Patrioten. Berlin 1921: Malik
Fig. 12.3 Marcel Breuer: Stahlrohr-Sessel »Wassily«. Designed 1925
Fig. 12.4 Laszlo Moholy-Nagy: Malerei Fotografie Film. Bauhaus Bücher 8. München 1927: Albert Langen

References

Adorno, Theodor W. 1970. Ästhetische Theorie. Gesammelte Schriften Band 7, Frankfurt/M.: Suhrkamp.

Avgikos, Jan. 1995. Group Material Timeline: Activism as a Work of Art. In *But is it Art? The Spirit of Art as Activism,* Nina Felshin (ed.), Seattle: Bay Press.

Bourdieu, Pierre. 2015. Manet. Eine symbolische Revolution. Aus dem Französischen von Achim Russer und Bernhard Schwibs. Berlin: Suhrkamp.

Breidecker, Volker. 1996. »Ferne Nähe«. Kracauer, Panofsky und »the Warburg tradition«. In *Siegfried Kracauer–Erwin Panofsky, Briefwechsel* Volker Breidecker (ed.), 129–226. Berlin: Akademie.

Bürger, Peter. 1984. Theory of the Avant-Garde. Translated by Michael Shaw. Minneapolis: Uniceresity of Minnesota Press.

Crimp, Douglas. 1993. On the Museum's Ruins. Massachusetts–London: MIT Press.

Crimp, Douglas. 2003. Gran Fury talks to Douglas Crimp. Artforum April 2003, 70–72.

documenta Retrospective. n. d. https://www.documenta.de/de/retrospective/#. Accessed: 25 September 2019.

Elias, Norbert. 1996. The Germans: Power Struggles and the Development of Habitus in the Nineteenth and Twentieth Centuries. Translated by Eric Dunning and Stephen Mennell. Cambridge: Polity Press.

Fitzke, Kirsten. 2015. Auf dem Weg zur documenta. Die Wochenzeitung DIE ZEIT und ihr Autor Werner Haftmann spiegeln und gestalten Positionen bildender Kunst in Westdeutschland. In *Beziehungsanalysen. Bildende Künste in Westdeutschland nach 1945,* Gerhard Panzer, Franziska Völz and Karl-Siegbert Rehberg (ed.), 151–171.Wiesbaden: Springer VS.

Germer, Stefan. 1992. Documenta als anachronistisches Ritual. Texte zur Kunst Nr. 6, 49–63.

Geschichte der documenta. n. d. https://www.kassel.de/kultur/documenta/geschichte/. Accessed: 21 February 2019.

Gober, Robert. 2011. Gran Fury. Im Katalog zur »Ausstellung Gran Fury: Read My Lips«, 80WSE Gallery der New York Universiy, 31.01.–17.03.2012. 11–18.

Grosz, George and Wieland Herzfelde. 1925. Die Kunst ist in Gefahr. Berlin: Malik.

Guerrilla Girls. 1995. Confessions of the Guerrilla Girls. New York: HarperCollins.

Haftmann, Werner. 1955a. Einleitung. In Katalog der »documenta«, 15.07.–18.09.1955. 15–25. München: Prestel

Haftmann, Werner. 1955b. Über das moderne Bild. Rede, gehalten zur Eröffnung der Ausstellung »Documenta, Kunst des XX. Jahrhunderts«. Frankfurter Allgemeine Zeitung vom 30.07.1955.

Hieber, Lutz. 2015. Politisierung der Kunst: Avantgarde und US-Kunstwelt. Wiesbaden: Springer VS.

Hieber, Lutz. 2017. Hippie-Aktivitäten. Im Katalog zur Ausstellung »San Francisco 1967«, Museum Folkwang Essen, 09.06.–03.09.2017, 17–24. Edition Folkwang/Steidl: Göttingen.

Hinrichsen, Jens and Saskia Trebing. 2020. Report Documenta 1955. monopol Magazin für Kunst und Leben, Februar, 46–55.

Hirzel, Stephan. 1955. Zur Architektur von 1905–1955. In Katalog der »documenta«, 15.07.–18.09.1955. München: Prestel. o. S.
Huelsenbeck, Richard (ed.). 1920. Dada Almanach. Berlin: Erich Reiss.
Kuhn, Thomas S. 2012. The Structure of Scientific Revolutions. Fourth edition. Chicago: The University of Chicago Press.
Leonard, Zoe and Anna Blume. 1997. Zoe Leonard Interviewed by Anna Blume. Im Katalog zur Ausstellung »Zoe Leonard«, Wiener Secession, 23.07. –14.09.1997. 7–23.
Lorch, Catrin. 2017. Mittelpunkt Provinz. Süddeutsche Zeitung vom 08.06.2017.
Moholy-Nagy, Laszlo. 1927. Malerei Fotografie Film. Bauhaus Bücher 8. München: Albert Langen.
Moore, Alan W. 2011. Art Gangs: Protest & Counterculture in New York City. New York: Autonomedia.
Nechvatal, Joseph. 2016. Reproduction – and Non-Reproduction – of Multitudes: Up With People. D/RAILED MAG. http://www.drailedmag.com/2016/08/22/reproduction-and-non-reproduction-of-multitudes-up-with-people/. Accessed: 12 April 2019.
Panofsky, Erwin. 1947. Style and Medium in the Motion Pictures. Critique: A Review of Contemporary Art, 1(3), 5–28.
Panofsky, Erwin. 1978. Drei Jahrzehnte Kunstgeschichte in den Vereinigten Staaten. Eindrücke eines versprengten Europäers. In *Sinn und Deutung in der bildenden Kunst*, Erwin Panofsky, 378–406. Köln: DuMont.
Panzer, Gerhard. 2015. Wie der Phönix fliegen lernte: Beziehungen aus der Asche. Beziehungsanalysen der Künstlergruppe »Hessische Sezession« (1946–1949). In *Beziehungsanalysen. Bildende Künste in Westdeutschland nach 1945*, Gerhard Panzer, Franziska Völz and Karl-Siegbert Rehberg (ed.), 61–88. Wiesbaden: Springer VS.
Rosebrock, Tessa Friederike. 2015. Anmerkungen zu Unterlagen aus dem Arbeitsausschuss: Kurt Martins Dokumente zur documenta I. In *Beziehungsanalysen. Bildende Künste in Westdeutschland nach 1945*, Gerhard Panzer, Franziska Völz and Karl-Siegbert Rehberg (ed.), 199–226. Wiesbaden: Springer VS.
Stather, Martin (ed.). 2014. Katalog zur Ausstellung »Art & Activism: Kunst und politischer Aktivismus in New York, Sammlungen Hieber/Theising«. Mannheim: Mannheimer Kunstverein.
Theising, Gisela. 2008. Im Katalog zur Ausstellung »Break the Rules, Sammlungen Hieber/Theising«, hg. von Martin Stather. Mannheim: Mannheimer Kunstverein, 49–68.
Wingler, Hans M. 1975. Das Bauhaus. Bramsche–Köln: Gebr. Rasch–DuMont Schauberg.

The Aesthetic Subversion of Cultural Posters—Examples from the Former People's Republic of Poland

13

Karolina Kempa

Abstract

Like in all socialist countries, art was also expected to serve the state in the People's Republic of Poland. The doctrine of socialist realism was predominant, especially in the Stalinist era. However, in Poland, an aesthetic (counter) movement established itself in a niche area of art, namely the Polish School of Posters, whose artists designed above all cultural posters. These artists managed to use aesthetic means to design an antitype to the predominant semantics of Socialist Realism. This allows one to localise them in the field of aesthetic subversion. Furthermore, numerous posters are in their very own way aesthetically linked to the opposition movements against the Polish regime, in particular KOR (Komitet Obrony Robotników, Workers' Defence Committee) and the Solidarność movement, from the end of the 1970s onwards. Symbolic and iconographic moods and subject matters from the period in which the posters were created come together in compact form in the posters. This work analyses selected exemplary cultural posters designed by two of the best-known poster artists in Poland: Jan Lenica und Franciszek Starowieyski.

Keywords

Polish culture poster · Polish School of Posters · Aesthetic subversion · Socialist realism · People's Republic of Poland

K. Kempa (✉)
Leibniz Universität Hannover, Hannover, Germany
e-mail: karolina.kempa@zew.uni-hannover.de

L. Gaupp et al. (eds.), *Arts and Power*, Kunst und Gesellschaft,
https://doi.org/10.1007/978-3-658-37429-7_13

Subversion and hegemony are dialectically interrelated. Without hegemony there would be no need for subversion. If one understands hegemony as the dominance of a class or group, then subversion can be described as the activity of a person or groups aimed at destabilising the existing conditions in society (Fuchs-Heinritz et al. 1994). Dick Hebdige—following Stuart Hall—describes the term hegemony as a situation "in which a temporary alliance between certain groups within society is able to exert 'total social authority' over other subordinate groups" (Diederichsen et al. 1983, p. 20). This hegemony is not permanent and stable. Rather, it must be gained and reproduced in an ongoing process and can only be sustained as long as the dominant group can "succeed in framing all competing definitions within their range" (Hall quoted in Diederichsen et al. 1983, p. 20).

Totalitarian political systems must also permanently maintain their hegemonial status and they do this in different ways in different social functional areas. How fragile their status can be is shown by the example of the People's Republic of Poland (1944–1989). The socialist state apparat began to falter at the latest when the Solidarność movement formed in alliance with the Catholic Church, and then collapsed in the face of the movement's persistence (see below). However, reducing resistance against the system to acts of political protest alone would be falling short of reality. Oppositional and subversive forces were always there at other levels, and also in art. In Polish film, for instance, there are various examples that still enjoy cult status to the present day. One such film is Andrzej Wajda's *Człowiek z marmuru (Man of Marble),* which was made in 1976. The film portrays the exploitative and violent behaviour of the socialist rulers and party functionaries towards the civilian population. In the world of music, the songs of musician and poet Jacek Kaczmarski deserve special mention. Many of these became anthems of the Solidarność movement, for example *Mury (Walls).*

These and numerous other examples clearly show how the content of works produced in the field of art was linked aesthetically in an alliance with the trade unions and with the opposition movement in Polish society.

Nevertheless, there were also forms of aesthetic subversion that were more subtle, because they were more indirect. These includes Polish cultural posters, which I would like to present here.

In doing so, I will analyse three typical examples of such posters and look at how they can be interpreted as aesthetically subversive within the context of the social and political framework conditions of the period in which they appeared. For better understanding, I would like beforehand to provide a brief rundown of the most important socio-political events in the People's Republic of Poland and, following that, outline the Socialist Realist doctrine about art in order, ultimately, to present my own analytical model and then examine the posters against this background.

13.1 Relevant Socio-Political Events in the People's Republic of Poland

Since its foundation in 1944, the People's Republic of Poland had repeatedly experienced heavy protests against the state apparatus by different social groups. The most important critical developments and events were:

3 June 1956: Demonstrations in Posen by several thousand workers protesting against bad working and living conditions. The protests are violently suppressed.

3 March 1968: In Warsaw, students protest against the banning of a performance of the theatre play *Dziady*, in English *Forefathers' Eve,* by playwright Adam Mickiewicz and directed by Kazimierz Dejmek. The revolts against state censorship spread throughout all of Poland and are met by severe repression by the state powers. On 28 March 1968, protests at the University of Warsaw end in students being arrested, ex-matriculated and entire faculties are dissolved (Arndt 2007, p. 48).

16 December 1970: In Danzig, dock workers go on strike in protest against increasing meat prices. The demonstrations spread to other coastal cities including Gdynia, Szczecin and Elblag and are brutally suppressed.

25 June 1976: Drastic price increases for basic foodstuffs set off rioting in Radom and Ursus in Warsaw. On 3 September 1976, intellectuals set up KOR (Komitet Obrony Robotników, Workers' Defence Committee).

17 September 1980: The Solidarność trade union is formed. This is preceded by strikes on the premises of the Lenin Dockyard in Gdańsk.

13 December 1981: General Wojciech Jaruzelski declares a state of war. The measures imposed include banning Solidarność and imprisoning most of its active members.

6 February 1989: Roundtable talks begin in Warsaw and, on 17 April, Solidarność is legalized again.

11 April 1990: The Sejm (Polish parliament) decides unanimously to do away with censorship (all Dates from Szymanski 2012).

The posters I have in mind must be examined against the background of these events. As I will show, the posters give subtle expression to the mood of the times. They correlate—sometimes subliminally, sometimes obviously—with the topics and events of the respective period in which they were produced, which is why their political dimension must also be examined.

13.2 Socialist Realism as an Art Doctrine in the People's Republic of Poland

Besides historical events in the People's Republic of Poland, another circumstance is of central importance for my analysis, and that is the significance of art. Art plays a particularly important role in totalitarian political systems (Türk 2006, p. 145). As such, it must be said that in Poland, in the years between 1944 and 1989, the state mostly pursued a strict cultural policy. Like in all socialist states, Poland also had a censorship authority that strictly monitored which works were to reach the public and which not. The so-called "Black Book of Polish Censorship" contains numerous original documents from the censorship offices in Poland that describe the censorship process in detail (Strzyżewski 1978).

In addition to the censorship authorities, which examined all cultural works, artists had to face another strong opponent at that time—the Socialist Realism art doctrine, the aim of which was to make art serve party political goals. Joseph Stalin is supposed to have said at a secret meeting in the apartment of writer Maxim Gorky on 26 October 1932: "If the artist wants to show life as it really is, he must of necessity see and show the development towards socialism. That is what socialist art, Socialist Realism must do" (translated by the author, quoted in Christ 1999, p. 24, footnote 31). This statement already implicitly shows what lies at the heart of Socialist Realism as well as its goal. According to it, art is assigned the job of taking on board the concept of socialism and being a servant to it. According to this understanding, people were to be re-educated or educated in line with the "spirit of socialism".

It is true that Socialist Realism in Poland only enjoyed a brief heyday in the 1950s (Ostrowski 2011, p. 538). Nevertheless, its language of form and its specific aesthetics were maintained, in particular by the party, for the purposes of its own self-aestheticization. This is ultimately proved by looking at the poster for the 1st of May, Workers' Day (see Fig. 13.1, 13.2, 13.3). The Socialist-Realist way the Workers' Party presented itself in public hardly changed at all over the years after the initial years of socialist rule. Its main motifs remained images of workers and peasants as the central motifs of socialism, industrial landscapes as symbols of progress as well as generally optimistic and happy-looking visual images (Damus 1981). A series of keywords such as "peace", "work", "socialism", "rebirth", "pride" and "nation" joined with the motifs being presented to give clear messages that were easy to understand. They were not aimed at a specific "expert audience" but intended to reach the "people in its entirety". They were never "provocative"—either in the sense of an "innovative" design modality

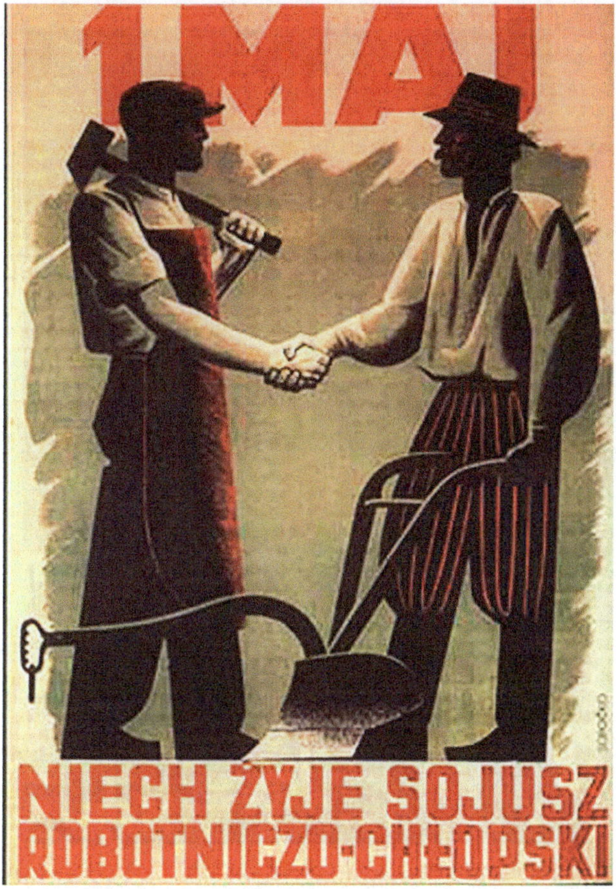

Fig. 13.1 Konstanty Maria Sopoćko: Niech żyje Sojusz robotniczo-chłopski. (Long live the Alliance of Workers and Peasents. Poster for International Workers' Day). 1948 ©

or in terms of the subject matter. They remained stuck in the traditional. From a design perspective, it is noticeable that the typography was always good and easy to decipher and was designed to be easy to read.

Taking this into consideration, one important question arises, namely how could posters that obviously did not fit in with the hegemonial way of seeing things emerge and be circulated legally despite censorship and the prevailing

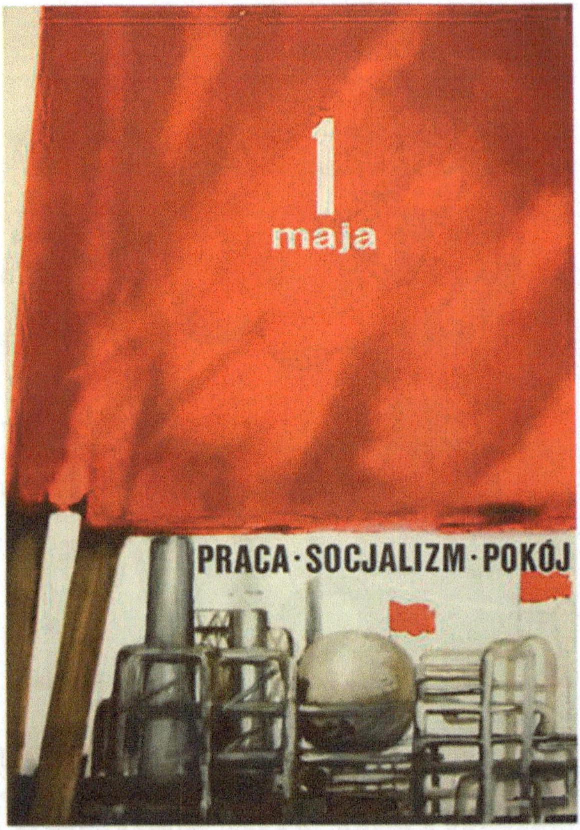

Fig. 13.2 Maciej Urbaniec:1 maja. Praca. Socjalizm. Pokój. (1. May. Work. Socialism. Peace. Poster for International Workers' Day). 1964 ©

art doctrine? This has to do on the one hand with the fact that different genres of art were subject to different levels of inspection depending on what cultural relevance they were hierarchically attributed. As such, in the Stalinist era after the Second World War, architecture ranked top place in this respect, followed by sculpture. By comparison, painting was not held in very high esteem and led an isolated existence in museums and archives (Szydłowska 2008, pp. 196–197). Posters were held in the lowest esteem which, according to Agata Szydłowska, had an (unintended) positive effect on the development of Polish poster design:

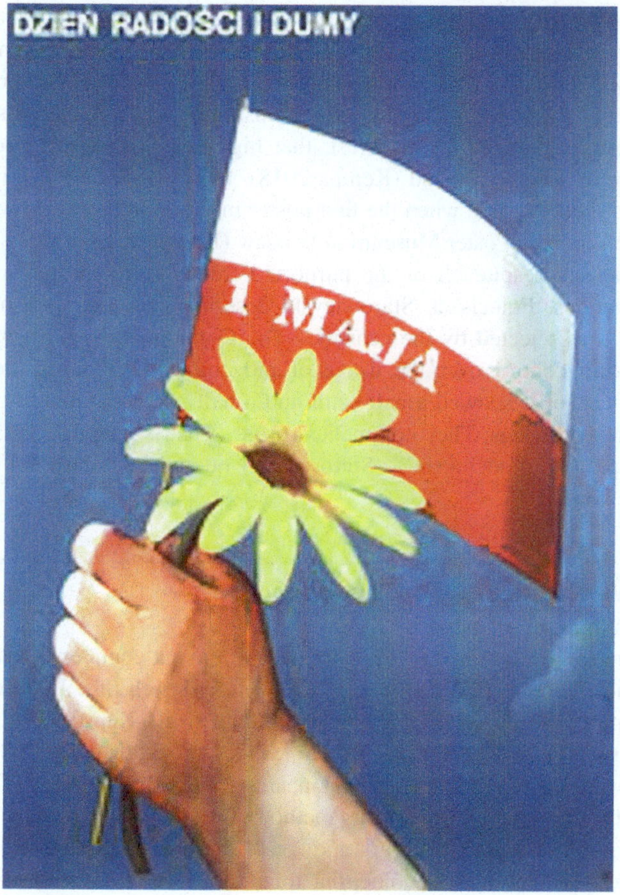

Fig. 13.3 Gorka Wiktor: Dzień radości i dumy. 1. Maja. (Day of Happiness and Pride. 1. May. Poster for International Workers' Day). 1978. 70,7 × 100 ©

> The least respected art in the communist world was printmaking, including poster art. This resulted in the artists in this field having the greatest freedom, a good example of which can in fact be seen on the fact that Henryk Tomaszewski's and Józef Mroszczak's studios at Warsaw's Academy of Fine Arts met with an almost complete lack of interest on the part of the authorities. (Szydłowska 2008, p. 197)

And yet this low level of respect and the disinterest this entailed changed in the 1960s when it became obvious how successful polish posters were becoming internationally: "The multiplicity of talents caught the attention of the authorities, who patronized poster art and made it a Polish export to the international art market" (Kurpik 2008, p. 44). From then on, the regime began taking the credit for the success of the posters and started demonstrating towards the outside world, with its focus especially on the West, that high-quality cultural products were being made in socialist Poland (Kempa 2018). Polish poster art also received an additional tribute in 1968 when the first poster museum in the world was opened, namely the Wilanów Poster Museum in Warsaw (Kempa 2018).

Nevertheless, despite all of the purported freedoms for poster artists, there were restrictions. Franciszek Starowieyski describes, for example, how one of his posters was rejected by the censors: "The Communist Regime was stupidly obsessed with control over the usage of red colour. One of my designs was rejected once, for it showed a half-scull and a half-chair (painted in green) as if walking on its red legs. They commented on it: 'That is how reactionaries imagine the approach to socialism'" (Starowieyski quoted in Dąbrowski and Kulon 2003, Foreword, p. 5).

13.3 Methodical Approaches

The posters to be examined in the next section have a complex visual character and it is necessary to find a systematic way to approach this. In an essay titled "Ikonographie und Ikonologie. Eine Einführung in die Kunst der Renaissance" (1978),[1] Erwin Panofsky develops a three-levelled model that is not only analyses visual images, but also identifies their meaning within the context of social and historical factors. The strength of his method also lies in the constant inter-referentiality between the concrete work and the theoretical-methodical model. Panofsky's model is therefore not a blueprint for interpretation determined by content but is rather a way of approaching something reflected in its application, and analyses how individual works are embedded in society, in history and in ideology.

His model proceeds on three levels, beginning with the pre-iconographical description, which aims to identify or describe the objects in a visual image without interpretation. In the second step an iconographic analysis follows, which defines and classifies the objects of the visual image within the context of a

[1] English: Iconography and Iconology. An Introduction to the Art of the Renaissance, (Panofsky 1978).

cultural tradition (Panofsky 1978, p. 41). Panofsky's third level is iconological interpretation. Iconology attempts to convey the interrelationship between works of art and the conditions and contexts in which they are created. By doing so, it focuses not only on works of art and their aesthetic form, but contextualizes these with the general conditions in society that, above all, have to do with a certain discourse as well as the knowledge and beliefs of a historical period through which the work of art becomes a historical episteme (Panofsky 1978, pp. 48–49). Panofsky's model has mainly proven to be of value in the iconological interpretation when a model for this also has to be found. As such, one cannot implicitly derive an episteme or even a dominant ideology from the posters. It is possible, however, to relate the work to concrete historical, social and political events and to interpret the work within this relationship (Kempa 2018, pp. 127–129).

However, the posters require a special approach when analysing them due to their special mediality. This is necessary, because the posters refer to other media, in the case of Polish cultural posters to theatre performances, operas and films which, for their part possess a medial complexity made up of topic, language, acoustic, visuality and aesthetics (Kempa 2018, pp. 131–133). Complementing Panofsky's model, the intermediality of the posters is also analysed accordingly. The term intermediality, which is analogous to the term intertextuality used in literature, looks at the ways in which different media relate to one another (Rajewsky 2002; Wolf 2004). The posters pick out aspects and bring them together in concentrated form resulting in a complex mediality. This added methodological aspect makes sense as it makes it possible to decipher the posters, which are otherwise difficult to understand, in a systematic manner. While the intermediality explains the inner relationship of the posters to another work, Panofsky's iconology, modified here, allows us to grasp the relationship to external social events and discourses.

13.4 Analyses of Polish Cultural Posters

The so-called Polish School of Posters enjoys a strong international reputation as evidenced by numerous catalogues, international monographs and anthologies, as well as the holdings of important museums such as MoMa (Museum of Modern Art in New York; Kempa 2018, p. X–XI). The Polish School of Posters emerged as a new movement in 1956 in post-war Poland (Kurpik 2008, p. 42). It has since produced many internationally renowned poster artists. I would like to discuss two of them: Firstly, Franciszek Starowieyski (1930–2009), who was also active as a painter, illustrator and stage designer. He also worked outside of Poland, for example for German clients such as the Städtische Bühnen Münster. In the mid-1980s, Starowieyski was a guest lecturer at the University of the Arts (HdK) in West Berlin, and then from 1993 he taught at the European Academy of Art in Warsaw (Schubert 2008, p. 208).

The second poster artist is Jan Lenica (1928–2011), who not only designed posters, but was also a film maker (animated films), caricaturist and stage designer. He began working as an illustrator for the satirical weekly *Szpilki* in 1945 while still an architecture student. In 1950 he became a graphic editor there. After several periods abroad, Lenica finally moved to Paris in 1963. In 1974 he taught as a guest lecturer at the Carpenter Center for the Visual Arts at Harvard University in the United States. Between 1979 and 1985 he was a lecturer for animated film at the Kunsthochschule Kassel. He then moved to Berlin, where he worked at the Berlin University of the Arts until his death (Dzikowska 2011, p. 120).

What both artists have in common is that they worked in the field of cultural posters. They designed posters for films, operas and plays. In the following, I analyse three examples of their posters.

I start with a poster by Franciszek Starowieyski for Jonasz Kofta's 1979 play *Wojna Chlopska (Peasants' War)* (see Fig. 13.4).

This poster clearly runs counter to our general understanding of socialism and its associated imagery. Examined from an iconographic, intermedial and socio-critical perspective, the poster's subversive tendencies become clear.

The poster is entirely black and white. In the centre a hybrid creature is depicted. It consists of a male lower body and a (provisionally) indeterminate upper body. The left leg, which is moving forward energetically, is crippled and from the calf down replaced with a prosthetic wooden leg. The muscular lower body appears powerful and martial despite its disabilities. The upper part of the body symbolically supports this martial effect. It is a mass of pointed, dangerous spikes protruding in all directions. The name of the work and the obligatory information (title, location of the performance) are also moved into the far top background of the poster. The typography of this text is faint, which makes the figure all the more impressive and present in the foreground.

The figure is easier to understand if we examine the play by Jonasz Kofta, which was performed in Opole in 1979. The play takes place at the time of the German Peasants' War in 1525. The first act introduces an impoverished group of peasant farmers. They heatedly discuss their troubles (poverty, bad harvest) and their anger towards their feudal lords. They make the decision to go out into the world and join the fight for "justice". In the second act they go to battle against their feudal lords. The final act is a bloody battle with members of the nobility in which all the men die.

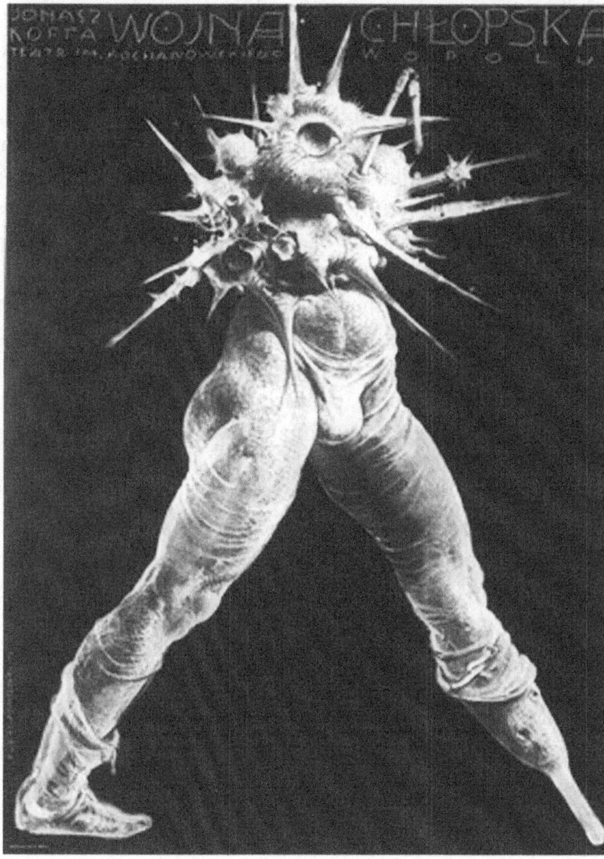

Fig. 13.4 Franciszek Starowieyski: Wojna chłopska (Peasants' War. Drama by Jonasz Kofta). 1979. 94,5 × 66,5 ©

Starowieyski's aesthetic density in the poster now takes shape against the backdrop of the piece. The belligerent figure, who, despite disability (wooden leg), energetically strides forward with his weapon-like upper body, stands symbolically for the peasant farmers in the play who are marching to war to fight for justice. The peasants are crippled to a certain extent (physically and mentally) by the circumstances of their time (feudal dependence), but they are also determined to fight.

But in Kofta's play, the peasants are by no means only depicted as warlike and courageous. They are also vulgar (they drink excessively and make offensive jokes), barbaric, brutal, unreflective, impulsive, uncontrolled. These aspects allow the poster to be seen in a different light. Thus, with its sparse clothing, prosthesis but no head, the figure appears almost frightening and threatening. The aspect of the vulgar is present in the genitals of the figure, which are visibly emphasized in the poster. They are only covered by a small piece of fabric. This gives the figure a libidinous quality. The ambivalence of the peasant characters in the play is thus transferred by Starowieyski in a complex fashion.

If one compares this poster to the naively coloured May Day posters discussed earlier, this poster clearly distinguishes itself from those starkly with its gloomy and dark atmosphere. Starowieyski's poster evokes a range of emotions in the viewer. It is threatening, frightening, confusing, intriguing, stimulating, repulsive, and so forth. All these qualities are unimportant or undesirable in the context of socialist-ideological social utopia.

When we consider the historical-social context in 1979, when this poster was created, another dimension of meaning becomes apparent. In socialism the peasant serves beside the worker as a figurehead. The image of the peasant is thus ideologically charged. At the same time, the image of the peasant in the late 1970s occupied a socially romantic status among the Polish population. The peasant here was included in the so-called "canon of national mythology of resistance" (Mędrzecki et. al. 2003, p. 163) against the system. The culture of the peasant was idealised yet his glorification under socialism was also opposed. The image of the peasant was characterised by a focus on the areas of family, faith and their associated values. The culture of the peasant was mystified as sacrosanct and primal. That is, it was stylised as the last fortress against the socialist reformation of the regime. At the same time, real resistance against the regime was forming among farmers in the 1970s. It succeeded in achieving representation of its interests on the political stage. Like the movement which began among trade unions, farmers also used the name *Solidarność*.

The poster by Starowieyski as well as the play by Kofta were created precisely at this time. Posters and theatre thus represent places to engage in debate about peasant identity and resistance. Both media paint an ambivalent picture of the martial peasant. On the one hand, Starowieyski demonstrates his readiness to fight against oppression by feudal rule, which can be transferred to the politics of the period in 1979: here, too, farmers are in fact opposed to the system's "quasi-feudal rule". The introduction of production cooperatives and the expropriation of farmers' properties under socialism are seen as analogous to feudalism. On the other hand, both the poster and the play diminish the heroic image of the peasant.

The barbarous, the headless, that is, the senseless, the unthinking initiation of an indeterminate war seems lackadaisical, thoughtless, and, so to speak, aimless. At the same time, the damaged image of the peasant undermines the heroic utopia that socialism imposes on it.

Viewers may make their own interpretation of the poster according to their individual perspective. The poster's open structure invites the attentive and willing observer to reflect, to negotiate various viewpoints. It does not commit to any certain single narrative or clear message, as typical May Day posters strive to do. Rather, it subverts the idealised clarity and simplicity of the mythological exaltation of peasants as well as the idealisation of their function in the propagated socialist workers' and peasants' state. The strength of Starowieyski's poster is its aestheticisation of ambivalences.

In my next example, too, these semantic and iconographic connections can be found between the poster and the advertised work and also the events of the time of its creation (see Fig. 13.5).

The 1977 poster by Jan Lenica promotes Janusz Majewski'*s* film *Sprawa Gorgonowej (The Case of Gorgonowa)*. The poster's iconography is not difficult to pin down. It depicts the classic attributes of Lady Justice: in the right hand the sword (symbol of punishing judgement), in the left hand scales (symbol of weighing up right and wrong), the blindfold (mocking the blindness of Lady Justice, which has represented impartiality since 1520 in Germany; Riese 2007, p. 220). But the figure is, and this is crucial, not female, that is, it is not the classical personification of justice as Lady Justice, but an abstract, elusive creature of some kind. It has eight aggressive and threatening mouths, some of which are screaming. While in classical representations, Lady Justice is depicted in a particularly "sober" fashion to emphasize the reasoned and rational foundations of justice, the creature in Lenica's poster is an emotional and depersonalised entity that is blindly shouting out its judgements in all directions. The illustration thus acts as a mockery of the traditional figure of Lady Justice. It conveys an impression of aggressive arbitrariness driven by strong emotions.

The film is based on a true story of a scandalous murder trial held in Poland between 1931 and 1933. A woman named Rita Gorgonowa was tried for murder. She was accused of having murdered her fiancé's daughter with an ice pick. In 1932 she was found guilty and sentenced to death. Gorgonowa's case was reopened and in 1933 she was sentenced to eight years in prison. Her prison term ended early on account of the outbreak of war in 1939. Gorgonowa always insisted on her innocence and there was no clear evidence that she did commit the crime. The "Gorgonowa Case" became one of the most famous trials of the Second Polish Republic.

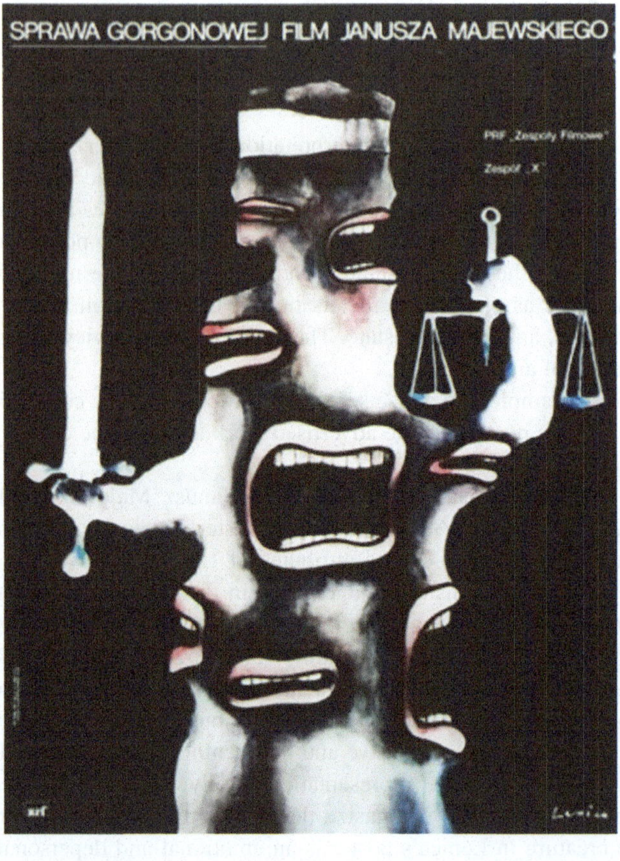

Fig. 13.5 Jan Lenica: Sprawa Gorgonowej. (The Case of Gorgonowa; Polish Film by Janusz Majewski). 1977. 94 × 68 ©

In his film, director Majewski focuses primarily on the court proceedings. He emphasises the arbitrary and contradictory nature of the individual court hearings and judgements—hence the Lady Justice in Lenica's poster, screaming in all directions. And the poster fits almost seamlessly with the creative context of the time. The years 1976/1977 can be seen as an important turning point in the political history of the People's Republic of Poland. The Polish opposition movement KOR was officially formed in these years. The committee was created after violent protests on 24 June 1976 against the dramatic price increases of Edward Gierek's regime for important staple foods. The response of the government included the arrest of hundreds of demonstrators. Thousands also lost their jobs (Arndt 2007, p. 52). The KOR legally supported the detainees, repeatedly criticising the fact that the judiciary decided unilaterally and always to the detriment of the accused. Against this background, Lenica's poster is also indicative of a morose mood regarding the state apparatus and its unjust, repressive and aggressive judiciary.

The best-known poster by Jan Lenica, which is for Alban Berg's opera *Wozzeck*, not only refers with its expressive design to the opera but also the historical context of the time (see Fig. 13.6).

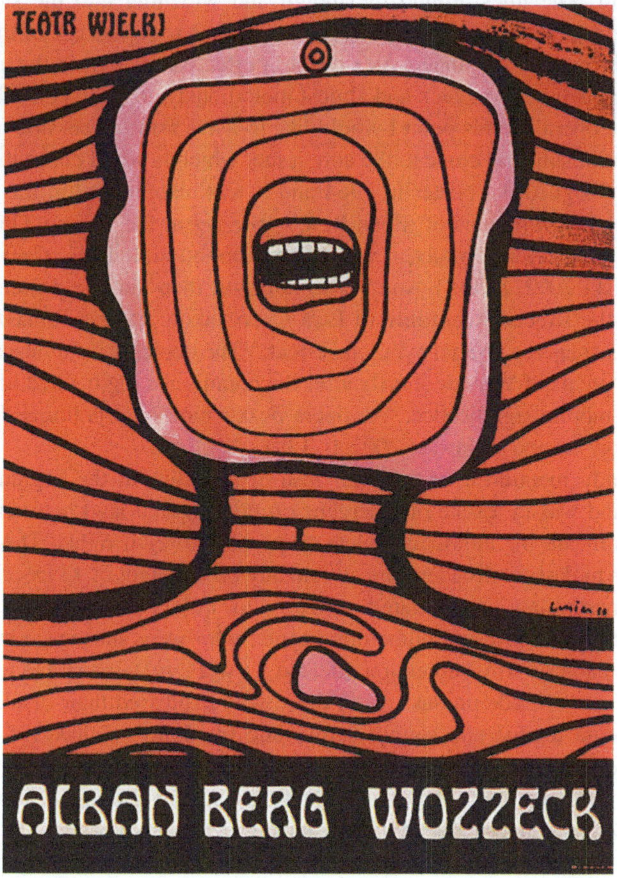

Fig. 13.6 Jan. Lenica: Wozzeck. (Opera by Alban Berg). 1964. 96,8 × 67,2 ©

The poster depicts an abstract portrait of a head with a gaping mouth in the centre, a badge on its forehead and a kind of heart on its chest. Throughout the background are different wavy stripes of red divided by black lines. The figure itself is made of the same background stripes, and here too the black lines create a special effect. The names of the composer and the play are situated at the bottom of the picture in the partitioned area and in Lenica's typical typography, which is based on the Art Nouveau typography by Hector Guimard. At the top left of the picture border, we see the name Teatr Wielki, (the Big Theater in Warsaw).

This poster is straightforward in terms of its choice of subject and design. The power of the poster lies primarily in its colour design, which is expressive, loud and forceful. The lines around the mouth and widening into the surroundings appear to evoke shock waves or a blast of sound, as if the energy of the cry were vibrating throughout the image. As a result, it effectively captures the tragic figure of the simple soldier Wozzeck and almost melodiously compliments Alban Berg's twelve-tone opera. At this point I would like to leave aside discussion of the intermedial relationship and come to the core issue, which is why I chose this poster.

Lenica designed the poster in 1964 and it was also printed in the same year. However, the performance of Alban Berg's *Wozzeck* at the Warsaw Opera did not take place in this year. In fact, it was not performed until 1984 (Czerwiakowska and Kujawski 2002, p. 138). And it was not until then that the posters were displayed on the streets. According to Ewa Czerwiakowska and Tomasz Kujawski (2002), this created a special symbolic effect: "The depiction of a face and an outcry on the streets of Warsaw shortly after the communist regime had lifted martial law had symbolic significance. 'An opera poster took on a political dimension'" (Czerwiakowska and Kujawski 2002, p. 139).

As already mentioned in the brief historical outline at the beginning of this essay, Poland was in a state of emergency (under martial law) between 1981 and 1983. This meant the almost complete suspension of civil rights. The aim of the regime was to destroy dissident forces. Already on the night of 13 December 1981, in which martial law was declared on the radio, militia patrols broke into homes with crowbars and arrested numerous people, who were subsequently housed in internment camps. "This repression affected … workers, farmers and intellectuals, representatives of trade unions …, all those who expressed their displeasure with the totalitarian regime and sought democratic freedoms" (Bujak 1984, pp. 120–121). There were unacceptable sanitary problems in these detention camps as well as (unjustified) acts of violence, a prohibition against communication with the outside world and non-transparent decisions to detain individuals. The number of detainees reached 10,131 by December 1982. In addition, street demonstrations were smashed, the excessive use of smoke bombs and batons by the authorities resulted in hundreds of victims, some of whom died (Bujak 1984, pp. 120–142).

If Lenica's Wozzeck poster is considered against the background of these events, which were only one to two years earlier and which had burned themselves into the collective memory of the suffering population, indeed a new dimension of meaning is revealed that goes beyond the poster being simply a successful opera poster. The expressive cry on the poster literally becomes the metaphor of a deeply frustrated civil society whose rebellion was severely punished. In comparison, the May Day poster of 1983, that is, in the same year martial law was lifted, strikes one as mockery (see Fig. 13.7). It reads: "Pokój Praca Odrodzenie" ("Work, peace, rebirth").

13.5 Conclusion

Numerous Polish cultural posters are not just successful works of art. They can be read as works of aesthetic subversion in two ways. Although they apparently only promote cultural artefacts, they contain implicit references to socio-political events and sentiments that can be determined in the context of their origin by means of an iconological interpretation following Panofsky. Thus, long before the overthrow of the Socialist apparatus by the Solidarność movement, they subtlety express a social tension that can be recognised by the attentive observer. On closer inspection, the underlying purpose of the posters is not to advertise other works, but rather to transcend this purpose by referring to politics without directly articulating themselves in a political fashion.

Furthermore, their specific iconography, composition and design stand in striking contrast to hegemonic aesthetic concepts, such as those formulated by the socialist state system with Socialist Realism and whose formal language and style of design the regime preserved until its collapse. The posters by Starowieyski and Lenica represent an aesthetic counter-design to the aesthetics of the May Day posters discussed at the beginning. And they therefore articulate—via the aesthetic—an alternative to the political doctrine that favours the unambiguous, the self-affirmative and the euphemistic in order to stabilise its own status. The cultural posters in their complexity, to put it in a nutshell, represent an aesthetic counterproposal to the political propaganda of the socialist regime. The special quality of the posters is that they do not draw upon the means of counter propaganda. They are rather recruited for the political fight via aesthetic means.

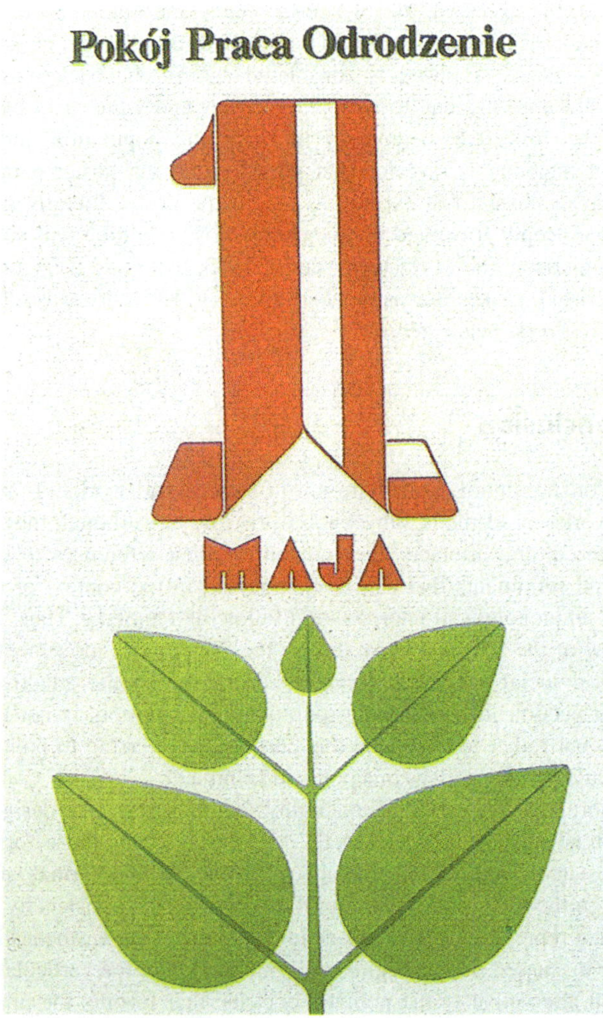

Fig. 13.7 Maciej Hibner: Pokój Praca Odrodzenie. (Work, Peace, Renaissance. [K. K.]. Poster for International Workers' Day). 1983 ©

References

Arndt, Agnes. 2007. Intellektuelle in der Opposition. Diskurse zur Zivilgesellschaft in der Volksrepublik Polen. Frankfurt am Main: Camous Verlag GmbH.

Bujak, Zbigniew/ Hardek, Wladyslaw/ Lis, Bogdan und andere. 1984. Die Menschen- und Bürgerrechte in der Volksrepublik Polen während des Kriegszustandes (13. Dezember 1981–31. Dezember 1982). Mainz: Hilfskomitee Solidarność.

Christ, Thomas. 1999. Der Sozialistische Realismus. Betrachtungen zum Sozialistischen Realismus in der Sowjetzeit. Basel: Wiese Verlag.

Czerwiakowska/ Kujawski, Ewa and Tomasz . 2002. Jan Lenica. Labirynt. A Labyrinth. Poznań: Muzeum Narodowe w Poznaniu.

Fuchs-Heinritz, Werner/ Lautmann, Rüdiger/ Rammstedt, Otthein/ Wienold, Hanns. 1994. Lexikon zur Soziologie. 3. Opladen: Westdeutscher Verlag GmbH.

Dąbrowski, Piotr and Agnieszka Kulon. 2003. Franciczek Starowieyski. Plakaty. Retrospektywa. Posters. Retrospective presentation. Z kolekcji Piotra Dąbrowskiego i Agnieszki Kulon. From the collection of Piotr Dąbrowski & Agnieszka Kulon. Olsztyn: Muzeum Warmii i Mazur.

Damus, Martin. 1981. Sozialistischer Realismus und Kunst im Nationalsozialismus. Frankfurt am Main: Fischer Taschenbuch Verlag GmbH.

Dzikowska, Elżbieta. 2011. Artyści mówią. Wywiady z mistrzami grafiki. Warszawa: Rosikon Press.

Diederichsen, Diedrich/ Hebdige, Dick / Marx, Olaph-Dante. 1983. Schocker. Stile und Moden der Subkultur. Reinbek bei Hamburg: Rowohlt Taschenbuch Verlag GmbH.

Kempa, Karolina. 2018. Polnische Kulturplakate im Sozialismus. Eine kunstsoziologische Untersuchung zur (Be-)Deutung des Werkes von Jan Lenica und Franciszek Starowieyski. Wiesbaden: Springer VS.

Kurpik, Maria. 2008. Czas, Twórcy, Plakat. Time, Artists and the Posters. In Plakaty w zbiorach Muzeum Plakatu w Wilanowie. Posters from the Poster Museum at Wilanów collection, ed. Kurpik, Maria and Szydłowska, Agata, 26–51. Warszawa: Muzeum Narodowe w Warszawie.

Mędrzecki, Włodzimierz/ Rudnicki, Szymon/ Żarnowski, Janusz. 2003. Społeczenstwo polskie w XX wieku. Warszawa: Instytut Historii PAN.

Ostrowski, Jan K. 2011. Wielka encyklopedia malarstwa polskiego. Kraków: Wydawnictwo Kluszczyński.

Panofsky, Erwin. 1978. Sinn und Deutung in der bildenden Kunst. Köln: DuMont-Taschenbücher.

Rajewsky, Irina O. 2002. Intermedialität. Stuttgart: UTB GmbH.

Schubert, Zdzisław 2008. Mistrzowie plakatu i ich uczniowie. Poster masters and pupils. Warszawa: Predsiębiorstwo Wydawnicze Rzeczpospolita SA.

Riese, Brigitte. 2007. Seemanns Lexikon der Ikonografie. Religiöse und profane Bildmotive. Leipzig: Seemann Verlag in der Seemann Henschel GmbH & Co. KG.

Strzyżewski, Tomasz. 1978. Czarna księga cenzury PRL. Black book of Polish censorship. London: AN EKS.

Szydłowska, Agata. 2008. Plakat Okresu Stalinowskiego w Polsce: od Socrealizmu do Popkultury. Posters of the Stalinist Period in Poland: from Socialist Realism to Pop Culture.

In Plakaty w zbiorach Muzeum Plakatu w Wilanowie. Posters from the Poster Museum at Wilanów collection, Kurpik, Maria, and Szydłowska, Agata, 196–201. Warschau: Muzeum Narodowe w Warszawie.

Szymanski, Berenika. 2012. Theatraler Protest und der Weg Polens zu 1989. Zum Aushandeln von Öffentlichkeit im Jahrzehnt der Solidarność. Bielefeld: transcript Verlag.

Türk, Klaus. 2006. Arbeitsdiskurse in der bildenden Kunst. In Bilder als Diskurse – Bilddiskurse, ed. Maasen, Sabine, Mayerhauser, Torsten, and Rengggli, Cornelia, 142–180. Weilerswist: Velbrück Wissenschaft.

Wolf, Werner. 2004. Intermedialität. In Metzler- Lexikon Literatur- und Kulturtheorie: Ansätze – Personen – Grundbegriffe, ed. Nünning, Ansgar, 296–297. Weimar: Metzler.

"The Dead are Coming". Contemporary Interventionist Art, Political Beauty, and the Power of Reason

14

Aude Bertrand-Höttcke and Matthias Kettner

Abstract

Have artistic interventions in public space run out of power? Following a first wave in the 1970s, public interventionist art re-emerged in the mid-90s outside the mainstream of the art world as the method of choice for artists with strong intentions to induce real changes in society. Most of them have very little impact since they either address very specific milieus that are confined to local communities or they get absorbed by the self-enclosed art world. In the present chapter, we ask to what extent the strong media resonance generated by the 2015 intervention art action "The Dead are Coming" by of the *Center for Political Beauty* (CPB, *Zentrum für politische Schönheit)* can be understood as by the quasi-argumentative power of this action. This intervention is especially interesting because it generated considerable coverage in the media and mobilised and polarised thousands of citizens as well as politicians, thereby making a noticeable impact on public policy in the context of what became framed as Germany's "refugee crisis". We explain how we understand "discursive power", rhetorical forces and the "forces of good reasons" within the

A. Bertrand-Höttcke (✉)
University of Witten-Herdecke, Dortmund, Germany
e-mail: Aude.Bertrand-Hoettcke@uni-wh.de

M. Kettner
Universität Witten/Herdecke, Witten, Germany
e-mail: kettner@uni-wh.de

L. Gaupp et al. (eds.), *Arts and Power*, Kunst und Gesellschaft,
https://doi.org/10.1007/978-3-658-37429-7_14

framework of a discourse theory: Discursive power refers to forms of power of actors to deliberately change convictions of correctness through communicative means. Convictions of correctness are those opinions of persons of whose correctness they are convinced because they consider their reasons for these opinions to be so good that the persons concerned identify with these reasons personally, and also publicly.

Keywords

Interventionist art · Center for Political Beauty · Dramaturgical action · Communicative power · Good reasons · Philipp Ruch

14.1 Introduction

Despite many turns, frequent crises and proclaimed endings, contemporary arts remain a lively field of action, renewal, and innovation. However, and somewhat paradoxically, contemporary art can take many possible forms and pervade many aspects of everyday life, thus becoming ubiquitous and yet at the same time strangely unnoticeable. Are there forms of contemporary art that still possess enough power to influence, unsettle, irritate or even reorient public opinion, aside from epitomizing and decorating human and economic capital?

The paradox of ubiquity and vivacity paired with powerlessness and inconspicuousness holds true especially for forms of contemporary art aiming at politically relevant articulation in public space. Artistic practices and interventions in public space with a view to influencing public opinion oscillate between gestures of radical empowerment and helplessness as they make their way into the realm of art biennials, festivals and similar large public entertainment events, thus defusing their initial thrust of dissent and opposition.

Into a context of widespread but also harmless, complaisant and unobtrusive artistic practice fell a theatrical action gaudily entitled “The Dead are Coming” as startling as a declaration of war. “The Dead are Coming” was performed in public in June 2015 in Berlin. This token of interventionist art proved a remarkable exception in that it succeeded in generating an enormous and variegated amount of media coverage (with over 250 press articles and a lot of buzz on social media and mass media; Center for Political Beauty, n.d.). It sparked political controversy comparable to that of Christoph Schlingensief's millennial public performance *Ausländer Raus!* in Vienna.[1]

The *Center for Political Beauty* (CPB) is a hybrid constellation of theatrical performance, political activism and interventionist public art. It was founded 2009 by a group of artists under the lead of the philosopher and art activist Philip Ruch. In what follows, we focus only on one of 14 spectacular interventions that the CPB has put into effect so far. The highlight and main act of "The Dead are Coming" was undoubtedly the realistically enacted, or close-to-real, burial of a dead migrant Syrian woman whose corpse had been recovered and transported all the way from Sicily to central Berlin in the midst of the so-called migrant crisis of 2015. The Syrian woman had drowned in an attempt to cross the Mediterranean in order to make it into the European Union. This is how CPB framed the political rationale of their action:

> The victims of [Europe's] cordon sanitaire are buried in masses in the hinterland of Southern European states. They have no names. No-one looks for their relatives. No-one brings them flowers. The Center for Political Beauty took these dead immigrants from the EU's external borders right to the heart of Europe's mechanism of defense: to the German capital. Those who died of thirst or hunger at our borders on their way to a new life were thus able to reach the destination of their dreams beyond their death. Together with the victims' relatives, we opened inhumane graves, identified and exhumed the bodies and brought them to Germany.(Center for Political Beauty, n.d.)

The burial ceremony composed the centre, not the fringes, of this meticulously orchestrated performance, which climaxed in outrage in the public sphere. It would certainly be interesting to explore the whole range of symbolic and ritual ingredients that made this action such a powerful attractor of media attention. In this chapter, however, we want to concentrate on explaining the nature of the rhetoric forces that made CPB's intervention a powerful work of performative political art.

---

[1] In 2000, theater director and performance artist Christoph Schlingensief installed a container during the *Wiener Festwochen* in an allusion to the television show "Big Brother". Inside the container asylum seekers were placed. By voting, the audience could decide which participant had to leave the container and the country. The project was launched under the title *Foreigners Out! Schlingensief's Container*. Film-director Paul Poet documented the entire event. The "container" mentioned in the title is a polemical allusion to the containers that figure in the trashy TV show "Big Brother" for housing the candidates. For documentation cf. Lilienthal and Philipp (2000).

We interpret CPB's intervention as a series of theatrical speech acts. To perform a speech act is to *do* something with words (e.g. making a promise by uttering suitable words). Theatrical speech acts add a surplus of emotional significance to what the words literally mean: they artfully add expressive momentum to merely representational propositional content. Theatrical speech acts unfold their communicative force by virtue of a strategy at once aesthetical and ethical. People who are unconcerned about, and insensible to, frequent drownings of would-be immigrants at Europe's frontiers in the Mediterranean, are placed in a situation that makes it harder for them not to develop emotionally *and* morally strong attitudes and affects (e.g. moral outrage). The exposure to CPB's interventions as bystanders, as recipients of the extensive media coverage, or as participants in conversations instigated by the media coverage impinges on people's reflective self-understanding and their "spaces of reasons". In many ways, some effects are normatively desirable[2] while others are questionable. Arguably, a normatively desirable effect is the massive albeit momentary recapture of conscious attention for recurring fatal and tragic events. For these events, citizens in a democratic state have, and should not pretend not to have, some amount of civic co-responsibility, however infinitesimal this co-responsibility might appear to them. A normatively questionable effect, on the other hand, is that CPB's aesthetic strategy seeks to overwhelm by shock people's capacity for critical thinking and to unleash the rhetorical forces of awe. Awe may engender emotional empathy at best. Awe will hardly engender compassionate (active, problem-focused, change-motivating) empathy.

Turning to CPB's programmatic formulations concerning their mission, we find that creating states of overwhelming moral emotions is exactly what CPB leader Philip Ruch's proclaims as the aim of "political beauty". To Ruch, experiencing beauty must not only provoke "an earthquake in our existences", it should also be inseparable from action. Ruch's (and CPB's) call for "actions so morally right that they shine with light intensity" (Ruch 2015) is a call for a moral understanding of beauty in action which some critics (in particular Ullrich 2017) consider as morally totalitarian in that "political beauty" in Ruch's sense does not tolerate alternative and perhaps quite reasonable appraisals of how things are.

[2] By "normatively desirable," we mean desirable not merely in the way of subjective preferences but for normative reasons, i.e. for reasons that carry intersubjectively shareable convictions how one ought (or ought not) do something one actually does or could intend to do.

To put this point in terms of a controversy in political ethics of the late sixties about morality proper versus self-righteous moralism, one could characterize CPB's self-understanding as a case of hypertrophic moralizing, "hypermorality" in Arnold Gehlen's terms (Gehlen 1969).

Returning to our main question: Why is "The Dead are Coming" so interesting in relation to power and art? Nowadays many artists who intervene in the public sphere seek to give back power to people within given communities by using particular strategies meant to effect re-empowerment. Typical re-empowerment strategies include subjectivation, autonomisation, cooperation, and participation. As employed by advocates of interventionist art, such strategies often involve positive emotions and are deployed in projects that aim, e.g., at giving corroded communities a new sense of positive identity and pride. They might aim at reshaping decrepit inner-city neighbourhoods in a pleasant way (incidentally paving the way for gentrification); or at creating alternative social spaces for exchange and togetherness or inventing new entrepreneurial activities for making a living.

In contrast, CPS's "The Dead are Coming", although addressing a tragic problem of a specific group of people (refugees) and specific sets of political practices (border control), does not draw its attention-raising power from art-based empowerment strategies. In fact, the very impossibility of active participation by migrants and refugees, as symbolised via the corpse of a Syrian woman, and their exclusion from the site of the theatrical action, has been criticized and deplored (Lewicki 2017, p. 280). Some NGOs involved in coping with the refugee crisis in Germany, but also some largely unknown artist collectives with years of experience of working together with migrants in participatory and community-based projects, charged CPB with instrumentalising the pain and tragedy of fellow human beings for the benefit of CPB's own publicity and success as an artistic group. In this critical perspective, the manifest moral thrust of "The Dead are Coming" shrewdly masks the self-promotion that is (such critics hold) its latent meaning. Such radical criticism focuses narrowly—we think too narrowly—on CPB's arrogantly demonstrated moral heroism: CPB claims the moral high ground but has next to no real impact at all on the lives of those for whose fate CPB purportedly cares so much. Migrants and refugees were not part of the proceeding since merely their corpses were involved. In other words, they figured as alien and ominous objects in the hands of the European masters of ceremony. Their absence during the burial ceremony and the accompanying demonstration marks PCB's artists as those who are pulling the strings and setting the stage whereas migrants are used as mere stage props and do not even figure as walk-on actors.

From an alternative point of view, however, one can argue that the performance draws its theatrical poignancy from the practical, social and physical powerlessness of Europe-bound migrants and refugees whose fate is in the hands of rogue states, criminal human traffickers, and callous European governments. To vividly disclose a situation in which refugees are left with no power or dignity over their own identities, movements and bodies—epitomized in the dead Syrian woman whose corpse could only be identified, transported and decently buried by way of an extraordinary intervention of an artist group, no less than in the woman's surviving husband who was not allowed to attend the burial although he was in Germany at the time—*reveals* the absurdity and inhumanity of this situation to a silent, improperly informed, unconcerned, de-sensitized majority of German citizens. The theatrical action *discloses* our own failure as a society to respond humanely to human disaster at our border.

Creating via an artistic performance a fictional "as-if" situation in which *ex negativo* a more adequate and more humane way of managing the refugee crisis becomes imaginable is a critical practice of disclosure since it reveals that what we accept as normal and as the political status quo is, at least to a certain extent, the product of civil cowardice, of moral indifference, and of bad politics.

In Sect. 14.2, we look at the different steps and stages of the performance. In brief, we conclude that the performance successfully deployed strategies of de-invisibilisation and re-sensitisation against widespread public denial of the tragic deaths of refugees. In Sect. 14.3, we approach the communicative or discursive power of the action in terms of good reasons. In Sect. 14.4, we turn to Ruch's so called "aggressive humanism" (Ruch 2015, 2019), his second major concept alongside "political beauty". With reference to *moral virtue*, *beauty* and *humanism* we question from a philosophical and from an artistic/aesthetical perspective CPB's justification of their program. We formulate a threefold critique of CPB's normative self-understanding.

14.2 Performance, Methods, Strategies

14.2.1 A Performance in Several Acts

On rainy Tuesday June 16th 2015 at a cemetery situated in Berlin, members of the CPB organized the performance of a burial ceremony in conformity with Muslim ritual norms. In a semi-real way, the burial ceremony involved the real corpse of a Syrian woman, a real imam, and an audience of spectators and media reporters. The members of the audience were casually dressed whereas the performers were recognizable by their black outfit and their black painted faces. The family of the perished woman had authorized the staged burial beforehand but was prevented from attending in person. The dead woman's husband was temporarily living in a camp in Germany awaiting official recognition of his refugee status. He was not permitted to travel to Berlin. He was merely allowed to attend via video live stream. The burial setting included a red-carpet area with VIP-seats reserved for state representatives. This area was intentionally left empty in order to symbolically articulate their absence and guilty inaction. In the run-up to the burial's main act a number of difficulties had to be overcome: The anonymous corpse which was found stacked with numerous other dead bodies in Sicily had to be selected, identified, prepared for transport and then transported all the way to Berlin. The transport required cutting a lot of red tape in order to cross national borders.[3] As it turned out, only one cemetery in Berlin did not object outright to hosting the ceremony, and the performance remained at legal risk until a couple of days before the fixed date. On the next Sunday after the burial, thousands of people who had been notified via social media formed a funeral cortege in the neighbourhood of the German Chancellery. CPB had originally planned to bury hundreds of corpses of anonymous refugees on the lawn between the German Chancellery and Parliament. Small wonder the artist group did not obtain official permission for this proposal. Instead, people placed flowers and candles on the lawn and epitaphs with accusatory slogans (e.g. "borders kill", "murdered") thus marking the event as a fully real political demonstration in distinction to the funeral cortege's mere stage reality (since no fully real funeral had taken place). The demonstrators did not leave until the police intervened. Costs for the re-cultivation of the damaged lawn were estimated to run at several thousand Euro.

As soon as the staged performance was over, political and journalistic critique set in. Instead of art critics appraising a work of art, politicians and the mainstream press in and outside Germany voiced mixed opinions. Most commentators initially were not so much interested in whether the performance should count as a proper and legitimate token *of art.* Instead, they were interested in the fundamental ethical and political question raised as to how we Europeans deal with migrants in a context of crisis. There were only few aesthetically oriented

[3]Allegedly, the artist group on their way had to endure chicanery from Bavarian police on accusation of consumption of illegal drugs.

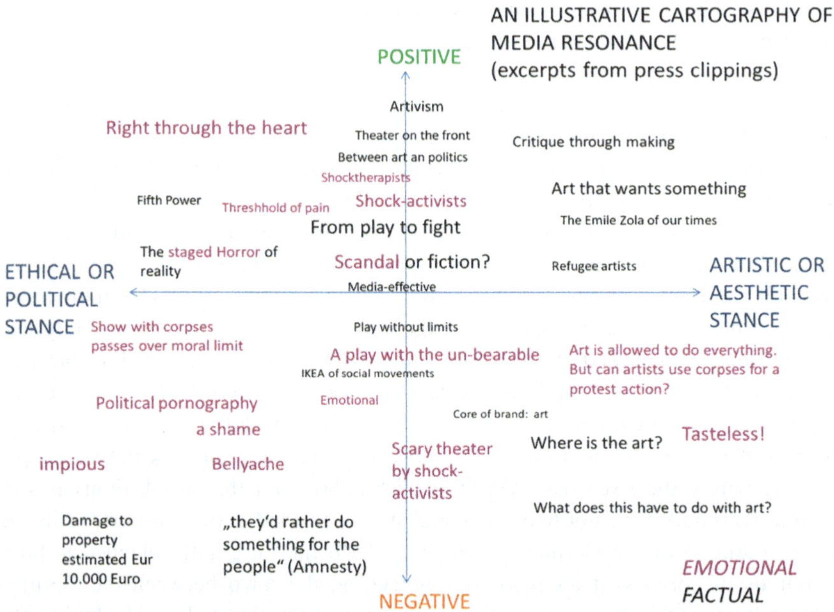

Fig. 14.1 Affective profile of press coverage of "The Dead are Coming"

comments. These comments came from outside professional journalism and politics, and they were unanimously in favour of the performance.[4]

We have analysed an extensive sample of newspaper clippings covering an interval of three months after the event took place. We now summarize our findings concerning media resonance in diagrammatic form. Figure 14.1 reveals a blatantly strong discursive divide along different affects and stances.

Figure 14.1 presents the result of a qualitative study by the authors. On the vertical axis we map the emotional quality of selected utterances, on the horizontal axis we distinguish two different normative perspectives ("stances") in which the respective utterances occur.

---

[4] With the notable exception of art historian Wolfgang Ullrich (2015) whose critique will be dealt with in Sect. 14.4 below.

14.2.2 Methods Between Activism and Art

What brought about such a remarkable media coverage? To be sure, CPB played an active role in feeding the press with information. They distributed their pre-prepared press kit (with revealing documentary photographs and information investigating the situation at European borders and in Sicily) at the right time, making clear to the public that the performance was embedded in an explosive context. CPB's press and social media campaign (press officer, Facebook campaign) was part of their artistic action but evidently goes far beyond conventional theatrical staging. However, all of this is no recipe for success in today's volatile economy of public attention. So, what additional methods did they employ? While we can identify numerous tactics in the orchestration of the performance, three methods played a major role in CPBs work: (fictional) overlapping, (historical) displacement and (conventional) activism.

Overlapping—Mingling reality with fiction. Mingling reality with fiction as a method plays a crucial role in CPBs action in creating a parallel reality. As Malzacher (2018, p. 330) underlined, CPB did not invent this method. Rather, it belongs to a set of methods that are characteristic for contemporary theatre.

Already prior to "The Dead are Coming", during the climax of the refugee crisis in the context of the Syrian war, CPB had launched in 2014 a deceptively "real" political campaign purportedly on behalf of the Ministry of Immigration seeking to save Syrian children. This campaign forced the German government to dissociate itself officially from what turned out to be a hoax. In CPB's 2015 performance the artists did not cause irritation while distorting reality through fiction (e.g. launching a mock-campaign and thereby producing a hoax), but by putting reality into fiction, as a real corpse was buried amidst a theatrical performance.

More than a mere "as if" installation, e.g. playing *as if* another reality were possible, CPB's 2015 performance shows our reality *as it ought to be but fails to be,* namely that each individual human being be treated with dignity and, if perished, be buried with propriety.

Historical displacement and "transfer emotion". For some, the intrusion of an unendurable reality into the fictional sphere was shocking. Reactions varied from rejection as "going too far", as indecent, or even as "political pornography".[5] Next to mingling fiction and reality, a displacement between now and then occurs as CPB implicitly draws parallels with significant historical times. The site

[5] Quoting *Süddeutsche Zeitung* journalist Sonja Zekri (Center for Political Beauty, n.d.).

for the symbolic memorial for perished refugees was chosen strategically for its neighbourhood to major political locations (between Chancellery and Parliament) and moreover for its symbolic link to German commemoration culture as incorporated in the close-by *Memorial to the Murdered Jews of Europe* with its symbolic gravestones. A "transfer emotion" is an emotion that is re-contextualised by artistic means (Beger 2016, p. 69). In this case, the transfer emotion is German citizens' shame for having failed to rise and stop the holocaust.

Conventional activism. Though refusing the label of "art activists" or "artivists", CPB did indeed use shock activism methods for mobilizing a critical mass of sympathising activists and other citizens, notably outside the art world. With over 10,000 declared attendants on Facebook alone, this critical mass radiated the performance' meaning and significance to a much wider audience and amplified its momentum. Social media circulation led to the performance being emulated in other cities and in other countries (from Basel to London), with people posting photographs of the symbolic graves which they had constructed. Obviously, such acts were acts of citizenship going well beyond artistic symbolism or mere playfulness.

14.2.3 Three Aesthetico-Political Strategies

CPB employed three strategies: de-invisibilizing the victims, re-sensitizing the citizens, and re-politicizing the debate accompanying the refugee crisis.

De-invisibilizing. Prior to CPB's performance, the scores of migrants who had drowned in the Mediterranean were reported in mainstream media coverage predominantly in quantitative terms, i.e. as death toll. Refugees tended to be presented in dehumanizing ways (corpses stacked, anonymised and made to disappear) and to be relegated to non-places consisting of borders, remote camps or mass graves. Interestingly, only a few months after CPB's performance a photo of a small Syrian boy found dead on the Turkish coast caught wide public attention and provoked strong public reactions that were followed by a change both in media coverage and in asylum policy across several European countries, particularly in Germany. CPB's performance succeeded in challenging and moving what Rancière calls the "distribution of the sensible" by presenting non-subjects as subjects. The audience was confronted with the horrible fate of an identifiable person and her relatives. In a certain sense, the person's identity and dignity were thereby restored.

Re-sensitizing. Through the confrontation with a determinate individual's fate, people could easily identify with her and her relatives. For many, this meant getting re-sensitized for a problem about which they were already objectively informed (through facts and figures in the media) and yet felt powerless, if not completely unconcerned. The performance was a shock, a wake-up call. It appealed to shame and guilt feelings that were both widespread and suppressed.

Re-politicizing. Since politicians were negatively but all the more explicitly involved in the performance by their conspicuous absence, the fate of migrants crossing the Mediterranean soon became a terrain for agonism in Chantal Mouffe's (2013) sense of the term: A political battlefield where opponents could directly issue objections, criticism, demands. For Mouffe and other advocates of political agonism this agitated mode of doing democracy is necessary for the vitality of democracy in the face of tedious expert consensus and a frustratingly limited ability to act, two tendencies of what Colin Crouch (2004) has termed "post-democracy".

14.3 The Communicative Power of Artistic Dramaturgical Action

How can we explain that a theatrical performance such as "The Dead are Coming" can have a powerful grip on our mind, especially on our imagination, and change the way we think and feel even though words and the power of speech play only a minor role in the total performance? Assume that for a moment at least we are affected somehow, perhaps touched emotionally. Does affect exclude reason? There is a long standing but preposterous philosophical prejudice that feeling (sentiment, emotion, "the passions") and reason (rationality, intellect) are radically opposed human "faculties" and that reason, in David Hume's famous but ill-conceived dictum, "is, and ought only to be the slave of the passions, and can never pretend to any other office than to serve and obey them" (Hume 1739/1975, p. 415). Fortunately, contemporary philosophy of mind, philosophical psychology, and cognitive science within the embodiment paradigm have done much to demolish the received dualistic view. They work towards non-dualistic views of complex entwinements and interactions of affectively meaningful and non-affectively meaningful intentional states (attitudes, contents, codes).[6]

[6] The observation that intentional states, affectively charged or neutralized, need not be conscious is another interesting point coming from psychoanalytic theory (cf. Kettner and Mertens 2010). For an illuminating deconstruction of the dualistic picture, see Nussbaum 2001.

14.3.1 Dramaturgical Action

Discarding both Humean and, on the opposite end, overly rationalistic accounts of humans as *animales rationales* clears the way for understanding how artistic dramaturgical action can have impact both on our reasoning and on our emotions. What is dramaturgical action? Sociologist Ervin Goffman (1978, p. 56) took up this term in order to refer to agents' power of governing the impression others get of them in their common interaction. This is how Jürgen Habermas in his *Theory of Communicative Action* (1984, pp. 85–86) characterizes Goffman's concept:

> The concept of dramaturgical action refers primarily neither to the solitary actor nor to the member of a social group, but to participants in interaction constituting a public for one another, before whom they present themselves. The actor evokes in his public a certain image, an impression of himself, by more or less purposefully disclosing his subjectivity. Each agent can monitor public access to the system of his own intentions, thoughts, attitudes, desires, feelings, and the like, to which only he has privileged access. In dramaturgical action, participants make use of this and steer their interactions through regulating mutual access to their own subjectivities. Thus, the central concept of "the presentation of self in everyday life" does not signify spontaneous impression behaviour but stylizing the expression of one's own experiences with a view to the audience.

Once we trace back the concept of dramaturgical action to its origin in performing arts and theatre analysis, it makes sense to understand dramaturgical *action* in analogy to a *text* (Ricoeur 1973), a text that is produced by an author-subject who skilfully augments the meaning of the text by employing suitable effective aesthetic techniques and selected strategies taken from practices that characteristically belong to theatre and other genres of the performing arts. Instead of employing such techniques and strategies for the regulation of mutual access *to their own subjectivities*, actors in original dramaturgical action employ techniques and strategies with a different intention. In original dramaturgical action the guiding intention of the author-subject is to enhance the rhetorical forces of all behaviours (linguistic, paralinguistic, mimetic, kinetic, etc.) that are relevant for the articulation of the *intrinsic meaning of a certain artistic composition* (a play, a narrative, a story, a ceremony, etc.) for all who figure in the composition's performance (as actors, as spectators, as viewers, as audience, etc.). The participants of dramaturgical action are situated within established institution (theatre, cinema, opera, etc.) specifically designed to bracket, re-frame or fictionalize the normative commitments of "real life".

14.3.2 The "Powers and Forces" of a Dramaturgical Performance

Conceptually, power has to be distinguished from force. Power is a complex notion (Haugaard and Kettner 2020) involving the notion of forces and of command or governance. Power in its conceptually central sense refers to the ability of an agent (or of suitably coordinated agents) to govern relevant (other) forces by forces that are the agent's own so as to effectively achieve something which the power-wielder(s) intend(s) to accomplish (Kettner 2018). There are of course man different kinds of power; but power is always of an individual or collective agent.[7] So the power we tend to attribute summarily to an extant artistic performance (e.g. "the dramatic power of *The Death are Coming*") really is a complex product of *the artistic power of the producers* of the performance (artists, ensemble, or in our case: Philipp Ruch and the CPB) who create, shape and design it as a concrete artistic composition. In other words, speaking of "the power of a performance" is an abridged way of speaking of a performance whose significant elements a dramaturgical agent has augmented so that we can experience their meaning as extraordinary and, in a sense, as touching or gripping (e.g. as moving us to tears or laughter, or thought).

Just as there are various kinds of power, there are also various kinds of force. An important class of forces are the so-called illocutionary forces (e.g. the force of an explicit command, the force of a promise, etc.) in classical speech act theory. Each distinctively illocutionary force of actual speech acts is a particular normative elaboration of certain conventions and the commitments such conventions imply. However, we can make our actual speech acts forceful in senses that are not, or at least not entirely, exhausted by conventions. Speech act theory as originally developed by philosophers John Langshaw Austin and John Searle with its sole focus on natural language communication marks such surplus forces off as "perlocutionary" (cf. Green 2017). Perlocutionary effects of an utterance go beyond the conventionally expectable effects of its wording. The term "perlocutionary" covers what speakers can intend to achieve with their utterances (e.g. *making your interlocutor very angry by* uttering an absurd accusation) over and above the conventionalized effects of the conventionalized illocutionary forces required for producing the utterance (an absurd accusation).

---

[7] This point is, of course, contested. We hold that postulates of "structural" power and similar non-agent views of power and power-relations confuse potentials of forces with actual power.

Speech act theory has been groundbreaking but remains too narrow in scope. Once we free speech act theory from its exclusive focus on linguistic utterances, we can see that *all* significant behaviour (i.e. all behaviour that is *meaningful* within the interpretative perspective of members of communities of communication) has a spectrum of what we might summarily dub *rhetorical* forces. Rhetorical forces of significant behaviours generally, like the illocutionary and perlocutionary forces of linguistic behaviour in particular, admit of governance, refinement and modulation by communicatively competent powerful agents. Our concept of rhetorical forces of significant behaviours generalizes speech act theory's concepts of perlocutionary and illocutionary forces and their respective effects. With this theoretical move, we can now conceive of art as involving *specifically aesthetic* rhetorical forces. Artistic power, then, involves the governance of a variety of specifically aesthetic rhetorical forces, forces that can be skilfully deployed in order to deliberately transfigure the ordinary meaning of significant elements in all kinds of behaviour, linguistic or other. The transfiguration of the ordinary meaning of significant behavioural elements can realize artistic purposes or other purposes. To illustrate: A non-linguistic corporeal gesture of grief, if artistically augmented by suitable dramatic means (facial expression, bodily movements) will not merely register in an audience as the information that someone feels grief; rather, the artistically augmented meaningful gesture is likely to prompt sympathy or compassion, to stimulate curiosity as to the why and what of the particular grief, and to catch episodic attention for how it fits into the unfolding overall story, even though its addressees are not "really" involved.

In the present essay we do not have enough space for elaborating our general concept of a spectrum of governable rhetorical forces. For now, we want to consider rhetorical forces only in their impact on persons' reasons viz. personal "spaces" of reasons. Significant behaviours, then, have determinate rhetorical forces in virtue of which these behaviours can serve to make a determinate difference in an interpretant's "space" of reasons.[8]

[8] Speaking of the "logical space of reasons" has become common parlance in contemporary pragmatist philosophy (Brandom 2000, Scharp and Brandom 2007). We put "space" in quotation marks since we think that the phenomenon's dynamic nature would be better captured by the metaphor of a field of force.

14.3.3 The Art of Bending the Fields of Force of Good Reasons

The crux of our discourse-theoretical good-reasons approach is this: In virtue of being mindful agents, persons possess what we can succinctly call their power "of good sense or reason" (Harris 1969, p. 621). We take this to mean that persons have power over, and have power by virtue of, the forces reasons have whenever they count as good reasons for someone, i.e., whenever persons recognize them as *their* good reasons. Reasons as good reasons can have, or can unfold, forces when appreciated by reason-sensitive persons ("reasonable evaluators"). In what does their goodness consist? It would be short-sighted to think of the goodness of reasons as reducible to their valuable role as justifiers. Good reasons serve us in more than one vitally important normative function. They serve us for (1) making things intelligible that would otherwise be inscrutable, (2) explaining why things occur the way they occur, and (3) justifying, i.e. rendering as acceptable what would otherwise be objectionable (*justification*) and inversely (*critique*). Reasons can carry sense-making forces, explanatory forces, justificatory forces, and still more. Unlike the illocutionary forces of speech acts, reasons' forces do not derive from conventions recognized as binding, but from convictions recognized as illuminating, as persuasive, as standing to reason, as right, or for short, as correct (*convictions of correctness*). Reasons and insight are intrinsically connected, speech acts and insight are not.

Persons have power of reason or good sense if, and to the extent that, they are capable of changing (or keeping from changing) their convictions of correctness in accordance with their insights. In other words: To possess power of reason or good sense essentially includes responsiveness to reasons, in particular to reasons that persons recognize individually *and* communally as their *good* reasons.[9]

Owing to what we may metaphorically call their "fields" of forces, one's good reasons are momentous for oneself, and so is changing one's evaluation of one's reasons ("changing one's mind about one's good reasons"). Changing (or keeping from changing) persons' recognition of their good reasons has practical consequences. It has consequences for *all* the behaviours whose making good sense

[9] Two caveats: (1) In the above formulation, "person" includes all persons who exercises good sense or have reason. (2) Recognition of some reason as so-and-so good may be bounded recognition: a reason can be a good reason in relation to some finite community of rational evaluators and a bad reason in relation to other some finite community of rational evaluators.

depends on their being related to adequate reasons within a community of communicatively connected persons sharing a "stock of common grounds" (Kettner 2017). That we share a stock of common grounds means that we have a common "space" of reasons which more or less all persons concerned would endorse as good reasons upon reasonable evaluation.[10]

Arguably, the rationally privileged method for governing the diverse forces of reasons is via disciplined argumentative discourse.[11] Here, interlocutors must operate with nothing but *argumentation* about validity claims. Participants in argumentative discourse rely on the authority of their insight when they try to reasonably evaluate particular reasons in comparison with other reasons with respect to their true value. Disciplined argumentative discourse is a dialogical cognitive practice that directs whatever powers the participants possess to the exercise of their "discourse power" through verbally articulated argumentative actions. So much for discourse proper. It is very important, however, to see that the forces of reasons are not only and exclusively susceptible to governance by discourse power. A "space" of reason, a field of forces of reasons, can be bent, deflected, disturbed and readjusted by rhetorical forces of all kinds of significant behaviours, for instance by dramaturgical action. In order to understand better how this works we must get clearer about the structure of good reasons:

The forces of reasons we identify with when we count reasons as "our good reasons" flow from sources located in four structural relations. Essentially, a reason is a thought that can pass in dialogue as an answer to a seriously raised why-question (Kettner 2012). Reasons as such bear determinable relations (1) to relevant facts, (2) to affective motivation, (3) to communities of communication whose members accept the respective reasons as adequate within a range of appropriate contexts; (4) to other reasons via conceptual (*inferential*[12]) relations of coherence constrained by consistency.

---

[10] To illustrate: if for more or less all my friends the fear of a Covid-19 infection has come to count as a good enough reason to carefully evade this topic in conversation, and if someone from "outside" is puzzled by how I talk round Covid-19 in conversation then I will perhaps justify with utter conviction my comportment by pointing out that "since we find the whole thing very disquieting" we would prefer not to talk about it at all.

[11] The normative notion of argumentative discourse as in "discourse ethics" (cf. Kettner 2006).

[12] That good reasons bear these four structural relations can partly be explained in inferential role semantic accounts of meaning, e.g. in Robert Brandom's inferentialism (Brandom 2000).

Let us assume that by watching the burial ceremony in "The Dead or Coming" gives me reason to think wholeheartedly thoughts such as *we owe some reverence to the deceased,* or, *it is really sad when a dead person cannot even be grieved for.* The dramaturgical performance is what prompts these thoughts which in turn might become my good reasons for *feeling* sad and for *resenting* the circumstances that prevent reverence to the deceased. These reasons, to repeat, stem from my experiencing a performance (i.e. certain significant behaviours within certain practices). There need not be interlocutors "giving" me these reasons by "telling" me what they think. No manifestly argumentative actions were involved and yet my set of good reasons, if you will: my personal "space" of reasons has changed a bit. The means involved were quasi-argumentative means: By letting myself be drawn into the meaning of theatrical actions augmented by dramaturgically powerful actors, and by exposing myself to their rhetorical forces I allow changes to occur in convictions of correctness and hence in my "space" of good reasons. It may turn out that now I have reason e.g. *to deplore the fate of those who go unmourned*, or reason *to object to those who do not give a damn about what we owe to the deceased*, and so on.

Note the significance of the aforementioned four structural relations of reasons: My thought (T) *we owe some reverence to the deceased* could not become my good reason (R+) for feeling pity unless it is true that someone died. Nor could T become R+unless I so identify with what I think that what I think disposes me to some behaviour guided by T. Nor could T become R+unless there are, or unless I think there are or should be, such-and-such others who would endorse and appreciate as proper and adequate my feeling pity on the strength of T once they knew that T constitutes my reason for feeling the pity I feel. Nor could T become R+unless T were such that we could bring T into the position of a premise, conjoin other premises and generate inferences involving T.

14.4 The *Center for Political Beauty*'s Aesthetic Intent: A Call to Action

The CPB artist collective has been prolific in commenting and justifying their own actions from a philosophical-ethical, an artistic-aesthetical, and a political perspective. Besides many press articles they also authored a manifesto (Ruch 2015) and more recently a publication on how to fight the spreading of right-wing narratives in Germany (Ruch 2019). Several irritating, willingly contradictory elements cast an ambiguous light onto the artist group. Tensions are already manifest in CPB's two most provocative key terms, "political beauty", and "aggressive humanism".[13] Notions of beauty have lost prominence in the conceptual repertoire of contemporary art. And notions of *polis* and politics imply collective action, coordination, deliberation, and legitimacy, nothing of which bears close associations with aesthetics in the modern sense of the term. "Political beauty" implies that art should be put in action in order to serve a higher cause than aesthetics, and that politics is this higher cause. CPB sees in politics the highest form of art. The flipside of this interpretation is a tendency to reduce art to an instrument of politics.

In CPB's parlance, the meaning of "beauty" centres not on aesthetic excellence but shifts to moral righteousness (or "moral beauty"), human rights heroism, and humanistic grandeur. Philip Ruch seeks to personify high-end humanism. Such aspirations for the moral high ground have attracted trenchant criticisms from many sides (Ullrich 2015), charging Ruch and the CPB with what amounts to hypermoralism. Even though this can hardly be denied, such criticism tends to overlook the specifically antique ideal of beauty that CPB's discourse seeks to reanimate: *kalokagathia*, a personified alignment between the beautiful and the good. This antique ideal relies on the conviction that outstanding human beings should elevate themselves to moral grandeur and perfection by living up to their virtues so that they shine with ethical as well as physical beauty. In other terms, following this ideal, beauty and moral virtue should be brought into personal unison by cultivating an exemplary virtuous life demonstrably by one's good and right deeds. Beauty becomes ethos in action.

Ruch's second surprizing concept, his "aggressive humanism", apparently conjoins contradictories: humankind as distinguished by reasonableness, benevolence, tolerance and peace-seeking, and on the other hand by a belligerent readiness to fight.[14] CPB's aggressive vocabulary and guerrilla-like methods have

---

[13] CPB describes itself as "an assault team that establishes moral beauty, political poetry and human greatness while aiming to preserve humanitarianism. The group's basic understanding is that the legacy of the Holocaust is rendered void by political apathy, the rejection of refugees and cowardice. It believes that Germany should not only learn from its history but also take action. The Center for Political Beauty engages in the most innovative forms of political performance art—an expanded approach to theatre: art must hurt, provoke, and rise in revolt. In one basic alliance of terms: aggressive humanism" (Center for Political Beauty About, n.d..). For observations on aggressive humanism *see* also Honnacker 2016.

[14] Quoting Ruch: "The term aggressive humanism merges two concepts that have commonly been deemed incompatible: European humanism and aggression. Occidental humanism was the epitome of human love, benevolence and friendliness. It vindicated the position of education, love and benevolentia with decidedly friendly means" (Zentrum für politische Schönheit 2015).

been criticized as pursuing the apparently noble cause of preserving humanism in ways on the verge of illegality, employing methods that range from hoaxes in the style of the *YES men* up to infringements of the law. Ruch (2019) advocates illegal methods such as refusing to pay taxes. He resorts a martial vocabulary as if Germany were at war, triggering in turn polemic reactions from right-wing critics who depict CPB as a terrorist organisation.

Art historian Wolfgang Ullrich (2017) denounces CPB's notion of beauty as totalitarian in aiming at affectively overwhelming a political collective, emphasizing a sense of belonging to a nation and conjuring up a sense of a historic mission. As Ulrich argues, beauty as holding a collective captive flies in the face of our enlightenment understanding of beauty as stimulating reflection and permitting a stance of intellectual consideration. For Kant and Schiller, if art has a function at all, its function is to support individuals on their diverse ways to freedom, emancipation and autonomy. Ullrich even goes as far as turning the moral table by pointing out some underlying similarities between CPB's beauty ideal and some strategies of visualisation right-wing activists use, groups whose anti-humanism CPB intends to oppose. Ulrich certainly has a point. However, given that CPB as artists cohere with contemporary performing arts rather than with conventional representation-based visual arts, Ullrich's frame of reference might not be fully appropriate in this respect.

Space does not permit us to discuss the interesting question of adequate and inadequate standards of criticism here. Briefly, in our view, an alternative to the stale opposition of art as emancipatory and art as ideologically manipulative would be to withdraw from the idea of beauty altogether, and to embrace instead recent theories of postmodern, post-dramaturgic theatre that illuminate theatre's capacity for tackling social reality. In our view, Florian Malzacher's (2018, pp. 321–322) analysis comes closer to CPB's normative self-understanding than Ulrich's. Malzacher is well acquainted with Christoph Schlingensief's ideas on drama and theatrical action including the essential role of communities in theatre: In order to be effective, political theatre should not so much present individual works of art to single individuals who can agree with other individuals with whom they share a *sensus communis*. Rather, political theatre should constitute arenas, spaces for exchange and debate. In a way, CPBs action "The Dead are Coming" can be considered to be participatory in that it reaches out into a collectively shared "space" of reasons. Its success as political theatre cannot be separated from the fact that journalists, critics, including art critic Ullrich himself, contributed to the artwork's unfolding significance via comment and critique.

We end this section by an illuminating comparison and contrast. CPB shares with Adorno and the first generation of the Frankfurt School of Critical Theory the conviction that the Holocaust was a moral and civilizational rupture and that this past implies for German citizens a forward-looking ethical co-responsibility for preventing in all future and by all means the reoccurrence of genocidal barbarism. As did Adorno, CPB holds that after Auschwitz all civilized culture and art, including modern art, have lost their purportedly de-brutalizing guarantees, if they ever had any. Whereas Adorno saw the problem in the fundamental impossibility to represent and depict the atrocity of genocide, and also in the intrinsically ambivalent nature of culture which bears the seeds both for humanism and barbarism, CPB is even more pessimistic: The problem is that despite thorough public information about, and sufficiently accurate representation in the mass media of, moral monstrosities and violations of human rights commitments by and large only few people really care. In Ruch's interpretation of contemporary "Western" culture, we live now in a context of consumerist complacency in which the very idea of humanism as an active struggle towards a world in which human rights would globally be realized is threatened at its roots. Therefore, Ruch and CPB think that they (we?) must abandon the propagation of humanism by merely persuasive means and critique. They are sceptical and even somewhat disdainful towards organizations like *Amnesty International*. For Ruch, signing petitions, publishing reports, organising political demonstrations is no longer efficient and sufficient for fighting inhumanity and breaking down the social-psychologically habituated walls of indifference and phlegm.

A good way to make sense of some of CPB's interventions is to see them as attempts to install in however transitory moments a better version of our world, taking it from what it deplorably is to what it should better be. If you will, this is the theatrical creation of a painfully fleeting ethical moment, fusing "what is" with "what should be". "The Dead are Coming" conveys a strong sense of liability and co-responsibility to the participating audience who, in the end, is encouraged to take action, albeit only in a vague and diffuse injunction.

14.5 Concluding Thoughts

From antiquity to the present, death, burials and funeral rituals have been a recurring theme in the arts. Therefore, it is not surprising that "The Dead are Coming" has been interpreted as a token within this typical tradition. As already mentioned, for some commentators (cf. Lewicki 2017, p. 285) the performance amounted to an audacious contemporary staging of *Antigone*. CPB's aesthetic-political program superficially looks like what Jacques Rancière (2016) describes as an "aesthetic regime", i.e. art works seeking to shift existing frontiers between what is admitted as showable and representable and what is not so. Looking closer, however, we find that PB's intervention reaches further not only in light of aesthetic representation (a performance with innumerable representations, documentations, and comments) but also in light of its political implications and its evocative message to the public (cf. Kester 1995).

In other words: CPB's decidedly ethical understanding of beauty and aggressive humanism cannot be interpreted solely within the aesthetic regime, as intelligently playing with the difference between art and non-art. A fuller understanding of CPB's intervention "The Dead are Coming" requires bracketing aesthetic considerations, immersing oneself in the "space" of reasons as this space is warped in determinate ways in virtue of this intervention, and confronting oneself with the theory of action and finally with its practice as well.

References

Beger, Karl-Leontin. 2016. Wir bergen moralische Größe aus der Geschichte. Das Zentrum für politische Schönheit und seine Strategien zur Generierung von öffentlicher Aufmerksamkeit und politischer Partizipation. m&z 3/2016, 64–77.

Brandom, Robert. 2000. Articulating Reasons: An Introduction to Inferentialism. Cambridge: Harvard University Press.

Center for Political Beauty. n.d. The Dead are Coming. https://politicalbeauty.com/dead.html. Accessed: 29 April 2020.

Center for Political Beauty About. n.d. About. https://politicalbeauty.com/dead.html. Accessed: 20 April 2020.

Crouch, Colin. 2004. Post-Democracy. London: Polity Press.

Gehlen, Arnold. 1969. Moral und Hypermoral. Eine pluralistische Ethik. Frankfurt: Athenäum.

Goffman, Erving. 1978. The Presentation of Self in Everyday Life. London: Harmondsworth.

Green, Mitchell. 2017. Speech Acts. The Stanford Encyclopedia of Philosophy Winter 2017 Edition. https://plato.stanford.edu/archives/win2017/entries/speech-acts/. Accessed: 10 September 2020.

Habermas, Jürgen. 1984. The Theory of Communicative Action. Volume 1. Reason and the Rationalization of Society. Boston: Beacon Press.

Harris, Errol. 1969. The Power of Reason. The Review of Metaphysics 22(4), 621–639.

Haugaard, Mark and Matthias Kettner. 2020. Theorising Noumenal Power. London: Routledge.

Honnaker, Ana. 2016. Absolute Helden? Anmerkungen zum "aggressiven Humanismus". Philosophie InDebate. https://philosophie-indebate.de/tag/zentrum-fuer-politische-schoenheit. Accessed: 29 April 2020.

Hume, David. 1975 [1739]. A Treatise of Human Nature (book 2, part 3, section 3) edited by L. A. Selby-Bigge, 2nd ed. revised by P. H. Nidditch, Oxford: Clarendon Press.

Kester, Grant. 1995. Aesthetic Evangelists: Conversion and Empowerment in Contemporary Community Art. Afterimage, 22(5), 5–11.

Kettner, Matthias. 2006. Discourse Ethics. Apel, Habermas, and beyond. In Bioethics in Cultural Contexts. Reflections on Methods and Finitude, ed. C. Rehmann-Sutter, M. Düwell, D. Mieth, 299–318. Berlin: Springer.

Kettner, Matthias. 2012. Gründe und Affekte. In Welt der Gründe. Deutsches Jahrbuch Philosophie Nr. 4, ed. J.Nida-Rümelin, E. Özmen, 444–454. Hamburg: Felix Meiner.

Kettner, Matthias. 2017. Argumentative Discourse: The Transcendental Starting Point of Apelian Discourse Ethics. In Transcendental Arguments in Moral Theory, ed. J. P. Brune, R. Stern, M. Werner, 325-347. Berlin: De Gruyter.

Kettner, Matthias. 2018. The Forstian Bargain. Overrationalizing the Power of Reasons. Journal of Political Power 11(2), 139–150.

Kettner, Matthias and Wolfgang Mertens. 2010. Reflexionen über das Unbewusste. Philosophie und Psychologie im Dialog. Göttingen: Vandenhoeck & Ruprecht.

Lewicki, Aleksandra. 2017. "The Dead are Coming": Acts of Citizenship at Europe's Borders. Citizenship Studies 21(3), 275–290. https://doi.org/10.1080/13621025.2016.1252717.

Lilienthal, Matthias and Philipp, Claus, 2000. Schlingensiefs Ausländer raus. Dokumentation. Frankfurt: Suhrkamp.

Malzacher, Florian. 2018. Aktivismus als Aufführung. Das agonistische Theater des Zentrums für Politische Schönheit. In Haltung als Handlung. Das Zentrum für Politische Schönheit, ed. M. Rummel, R. Stange, F. Waldvogel, 321–329. München: Metzel.

Mouffe, Chantal. 2013. Agonistics: Thinking the World Politically. London: Verso.

Nussbaum, Martha. 2001. Upheavals of Thought: The Intelligence of Emotions. Cambridge: Cambridge University Press.

Rancière, Jacques. 2016. Politik und Ästhetik. Im Gespräch mit Peter Engelmann. Wien: Passagen.

Ricoeur, Paul. 1973. The Model of the Text: Meaningful Action Considered as a Text. New Literary History, 5(1), 91–117.

Ruch, Philipp. 2015. Wenn nicht wir, wer dann? Ein politisches Manifest. München: Ludwig.

Ruch, Philipp. 2019. Schluss mit der Geduld. Jeder kann etwas bewirken. Eine Anleitung für kompromisslose Demokraten. München: Ludwig.

Scharp, Kevin, Brandom, Robert (eds.). 2007. In the Space of Reasons: Selected Essays of Wilfrid Sellars. Cambridge, MA: Harvard University Press.

Ullrich, Wolfgang. 2015. Das Erdbeben der Schönheit. DIE ZEIT Nr. 48. https://www.zeit.de/2015/48/philipp-ruch-kunst-politik-manifest-antimodernismus. Accessed: 29 April 2020.

Ullrich, Wolfgang. 2017. Die Wiederkehr der Schönheit. Über einige unangenehme Begegnungen. Pop-Zeitschrift. https://pop-zeitschrift.de/2017/11/07/die-wiederkehr-der-schoenheit-ueber-einige-unangenehme-begegnungenvon-wolfgang-ullrich07-11-2017/. Accessed: 29 April 2020.

Zentrum für Politische Schönheit. 2015. Aggressive Humanism. Of democracy's inability to generate great human rights activists. Medium. https://medium.com/@political-beauty/aggressive-humanism-bbff64cf4296. Accessed: 29 April 2020.

15 Musical Interface Agendas. Musical Appropriation via Technological Pre-configuration

Carsten Wernicke

Abstract

The use and implementation of software and digital-material controllers in conjunction with protocols indispensable for the process change the aesthetic experiences and capacities of musical and creative expression potentials in musical practices. What is first established as a constructed naturalness of individual music practices is challenged and disrupted by the digitality-specific multiplicity of the musical interface complex (hardware/software/protocol). The role of the empty-world hypothesis as a hidden agenda of these devices will be examined in more detail with the aid of a self-produced video documenting the use of the digital-material interface ROLI Seaboard by a musician playing in an "Arabic" idiom. Due to its dialectic framing character, the ROLI device will have to be described as a culturally expanding facilitator on the one hand and as a cultural capitalist artifact on the other.

Keywords

Musical interface · Interface design · Materiality · Video self-documentation · Empty world hypothesis

The study is based on data collected in the BMBF-funded (German Federal Ministry of Education and Research, November 2017–September 2021, funding code 01JKD1708B) research project "Musical Interface Designs: Augmented Creativity and Connectivity" (Subproject 2) at Leuphana University Lüneburg.

C. Wernicke (✉)
Leuphana University of Lüneburg, Lüneburg, Germany
e-mail: carsten.wernicke@leuphana.de

L. Gaupp et al. (eds.), *Arts and Power*, Kunst und Gesellschaft,
https://doi.org/10.1007/978-3-658-37429-7_15

15.1 Introduction

In recent years, novel digital-material musical interfaces have been introduced to the music market that enable musicians to transfer musical gestures directly from the grids or tactile surfaces of the interface to musical material with greater accuracy and sonic complexity. This was made possible above all by the introduction of a new data transmission standard with the aid of which the musical gesture, for example on a keyboard, can be displayed in more dimensions than was previously the case. In earlier decades (with the exception of a few special devices with only limited options for expression), it made no difference to data transmission whether the user's finger hit the key of a keyboard directly, whether a vibrato was added to the sound quality by tilting the finger from side to side or the tone was modulated by sliding the finger back and forth.

These musical interfaces are material because, unlike apps on tablets, they are not represented as emulations on screen surfaces in the form of front-end interfaces, but are actually still embodied by the use of tactile devices. In contrast to historical music instruments like the violin or the piano, they do not themselves function as producers of a sound when operated by a musician, but are instead embedded in a whole series of nested and interconnected interfaces that do not end at the casing of the device. For an audible sound to be produced at all, certain computational processes are necessary. These processes allow the musical content of the gesture performed on the controller to be ultimately converted into an electrical signal, which is then transmitted to a loudspeaker. This fact is also what makes the material interface digital. A complex interaction between all the material (hardware) and symbolic (software, protocol/API) interface layers (cf. Distelmeyer 2019) is necessary to produce an audible sound result.

The question now is to what extent this complex interplay can be contained within a concept that will help us explain how it is also possible to implement "non-Western" musical concepts using these musical interfaces, which are often packed with culturalistic pre-settings (Ismaiel-Wendt 2016, p. 7). To this end, we will examine to what extent the concept of the "empty-world hypothesis" (Simon 1981; Baecker 2018) can play a major role in this form of transculturality (Welsch 1997, 2017) and its specific limitations. These limitations are essentially based on a specific construction of naturalness made up of imagined user interface scenarios (Oudshoorn et al. 2004; Iivari 2006).[1]

[1] I would like to thank Victoria Simon for that last reference.

The fact that these limitations are manifested as technological conditions (Hörl 2011) reveals the precariousness of the digitality-specific trajectory between user spheres and their performative settings and musical interface complexes. What was once established as a constructed naturalness of individual music practices is challenged and disrupted by the digitality-specific multiplicity of the network of interfaces involved (hardware/software/protocol), which I will refer to as musical interface complexes, but is also reinstated for its own benefit when an empty-world situation actually occurs. Using a self-produced video clip documenting a musician playing in an "Arabic" idiom on the digital-material interface "Roli Seaboard" in combination with an artifact analysis of the Roli Seaboard and the distributed software bundle, which includes Roli Dashboard, Roli Equator and Ableton Live Lite,[2] it will be shown how it is possible, on the one hand, to transfer a "non-Western" musical concept to the musical interface (video self-documentation) and on how, on the other hand, restrictions are inscribed in the "MusickingThing" (Ismaiel-Wendt 2016) complex itself (artifact analysis). For while the digital-material interface opens up new realms of possibilities, the software environment in which the interface is embedded causes a certain form of cultural hegemonization to be rearticulated.

15.2 Interfaces, Devices, and Their Role as Designed Things

As a designed product, the digital-material interface is a tactile device with a dual nature: it is both thing and device at the same time (Ehn 2013; Jörissen 2014). From a technical perspective, it is what the design object embodies that ensures access to the functionality of the thing. As the result of design processes, it is always a thing that modifies the interactional space for its users and is prepared for unexpected use. It is polysemic in terms of aesthetics and cultural values, and can thus facilitate new ways of thinking and behaving (Ehn 2013, p. 82).

Designed things are also emulations due to how they are prepared. They fit into existing environments, simulating compatibilities, and bridging discrepancies in cultural codes without losing their specific autonomy and resistance, even though they will always be structurally linked to certain domains of knowledge

[2] The status of the software versions is the start of the data collection phase in the 2nd and 3rd quarters of 2018. Here, I refer only to the free and thus in this respect accessible software included in the bundle.

and design as music-making, supporting and enabling things. According to Dirk Baecker, things have certain modes of existence based on selections of elements which are connected to each other in a very specific way. However, this does not mean that the connections between the elements cannot be made in any other way. Certain distinctions can be made by making the connections visible, without excluding the possibility of alternative connections (Baecker 2018, p. 25).

Beyond that, it is not only about things, but also about interfaces. According to Alexander Galloway (2012), interfaces, as devices, are processual rather than material. The use and implementation of software and digital-material interfaces in conjunction with protocols that are indispensable for the musical process such as MPE[3] change the aesthetic experiences and capacities of musical and creative expression potentials in musical practices. They are entities that, together with human entities, make themselves available to each other as an interface (Knorr-Cetina 2007, p. 285) and characterize the transition to a "post-digital state" (Berry and Dieter 2015; Jörissen 2016; Stalder 2016; Baecker 2018) from both an individual and a collective perspective. This follows from the fact that the complex interplay between the various interface layers also realizes the manufacturer's intentional concealment of the interfaces between the system and the user (Kaerlein 2016, p. 31). To put it simply: if I have a well-made user interface, I as a user don't care which and how many interfaces lie between the actions I perform and the auditory and visual results.

The various states observable on the material level are only made possible in the first place by "interface effects" (Galloway 2012), in which various superimposed and yet autonomous interface layers inscribed in things become transformatively effective. These effects refer not only to the interfaces inherent to the thing itself, but also to interfaces of other things that can be structurally linked to it.

Basically, taking Lev Manovich (2001) as his point of departure, Galloway develops an almost paramaterialistic thesis of the primacy or exclusivity of software vis-à-vis Kittler's overemphasis of the role of hardware (Kittler 1993). The aim appears to be to derive a meta-effect for social structures and processes from the resulting interface effect. The concept of the interface is thus somewhat overstretched in certain respects. Galloway legitimizes this step by arguing that the

[3] MPE and its predecessor MIDI are protocols that can be used to standardize the transfer of musical values from one musical interface to another, such as MIDI-keyboards, synthesizers, and other controllers as well as personal computers. Codified in 2018, MPE (MIDI Polyphonic Expression) is a more advanced version of MIDI, and can directly and multidimensionally map a larger spectrum of musical gestures than MIDI can.

genuine materiality of interface layers as a constitutive feature does not necessarily have to be given in relation to the various media layers that make up informational artifacts.

15.3 Musical Interface Complexes and the Empty World

Musical interfaces, like all technical devices, instruments and interfaces, are "part of performative settings" with their own "media character" (Großmann 2013, p. 299). Software, digital-material interfaces, and their incorporated interface layers, however, are at odds with common practice definitions of musical instruments and their inherent constructions of naturalness, although the essential function of musical instruments is also fulfilled by these things: making music—whatever that may refer to. Particularly with regard to their being designed as interfaces, digital-material musical interfaces often imply a simplification of the actual music-making process, although paradoxically they themselves are constructed as musical black boxes that function as complex, technically interconnected algorithmic control and programming units. The answer to the question of the role and function of internal and external control units remains nebulous and undecided, since the respective black box usually oscillates between openness (e.g. protocols, data formats) and unity (also data formats, system architecture). Another reason for this is the aforementioned concealment of the interfaces, or as Galloway puts it: "New media hide as much as they show" (Galloway 2012, p. 61). They demonstrate a "free paternalism" towards the user that is typical of digital networks (Stalder 2016, pp. 228–229). Nevertheless, the complexity and nature of musical interfaces is not the only constitutive factor for the extent of this "paternalism"; the domains of knowledge available to the users and thus the performative setting mentioned above (Zirfas and Klepacki 2013; for more detailed design-theoretical implications see Jörissen 2014) are also contributing factors.

We are therefore not dealing with a simply completed artifact or thing, but instead with fluid music interface complexes consisting of different interface layers, interface tracks, and application scenarios. For the sake of clarity, however, we must resort to a simplified description of this phenomenon in order to clarify what such a musical interface complex might look like. This musical interface angle should correspond to the principle of simplicity (Baecker 2018, p. 183) and thus do justice to the "empty-world hypothesis" of designed things: this is based on the assumption that artificial systems or artifacts are always contingent on the use and benefit conceived by the designers. The actual field of action thus emerges only through actual use. The complexity of artificial systems or artifacts does lead to a hierarchization of the individual partial artifacts or system components. However, this is mitigated by the fact that "most things are only weakly connected with most other things; for a tolerable description of reality only a tiny fraction of all possible interactions needs to be taken in account" (Simon 1981, p. 221). To give a relatively simple example: in the case of our musical interface complex, this means that I neither have to understand all the functions of the individual complex layers, nor do all the layers have to be equally accessible in order for the system to function as understood by the user.

15.4 Method

The methodological framing of this study and the underlying research project is an artifact analysis according to Lueger and Froschauer (2018), which deals with the materiality of artifacts and their embedded meanings in everyday contexts. The artifact analyses will involve a multi-step procedure to describe the materiality, structures and contexts of the artifacts in question in order to reconstruct social meanings, involved actors and situated context analyses, and to generate structural and comparative analyses.

Data from self-documentation videos (Maleyka et al. 2018) created by musicians and instrumental teachers and acquired for the study are used for analysis, especially with regard to embedding the artifacts in the context of users' everyday lives (Schutz 1972; Soeffner 1996).The video analysis is based on the hermeneutic sociology of knowledge (Reichertz and Englert 2011; Reichertz 2018) and makes this data methodically accessible by analytically processing "moves" video instead of "stills" and, in addition to the action in front of the camera, taking the action of the camera itself and the concatenation of these two levels of action into account. At first glance, however, the implementation of the hermeneutic sociology of knowledge contradicts in principle the de facto symmetrical setting (Latour 1993, 2010) which underlies this study: this is because of the paradigm of the passive object inherent in the hermeneutic sociology of knowledge and therefore of the special (human) subject (Hitzler et al. 2020, p. 15). Furthermore, this dilemma also exists for the artifact analysis framing the study (Lueger and Froschauer 2018; Froschauer and Lueger 2020). Balbierz and Loon (2020), on the other hand, are in favour of abandoning anthropological primacy for the hermeneutic sociology of knowledge as well—not because this primacy has in any way been proven false or obsolete, but rather because the implementation of this premise, as a fact that cannot be circumvented by the method, weakens the claim to scientificity itself (Balbierz and Loon 2020, p. 495). Accordingly, for the purposes of sequential analysis, the actions to be considered in terms of inherent affordant content do not have to be at odds with the methodological setting. Instead, they indicate potentials for the transference of the methodological framework into contexts that are not genuinely anthropocentric.

15.5 Artifact-Analytical Descriptions

The following descriptions of the materialities are not exhaustive but are intended solely to provide a more detailed explanation of the case descriptions that will follow later and are supplemented by more specific aspects in the case descriptions themselves. They make a rough distinction between the three levels of protocol/hardware/software, whereby the laptop, including its hardware layers and wires required for the setting, cannot be taken into account at all and must remain a blind spot.

1. On the protocol level, the Roli uses the MPE protocol, which was not implemented as a new overall standard until 2018. As a more advanced version of MIDI, the former institutionalized standard for transferring controller data to a computer unit for conversion to sound, MPE offers a wider range of transferable controller information. With MPE, not only the pitch and velocity of the sound can be transmitted by pressing a key or pad on a controller, other gestures can also be transmitted, depending on the mapping. With a Roli Seaboard, the quality of the sound can be changed by moving one's finger from side to side and backwards and forwards on the key. Furthermore, it makes a difference not only how hard the key is pressed, but also how quickly it is released. It was of course also possible to modulate sounds with the aid of a Digital Audio Workstation (DAW, see below) before MPE. This, however, required the post-processing of the sound or the interposition of physical controllers or software plug-ins. Thus, it was not possible to edit the tone directly at the gestural source. The fact that data of this quality can also be processed by software means that the protocol must be implemented within the software structure.
2. The Roli is delivered with software that also enables the use of MPE. It comes with a software bundle which includes Dashboard software and a software called Equator. The most important configurations of the individual Roli components can be controlled and changed (mapping) with Dashboard. The Equator software works like a database of presets to translate MPE signals generated by using the Roli Seaboard keyboard into sound by using the

hardware resources of the computer required to run the software. Equator was specially designed for Roli products and also natively integrates MPE. Dashboard, which is also supplied, enables the mapping of the Roli Songmaker Kit (RSK) components (Fig. 15.1).

Another part of the software bundle is a license for the lite version of Ableton Live. Ableton Live is a Digital Audio Workstation (DAW). DAWs work like digital versions of music production studios. They can be used to record and edit entire songs. Compared to other DAWs, Ableton Live does not exclusively follow an established vertical track logic, but a modular structure instead. In addition, the software should not only enable the subsequent processing of samples, but also the manipulation of musical snippets in real time. Compared to the larger full versions, this lite version offers only eight tracks for processing samples and recording MPE signals, for example, in addition to the number of given sounds and samples. However, MPE is not embedded natively in this software, but must first be emulated into the environment by using several tracks in parallel.

3. Looking at the musical interface as hardware, it is now noticeable that the translation of the new musical-aesthetic possibilities offered by MPE has been implemented in a specific way on the surface level of the device. The Roli Seaboard features a metaphorical reference to a classical keyboard. At the same time, this similarity is breached by the use of silicone, which makes the chromatic keys, all of which can be perceived individually as elevations, more tactile and conveys an increased sensitivity. It also facilitates sliding via the tactile user interface.

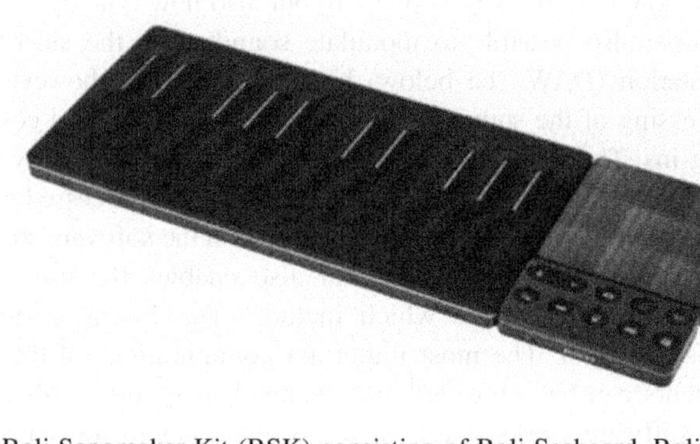

Fig. 15.1 Roli Songmaker Kit (RSK) consisting of Roli Seaboard, Roli Block and Loopstation (Jörissen et al. 2019)

Based on these descriptions, it is now necessary to ask what effects the constitution of the artefact has on its embedding in everyday contextual meanings for professional actors—especially for those of a "non-Western" background.

15.6 Implications of the Self-Documentation Videos

The following example illustrates how it is possible to make the empty world one's own. A self-documentation video is used for this purpose. A group of three study participants is shown in a DIWO (Do It With Others) situation. The third participant in the video is usually covered by the participant sitting on the chair. She is using a different musical interface complex, the Eigenharp TAU, which is part of the research project, but is not of interest here. The participant sitting on the chair on the left is playing his usual instrument, a microphoned violin, using an effect pedal to modulate its sounds. Sitting on the right handside of the sofa, there is an oud player with a RSK, who will be the focus of attention here (Fig. 15.2).

The aforementioned inability of the silicone surface to accurately represent exact chromatic tones in the sense of "Western" tonal systems is what opens up a

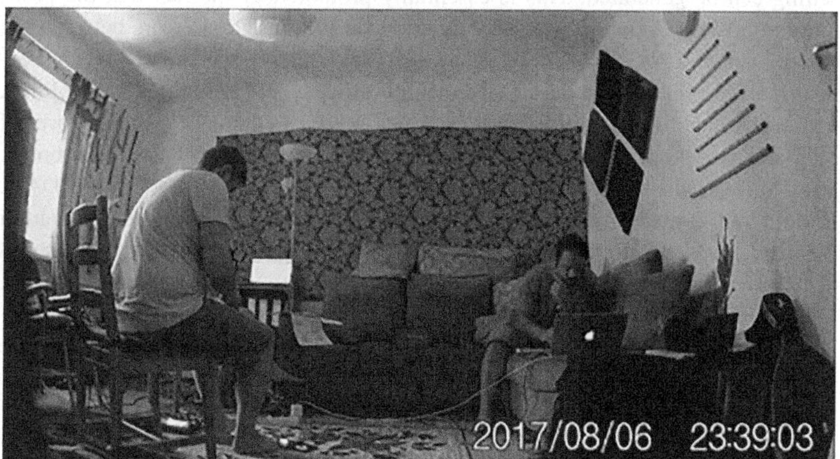

Fig. 15.2 Self-documentation video of a jam session with ROLI Songmaker Kit (wrong time stamp due to action cam default setting)

particular accessibility to the musical interface complex for the oud player. Even though the elevations of the silicone surface of the RSK Seaboard imply a reference to "Western" chromaticism, the lack of a clear physical separation between the chromatic centres of the key-like elevations facilitates the creation of non-chromatic tone sequences and melismas by touching the elevations slightly to one side of these centres and by sliding. This gestural implementation of basic concepts of the music practiced in Arabic-speaking regions, among others, whose characteristics include the use of quarter tones (measured by the chromatic notation system or similar) with the RSK is essentially due to the use of MPE protocol. Only by translating the gestures into MIDI data that is more complex than in the previous standard does a multi-dimensional acceptance and translation become possible, one which does not require additional simulation via an external controller or internally, via mouse movements.

15.7 Discussion

The actual keyboard logic of the Roli is extended beyond its limits in the empty-world space; "Arabic" musical parameters such as quarter tones become part of the concept, and a unity between hardware, software, protocol and performative setting can be generated. This is essentially promoted by the fact that the problem of software logic in this case does not lead to translation errors between the oud player and the musical interface complex. The connection between Seaboard, USB cable, Dashboard and Equator software offers the user a simplified, non-critical pathway into the use of musical interface complexes. In this form of plug-and-play logic, low-threshold performative settings also become resonant. The Roli can also be used as a standalone without being embedded in a DAW. In doing so, it also provides the opportunity for cultural transversality (Guattari 2000, p. 43; Braidotti 2014, pp. 96–97). As the example shows, this also applies to hardware such as protocol and thus to the entire musical interface complex. Self-optimization in the sense of embedding one's own practice in the information logic provided by the musical interface complex is thus made possible. The informational interface is also brought back into the performative logic of the oud player, intertwining the technosphere, sociosphere and individual spheres.

But the empty world is also a user experience instrument for the brand manufacturer. In this sense, the empty-world agenda does not overcome the capitalist logic of commodification. Rather the opposite is also(!) the case, since commodification as capitalist "Landnahme" (Dörre 2011) is facilitated by the empty world. Not only does this conceal the computicity of the post-digital artifact (Kaerlein

2016), the agenda itself is also masked by its semantic indecision. The Roli is thus simultaneously both the facilitator and the beneficiary of the empty world that has been made available, because the Roli requires the empty world in order to pervade the musical practices of musicians playing in an "Arabic" idiom, similar to Guattari's lagoon-invading algae in *The Three Ecologies* (Guattari 2000, p. 43).

The fact that the Roli does not detach itself from the commodifying logic is also shown in the paradoxical practice of selling the Roli along with a software that is not MPE-capable. The culturalistic expansion narrative that is actually to be inscribed is thereby corrupted. Because Ableton Live, as a non MPE-enabled DAW, can only be very laboriously simulated or emulated, MPE music, like that of the "Arabic" musician, remains ephemeral, while a clear primacy of chromatic "Western" music is conceded, since it is considered to be 'recordable'. The entire musical interface complex constitutes, according to a distinction borrowed from Braidotti by Foucault, a power that is both productive and restrictive (Braidotti 2014, p. 32). The software is thus an expression of hegemonic norms in the sense used by Victoria Simon (2018). This corresponds less to the users' intentions, and rather to those of the (mostly "Western") producers, who imagine their target users as being similar to themselves (Oudshoorn et al. 2004; Iivari 2006), thus producing specific constructions of naturalness made up of imagined user interface scenarios. With this bundle, "Western" ideologies are inscribed into the complex which is imbued with a certain "Western" idea "of musical practice, control over a musical instrument through embodied knowledge of musical technique" (Simon 2018, p. 3). What is subjectively possible for the artist is administered as capitalist subjectivity (Guattari 2000, p. 50) by techno-social shaping mechanisms.

15.8 Summary

The aim of this paper is to illustrate that these interactions and symmetries between users and, depending on the level of description, interface layers, devices or things are shaped by simultaneously productive and restrictive powers. This constitutes the necessity of thinking along the lines of different interface layers as well as the entire complex. Understanding situational use means becoming aware of ephemeral flashes of emergent interconnections between layers.

As the comparatively restricted insight into the data material is already able to show, the changes in aesthetic experience brought about by musical interfaces call into question perception itself:

> At a time when machines are being upgraded in terms of sensory and motor quality and one not only wishes to know how this works, but also, conversely, comes to the realization that artificiality, or artificial self-production, is already part of every human being, then the issue here is nothing short of human perception itself. Those elements that come together as organism, brain, consciousness, technology and society might also be arranged quite differently. The machines are doing it. And people imitate it, but on both sides of the boundary, on the side of technical self-optimization and on the side of maintaining a robust or even fragile stubbornness. (Baecker 2018, p. 130; my translation)

The distinction between the inside and outside of what is artistically possible with the device can be made by the user him or herself, provided that his or her imagination of the empty world resonates with the empty world of the device. At the same time, the empty world aids and abets the commodification of "non-Western" musical material, which has been difficult to anticipate by non-MPE musical interfaces so far. The empty world is thus not only a facilitator for the use of digital-material musical interfaces by "non-Western" musicians, but at the same time also a facilitator of musical interface´s own transference of the interface concept into another cultural context, enabling it to pervade a not yet overrun habitat.

References

Baecker, Dirk. 2018. 4.0 oder die Lücke die der Rechner lässt. Leipzig: Merve.

Balbierz, Sandra and Joost van Loon. 2020. Ego—Anthropo—Sozio. Zur Verortung des Prinzips der Generalisierten Symmetrie in der Hermeneutischen Wissenssoziologie. In Kritik der hermeneutischen Wissenssoziologie, ed. Ronald Hitzler, Jo Reichertz, and Norbert Schröer, 485–497. Wiesbaden: Springer VS.

Berry, David M. and Michael Dieter (ed.). 2015. Postdigital Aesthetics. Basingstoke: Palgrave.

Braidotti, Rosi. 2014. Posthumanismus Leben jenseits des Menschen. Frankfurt/M.: Campus.

Distelmeyer, Jan. 2019. From Object to Process. Interface Politics of Networked Computerization. Artnodes 24: 83–90. http://dx.doi.org/https://doi.org/10.7238/a.v0i24.3300

Dörre, Klaus. 2011. Capitalism, Landnahme and Social Time Régimes: An Outline. Time & Society 1: 69–93.

Ehn, Pelle. 2013. Partizipation an Dingen des Designs. In Wer gestaltet die Gestaltung? Praxis, Theorie und Geschichte des partizipatorischen Designs, ed. Claudia Mareis et al., 79–104. Bielefeld: transcript.

Froschauer, Ulrike and Manfred Lueger. 2020. Materialisiertes Wissen: Ein hermeneutischer Blick auf die Organisationswelt. In Kritik der hermeneutischen Wissenssoziologie, ed. Ronald Hitzler, Jo Reichertz, and Norbert Schröer, 312–325. Wiesbaden: Springer VS.

Galloway, Alexander R. 2012. The Interface Effect. Cambridge: Polity.

Großmann, Rolf. 2013. 303, MPC, A/D. Popmusik und die Ästhetik digitaler Gestaltung. In Performativität und Medialität Populärer Kulturen, ed. Marcus S. Kleiner, and Thomas Wilke, 299–315. Wiesbaden: Springer.

Guattari, Félix. 2000. *The Three Ecologies*. London, New Brunswick/NJ: Athlone.

Hitzler, Ronald, Jo Reichertz, and Norbert Schröer. 2020. Zwei Jahrzehnte Hermeneutische Wissenssoziologie – Ausbau und Differenzierung. In Kritik der hermeneutischen Wissenssoziologie, ed. Ronald Hitzler, Jo Reichertz, and Norbert Schröer, 9–24. Wiesbaden: Springer VS.

Hörl, Erich. 2011. Die technologische Bedingung. Zur Einführung. In Die technologische Bedingung: Beiträge zur Beschreibung der technischen Welt, ed. Erich Hörl, 7–53. Berlin: Suhrkamp.

Iivari, Netta. 2006. 'Representing the User' in Software Development - a Cultural Analysis of Usability Work in the Product Development Context. Interacting with Computers 18(4): 635–664.

Ismaiel-Wendt, Johannes. 2016. post_PRESETS. Kultur, Wissen und populäre Musikmach-Dinge. Hildesheim: Olms.

Jörissen, Benjamin. 2014. Bildung der Dinge: Design und Subjektivation. In Subjekt Medium Bildung, ed. Benjamin Jörissen, and Torsten Meyer, 215–233. Wiesbaden: Springer.

Jörissen, Benjamin. 2016. Hegemoniale Ästhetiken und ästhetische Gegenstrategien—Kulturelle Bildung in der postdigitalen Kultur [pre-print]. In infodienst 120. https://joerissen.name/joerissen-benjamin-2016-preprint-hegemoniale-aesthetiken-und-aesthetische-gegenstrategien-in-der-postdigitalen-kultur/. Accessed: 27 December 2019.

Jörissen, Benjamin, Michael Ahlers, Martin Donner, and Carsten Wernicke. 2019. Musikalische Interface-Designs: Augmentierte Kreativität und Konnektivität (MIDAKuK). In Forschung zur Digitalisierung in der Kulturellen Bildung, ed. Benjamin Jörissen, Stephan Kröner, and Lisa Unterberg, 195–214. München: kopaed.

Kaerlein, Timo. 2016. Intimate Computing. Zum diskursiven Wandel eines Konzepts der MenschComputerInteraktionen. Zeitschrift für Medienwissenschaft 15: 30–40.

Kittler, Friedrich. 1993. Es gibt keine Software. In Draculas Vermächtnis by Friedrich Kittler, 225–242. Leipzig: Reclam.

Knorr Cetina, Karin. 2007. Postsoziale Beziehungen. In Kulturen der Moderne, ed. Thorsten Bonacker, and Andreas Reckwitz, 267–300. Frankfurt/M.: Campus.

Latour, Bruno. 1993. We Have Never Been Modern. Cambridge: Harvard UP.

Latour, Bruno. 2010. Eine neue Soziologie für eine neue Gesellschaft. Berlin: Suhrkamp.

Lueger, Manfred and Ulrike Froschauer. 2018. Artefaktanalyse. Wiesbaden: Springer.

Manovich, Lev. 2001. The Language of New Media. Cambridge/MA: MIT.

Maleyka, Laura, Sascha Oswald, Holger Herma, and Michael Corsten. 2018. Filmische Selbstdokumentationen als Datenmaterial. In: Handbuch Qualitative Videoanalyse, ed. Christine Moritz and Michael Corsten, 719–733. Wiesbaden: Springer VS.

Oudshoorn, Nelly, Els Rommes, and Marcelle Stienstra. 2004. Configuring the user as everybody: Gender and Design Cultures in Information and Communication Technologies. Science, Technology, & Human Values 29(1): 30–63.

Reichertz, Jo. 2018. Die Kunstlehre der wissenssoziologisch- hermeneutischen Videointerpretation. In Handbuch Qualitative Videoanalyse. ed. Christine Moritz, and Michael Corsten, 101–119. Wiesbaden: Springer VS.

Reichertz, Jo and Carina Jasmin Englert. 2011. Einführung in die qualitative Videoanalyse. Eine hermeneutisch-wissenssoziologische Fallanalyse. Wiesbaden: Springer VS.
Schutz, Alfred. 1972. Common Sense and Scientific Interpretation of Human Action. In Collected Papers I: The Problem of Social Reality, ed. Maurice Natanson, 3–47. Dordrecht: Springer.
Simon, Herbert. 1981. The Science of the Artificial. Cambridge/MA: MIT.
Simon, Victoria. 2018. Guided by Delight: Music Apps and the Politics of User Interface Design in the iOS Platform. Television & New Media, 1–15. https://doi.org/10.1177/1527476418794634
Stalder, Felix. 2016. Kultur der Digitalität. Berlin: Suhrkamp.
Soeffner, Hans-Georg. 1996. The Order of Rituals. The Interpretation of Everyday Life. New Brunswick, NJ: Transaction.
Welsch, Wolfgang. 1997. Transkulturalität. Zur veränderten Verfassung heutiger Kulturen. In Hybridkultur: Medien, Netze, Künste, ed. Irmela Schneider, and Christian W. Thomson, 67–90. Köln: Wienand.
Welsch, Wolfgang. 2017. Transculturality—the Puzzling Form of Cultures Today. In Perspectives on a 21st century comparative musicology. Ethnomusicology or Transcultural Musicology? ed. Francesco Giannattasio and Giovanni Giuriati, 30–49. Udine: Nota.
Zirfas, Jörg and Leopold Klepacki. 2013. Die Performativität der Dinge. Pädagogische Reflexionen über Bildung und Design. Zeitschrift für Erziehungswissenschaften 16: 43–57.

Who Plays? an Analysis of Musicians' Global Mobility in the World Music Market

16

Glaucia Peres da Silva

Abstract

This paper discusses musicians' global mobility using the example of the World Music market. It draws on theoretical debates on migration and mobility, focusing on Glick Schiller and Salazar's (Journal of Ethnic and Migration Studies 39(2): 183–200. 10.1080/1369183X.2013.723253, 2013) approach to regimes of mobility, in order to investigate how global mobility becomes part of musicians' careers within unequal fields of globe-spanning power. An empirical analysis of migration history of musicians who performed at WOMEX–The World Music Expo between 2008 and 2017 reveals that a high share of migrant musicians (18%) are active in the World Music market. Nevertheless, it is the non-migrant touring musicians who seem to sustain this market as a transnational space beyond the boundaries of nation-states and ethnic groups. Most of them live in Europe, or, more specifically, in Spain, France, Brazil, the UK, and also the USA, which implies clear power imbalances within the World Music market.

Keywords

Globalization · Mobility · Musician · World music · Market · Transnational space

G. Peres da Silva (✉)
Universität Tübingen, Tübingen, Germany
e-mail: glaucia.peres-da-silva@uni-tuebingen.de

L. Gaupp et al. (eds.), *Arts and Power*, Kunst und Gesellschaft,
https://doi.org/10.1007/978-3-658-37429-7_16

16.1 Introduction

Research on globalization has evolved around the concepts of connectivity and flows, emphasizing the integration of geographically distant areas through social relations that span the globe and form one single structure (Giddens 1990; Harvey 1989). In this context, capital flows in finance markets, commodity flows in new modes of capitalist production, information flows through new communication technologies, and flows of people around the globe have attracted the attention of scholars from various disciplines (Appadurai 1990; Urry 2012). Specifically, with respect to flows of people, a new interest in global mobility has emerged, which seeks to elucidate the relationship between mobility and stasis. This has, in turn, fostered the development of migration studies. As Favell et al. (2007) argue, while the first generation of studies on global mobility was "macro in its scope and argumentation" (Favell et al. 2007, p. 15), a second generation of research has highlighted migrants' agency, resistance, and freedom to form transnational networks. More recently, a third wave of social research in this area has focused on mobile and networked populations, which aim to dissolve the concept of society and its connection to national states. Building on this, Glick Schiller and Salazar (2013) maintain that the current discourse on global mobility contributes "to an emerging theory of society as globally but unequally relational" (Glick Schiller and Salazar 2013, p. 185). In this sense, society is conceived as "various circuits of interconnection [that] are part of the ordinary structuring of human sociability" (Glick Schiller and Salazar 2013, p. 185). In order to understand these "circuits", Glick Schiller and Salazar suggest addressing "the dynamics between mobility and stasis within unequal fields of globe-spanning power" (Glick Schiller and Salazar 2013, p. 187) with a focus on regimes of mobility. This perspective has the advantage of including both macro structures of international regulatory and surveillance systems and micro-level disputes over categories of similarity vs. difference and belonging vs. foreignness strangeness in different forms of governmentality.

The present paper discusses the global mobility of musicians by taking a mobility regime perspective (Glick Schiller and Salazar 2013). While migration studies have so far focused primarily on "the lower end of the labor market … in terms of minority race, ethnicity or culture" (Favell et al. 2007), mobility studies have tended to study elite travelers, seen as cosmopolitan individuals who are open to difference (Glick Schiller and Salazar 2013). The mobility of artists in general and musicians in particular play a peripheral role in this literature. Musicians are difficult to pin down as a group because they are spread across social classes and represent diverse ethno-cultural backgrounds, and, as such, present a challenge to mobility studies. Additionally, their international tours have little in common with labour migration, although they do result in them living transnational lives (Guilbault 1996). Moreover, musicians do not constitute a professional group in the sociological sense (cf. Frederickson and Rooney 1990), which distinguishes them from highly skilled migrants. In view of these particularities, my study builds on Glick Schiller and Meinhof's (2011) suggestion of disregarding the migrant/native divide in the analysis of patterns of global mobility and focuses on power imbalances between musicians from different parts of the world.

The aim of this paper is to contribute to the debate on global mobility by exploring the following question: How does global mobility become part of musicians' careers within unequal power structures, mostly associated with geopolitical relations? By observing musicians' patterns of global mobility, it is possible to gain insights into the circuits of interconnection that allow certain groups to be more mobile than others. Comparing these circuits with the geopolitical macro structures that divide the world into the Global North and the Global South—terms that pertain to "an entire history of colonialism, neo-imperialism, and differential economic and social change through which large inequalities in living standards, life expectancy, and access to resources are maintained" (Dados and Connell 2012, p. 13)—makes it possible to verify the extent to which musicians' global mobility reproduces or breaks with these forms of structuring global sociability.

The musicians who are the subjects of this study do not come from specific nation-states or belong to specific ethnic communities, rather they move within the transnational space of the World Music market. A transnational space is constructed by stable movements of people and objects across the borders of nation-states (Pries 1996), interlocking different networks of social relations which reconfigure the boundaries of nation-states, without dissolving them (Glick Schiller 2010). The focus on the transnational space formed by the movement of musicians and their music in the World Music market allows me to avoid the problem of methodological nationalism (Wimmer and Glick Schiller 2002), which treats "social and historical processes as if they were contained within the borders of individual nation-states" (Glick Schiller and Salazar 2013, p. 185), since the borders of this transnational market do not coincide with the territorial borders of any nation-state. Thus, the World Music market can be considered a field of globe-spanning power, understood as a set of positions that give privileges to some or impede the movement of others, where the circuits of interconnection that structure global sociality can be observed. Analysing the movement of musicians in this market will also provide insights into its underlying power structure, since some of these musicians might be more mobile than others, not only due to the quality of their work, but also because of visa issues, etc.

This paper is divided into three parts. First, I give a brief overview of the literature on musicians' mobility. Then, I present the main characteristics of the World Music market. Lastly, I discuss some results of my empirical research on the careers of musicians in this market.

16.2 Understanding Musicians' Global Mobility

Although globalization makes us more aware of the movement of people across borders, an overview of the literature on the history of music reveals that musicians were rarely sedentary. If we look at developments in the "West", as described by Salmen (1960) in his study on travelling musicians in the Middle Ages, we can see that, since classical antiquity, musicians have taken on roles as court bards and entertaining minstrels, touring with political events or harvest festivals. Sedentariness particularly concerned those connected to the clergy and living in the cities. However, many of them still moved around as gleemen and minstrels, taking part in religious masses, trade fairs, military occasions, or events organized by the nobility. As they belonged to the lowest social class, although their music was appreciated, beyond their performance, their presence was unwelcome. Hence, they were impelled to constantly look for new performance opportunities in other regions. Even if they were hired by a noble, musicians at the time would be sent to tour in other regions in order to expand their repertoires and absorb new influences into their performances. Over the centuries, the institutionalization of musicians' activities in choirs, orchestras, and music academies led to the diversification of their professional fields. At the same time, this also increased opportunities for sedentariness (particularly of music teachers) and for combining periods of sedentariness with tours.

With the development of recording technology from the second half of the 19th century (first, of sound, and then of sound and image), musicians' mobility gained another quality. Musicians can now be mobile by broadcasting their music and image, even if they do not tour very often. Their fame and fans can be spread around the world and connect through the media, thus creating a translocal or virtual scene (Bennett and Peterson 2004)—a transnational space where musicians transit both physically and virtually. However, among musicians, there is no equality of opportunity for media exposure and transnational movement. As Glick Schiller and Meinhof (2011) illustrate, power inequality related to musicians' literacy, knowledge about copyrights, and contacts with members of the music industry are just some of the factors influencing musicians' global mobility. Nevertheless, musicians are still considered to be one of the most highly globally mobile professional groups.

Recently, the analysis of artists' mobility in general was reframed by new research on the creative industries. Lipphardt (2015) summarizes this perspective in three arguments. First, the mobility of those working in the creative industry enables freedom and independence. Second, it fosters intercultural exchange and cosmopolitanism. Third, it promotes creativity and innovation. By emphasizing the positive side only, Lipphardt argues that this perspective tends to obscure the precarious situation of creative professionals in the labour market and the pressure to be mobile in order to engage in new projects and keep working. Moreover, this literature refers to all different occupations in the creative industry, subsuming musicians under this category and so overlooking their distinctive features. Another problem concerns the availability of data that can be used to provide empirical evidence for an analysis of musicians' mobility. As Addison (2006) explains, there is little data on artists' geographical mobility and the prevalence of combining occupations in the cultural sector with commercial activities in other sectors. Moreover, artists may be registered as unemployed, while they are, in fact, still working. Thus, the artistic field is still under-researched in terms of artists' mobility and the effects of globalization on their careers. In order to shed new light on this topic, I will now turn to an analysis of the World Music market.

16.3 The World Music Market

World Music is a politically disputed concept. Originally used to refer to the search for new sources of musical creativity outside Europe with the aim of enriching European musical tradition (Franzen 2005), it has been characterized since its inception by a fundamental divide between "the West" and "the Rest" (Hall 1992). Nevertheless, ethnomusicology adopted the notion (though its use remains controversial) to question this very eurocentrism in musicology's concept of music and to offer an alternative scientific concept to refer to the musics (in the plural) of the world (Nettl 1985; Mendívil 2017). In jazz, World Music means the search for what is common to all human beings and encompasses the work of Yusef Lateef, John Coltrane, and Don Cherry (Berendt 1985). Yet, at the same time, World Music is criticized for perpetuating unequal power relations among musicians from different parts of the world in terms of copyright, control over the studio technology, and exoticism (Voswinkel 1985). In cultural studies, World Music is presented as non-Western music packaged as pop genre and marketed to

Western audiences (Frith 1991), which raises questions about cultural imperialism (Goodwin and Gore 1990). In the music market, World Music is used as a category to designate a special niche normally associated with the music of "the other", breaking with the "West-Rest" divide.[1] Since the 2000s, its use has been disputed (Peres da Silva 2017) and some market actors prefer the term "global music", while academics speak of World Music 2.0 (Burkhalter 2011). Nevertheless, the category World Music is still used in the market.

If we examine World Music as a category in the transnational market, it is possible to trace its origins back to the efforts of 11 independent labels in London in 1987 to establish this category in the record shops through a marketing campaign (Peres da Silva 2016). From this point on, diverse networks of social relations which crossed the borders of nation-states started to interlock with one another in order to build a transnational space for this emerging market. At first, record shops in England and in the USA agreed to create a special section dedicated to World Music. Then, in 1989, since the distribution of physical records was very unequal in this market, radio DJs linked to the *European Broadcasting Union* started the *World Music Workshop* and decided to create the *World Music Charts Europe* based on their radio shows (Fairley 1992). These charts were released in 1991. In the same year, the *Berlin Independent Days* event hosted a special festival of World Music under the label *Worldwide Music Days*, bringing everyone interested in this market together. With new iterations of this festival being held over the following two years, in 1993, organizers of World Music festivals decided to establish the *European Forum of Worldwide Music Festivals* (now the *Forum of Worldwide Music Festivals*) in order to create a forum for exchange about new acts and to plan joint tours in Europe. Together, all these labels, radio DJs, and festival organizers decided to create a special event exclusively for World Music, *WOMEX–The World Music Expo*, which took place for the first time in 1994. Since then, this event has become the central meeting point for representatives of the World Music market globally. It consists of a trade fair, concerts, conferences, network meetings, radio broadcasts, and film screenings, and it brings together around 2,700 professionals every year. In the same year, the book *World Music–The Rough Guide* (Broughton et al. 1994) was published. One of its editors, Simon Broughton, created the magazine *Songlines* in 1998, which

[1] From a European perspective, it could mean "non-European" music. However, from another geographical perspective, World Music could be European. In Brazil, for instance, Charles Aznavour's and in South Africa, Patrizio Buanne's albums are classified as World Music. In this market, the point of reference is always the culture of the person speaking.

alongside *fRoots* (published since 1979), is one of the main magazines specializing in World Music.

These market institutions were established on the basis of common values. An insight into these values can be acquired from the press release (fRoots, n.d.) for the 1987 marketing campaign, the criteria for being part of the charts (Fairley 1992), the articles published in the newsletters of the European Forum of Worldwide Music Festivals, and in the programs of the Worldwide Music Days and WOMEX.[2] Together, actors in the World Music market see themselves as representing the positive side of globalization: they are fostering cultural diversity and, at the same time, they are investing in a market to create economic sustainability for all artists and people involved in the business. Moreover, they respect all cultures, they comply with artists' copyrights, and their work in music studios and at concerts adheres to ethical standards, including rejection of all forms of racism. As they imported all types of music from around the world to present to Western audiences, they considered their work an inversion of the commercial relations that usually prevailed between the "West" and the "Rest", which was essentially focused on making Anglo-American pop musicians famous worldwide (Peres da Silva 2016).

Musicians were initially not aware that their work was being classified in Europe as World Music (Gillett 2009). The organization of the World Music market was in the hands of Western actors—labels, concert and festival organizers, journalists, and radio DJs—who were actively selecting who could and who could not enter this market. In this sense, musicians seemed to be powerless. Yet, they had to accept this classification of their work outside their home country if they intended to move within this emerging transnational space. Over the years, the stable reproduction of World Music market institutions has led to a certain crystallization of the technical and aesthetic features that characterize World Music—a preference for acoustic music, traditional instruments, etc. Musicians interested in being part of this market started adopting these traits, which, in turn, contributed to the formation of World Music as a music genre—a discursively constructed category (Wicke 2004) for "a set of artworks classified together on the basis of perceived similarities" (DiMaggio 1987, p. 441). However, musicians do face barriers to market entry (Le Coz 2019). Thus, their situation in this context can be interpreted as a powerless position in a trade-off between accepting a producer's view or not participating in this global market.

---

[2] This analysis did not take the content of the book *World Music–The Rough Guide* or the magazines *fRoots* and *Songlines* into consideration.

Nevertheless, musicians do have agency and are able to strategically enter and withdraw from the World Music market. The decisions they make concerning more or less global mobility are part of their careers, which are influenced by structural and institutional barriers to the World Music market, but also depend on which countries and world regions they are based in and would like to reach. In this study, I will present some preliminary results regarding musicians' mobility in the World Music market and thus shed some light on these issues.

16.4 Empirical Analysis of Musicians' Global Mobility

In order to identify which musicians move within the transnational space of the World Music market, this study takes participation at WOMEX as a starting point. WOMEX is the world's most important trade fair for World Music. The event involves actors from different countries and world regions meeting to generate business, network, and present their music. It gives musicians a unique opportunity to perform in front of a selected audience of market actors and to make contact with festival organizers, labels, and other relevant institutions that can help to shape their career paths. Besides its global centrality, WOMEX has the advantage of uniting most (if not all) different niches within the global World Music market. An analysis of the event thus provides us with an all-encompassing view of this transnational market.

This five-day event takes place in a different European city every year. The opportunities for artists to perform have increased since the event was first held in 1994. While the early years only featured evening Showcases (a term used to refer to short concerts of 30–45 min open only to the music industry and not the general public), nowadays WOMEX has seven different performance formats. The event begins with an Opening Concert, curated by the organizers (the WOMEX team together with the local organizers). The music program selected by the jury, which musicians can apply to be part of, consists of evening Showcases (five stages), Daycases performed during the day (one stage), and a DJ Summit, which takes place after the Showcases. Parallel to this, there are two formats for sponsored concerts, the programs for which might be designed by someone other than the trade fair organizers: off WOMEX (one stage) and Special Focus (one stage). In the former, concert slots can be sponsored, with the option of composing the program for an entire evening. In the latter, musicians from the region where WOMEX is being held that year are invited to present their work. To close the event, there is an Award Ceremony and, since 2008, there has also been an award for artists, with the winner being given an opportunity to perform (WOMEX 2019).

For the present analysis, we consider the WOMEX editions that took place between 2008 and 2017. Data about the participating musicians were collected from WOMEX programs, the catalogues distributed at the event, the musicians' websites, and from the press (interviews etc.). The analysis focuses on the migration history of the participants in order to acquire information about how mobile the musicians in the World Music market are. To this end, data on changes of domicile across national borders were collected (that is, every time a musician changed their residence address from one country to another). The results presented here are part of an ongoing study and, as such, do have some limitations. It thus follows that the analysis only provides us with an indication of musicians' transnational life.

In the period analysed, there were ten WOMEX editions with a total of 613 concert slots (an average of 61 concerts per year), in which 2,732 musicians participated. Of these, only 88 musicians (3.2%) performed more than once. This points to a very low number of repetitions within the event. Adjusting the data to exclude duplication and musicians for whom there was no available information about place of residence, the sample analysed consisted of 2,403 musicians.

In this study, the analysis of musicians' mobility (pertaining only to migratory movements) is based on the number of changes of domicile in relation to their place of birth. If a musician tells the public that he/she lives in one country and does not mention any other place of residence, he/she is considered a non-migrant. In all other cases, where more than one country of domicile is known, musicians are considered migrants. Our results show that 18% of our sample could be considered migrant, which appears high in comparison to the worldwide population of international migrants of just 3.4% (United Nations 2017). It is interesting to note that, over the years, the rate of migrant musicians performing at WOMEX has been consistently around an average of 18.2%. These figures indicate a high mobility rate among musicians in the World Music market, with one fifth of them changing their domiciles and thus being part of a transnational market. At the same time, the results show that musicians' mobility is not necessarily connected to migration, since the majority of musicians do not actually need to migrate to participate in the World Music market.

If we focus specifically on the group of migrant musicians, we observe that the overwhelming majority (89%) changed their domicile two or three times. Only a tiny group, consisting of 2% of all musicians analysed, reported being extremely mobile, with a history of four or more different places of residence. In addition, we note that just 6.2% of the migrant musicians (1.1% of our sample) can be considered cases of return migration, where the third country of domicile is the same as the first one. In this sense, it is not the change of domicile or the movement back and forth between the same two countries that fosters the creation of the transnational space of the World Music market. Rather, based on these findings, we can hypothesize that the configuration of this transnational market depends more on musicians who frequently tour than on those who are extremely mobile or move back and forth between just two countries.

Considering musicians' last place of residence before their performance at WOMEX shows that they come from 105 different countries. This means that more than half of all existing countries in the world[3] are represented at WOMEX. The countries that sent the largest number of musicians were Spain (208), France (173), Brazil (154), the UK (121), and the USA (100). Together, this group of countries accounted for 30% of the musicians at WOMEX. It is important to remember that this does not necessarily mean that musicians are citizens of these countries, but rather that they were living there at the time of their performance at WOMEX. An analysis of the continents these musicians come from reveals that 53% come from Europe, particularly from Western and Southern Europe. A total of 25% come from the Americas, especially from South America. 11% of the musicians come from Africa, primarily from Eastern and Western Africa, and 10% from Asia, predominantly Eastern and Western Asia. Oceania, Central and Southeast Asia, as well as Central, Northern, and Southern Africa are regions with almost no representation at WOMEX. These data depict a concentration of the World Music market in Europe. However, we should interpret the results of this analysis with caution, since WOMEX only takes place in Europe and is therefore more accessible to musicians based there. More data on other events and participants of the World Music market would enable us to create a more accurate picture of the whole market.

Thus, this analysis indicates that the circuits of interconnection in the World Music market, as represented at WOMEX, are concentrated in Europe and also offer opportunities for migrant musicians. At the same time, the global power structure of this transnational market excludes musicians from certain world regions who face more barriers to participation. The reasons for this exclusion are a potential subject for further researches. In this sense, our study suggests that musicians living in Europe enjoy a higher mobility in the World Music market than those in Oceania, Central and Southeast Asia, and Central, Northern, and Southern Africa, who appear to be more static in the context of the World Music market. Although these data do not allow for an analysis of the micro level of this regime of mobility, the macro structure depicted here points to clear power

[3] According to the United Nations, there are 195 countries in the world.

imbalances between musicians from different parts of the world. Moreover, the reproduction of this market globally seems to be more dependent on musicians' touring schedules than on their migration record. The touring musicians' routes are not arbitrary, but rather embedded in circuits comprised of interlocking networks of social relations, which cross the borders of nation-states and create a transnational space. Identifying how these routes are configured and the preferred circuits are the next steps in this research.

16.5 Final Considerations

With the aim of shedding new light on the under-researched topic of musicians' global mobility and the impact of globalization on their careers, this paper presented some exploratory results from a study that is currently being conducted. In light of the state of the art in the global mobility literature, we proposed an analysis of the circuits of interconnection that structure human sociality worldwide, focusing on a specific field of globe-spanning power: the transnational market for World Music. Embedding this market in the history of music in the "West" revealed that musicians were rarely sedentary. Thus, the effects of globalization on their mobility should be sought in other factors, not only in their movements across borders. The literature also points to the potential impact of the media and the creative economy on musicians' careers, but a lack of reliable data still hinders an accurate understanding of their global mobility. Against this backdrop, the current study aims to provide new insights on this topic.

Our analysis indicated that the transnational market for World Music, as represented at WOMEX (2008–2017), offers opportunities for migrant musicians, while it is probably sustained by the regular tours of musicians with no history of migration. Moreover, even if the structure of the event does not correspond with any specific geographical or political border, our results indicate privileged access to musicians living in Europe or, more specifically, in five countries: Spain, France, Brazil, the UK, and the USA. This demonstrates that the global power structure of the World Music market is not so different from other imbalances between the Global North and Global South. The expected high mobility of musicians coming from Global South could not be confirmed by our study. Despite the compression of time and space emphasized in some globalization studies, our study provides evidence that musicians from certain world regions have little access to transnational networks and are more sedentary. Further research would certainly foster a better understanding of musicians' careers and mobility in fields of globe-spanning power.

References

Addison, Tony. 2006. The international mobility of cultural talent. WIDER Research Paper, No. 2006/108. Helsinki: UNU WIDER.

Appadurai, Arjun. 1990. Disjuncture and Difference in the Global Cultural Economy. Theory, Culture & Society 7: 295–310. https://doi.org/10.1177/026327690007002017.

Bennett, Andy and Richard A. Peterson. 2004. Introducing Music Scenes. In Music scenes: Local, translocal and virtual, ed. Andy Bennett and Richard A. Peterson, 1-15. Nashville: Vanderbilt University Press.

Berendt, Joachim E. 1985. Über Weltmusik. Jazz Podium, March 24, 8–13.

Broughton, Simon, Mark Ellingham, David Muddyman Richard Trillo. 1994. World Music: The Rough Guide. New York: Penguin.

Burkhalter, Thomas. 2011. Weltmusik 2.0: Zwischen Spass- und Protestkultur. Norient. http://norient.com/academic/weltmusik2-0/. Accessed 8 November 2019.

Dados, Nour and Raewyn Connell. 2012. The Global South. Contexts 11(1): 12–13. https://doi.org/10.1177/1536504212436479.

DiMaggio, Paul. 1987. Classification in Art. American Sociological Review 52: 440–455. https://doi.org/10.2307/2095290.

Fairley, Jan. 1992. European world music charts. Popular Music 11(2): 241–244. https://doi.org/10.1017/S0261143000005055.

Favell, Adrian, Miriam Feldblum, and Michael P. Smith. 2007. The Human Face of Global Mobility: A Research Agenda. Society 44(2): 15–25. https://doi.org/10.1007/BF02819922.

Franzen, Stefan. 2005. Weltmusik. In Musiklexikon: In vier Bänden, vol. 4, ed. Harald Hassler, 737–740, 2nd edn. Stuttgart: Metzler.

Frederickson, Jon, and James F. Rooney. 1990. How the Music Occupation Failed to Become a Profession. International Review of the Aesthetics and Sociology of Music 21(2): 189–206. https://doi.org/10.2307/837023.

Frith, Simon. 1991. Critical Response. In Music at the margins: Popular music and global cultural diversity, ed. Deanna C. Robinson, Elizabeth Buck and Marlene Cuthbert, 280–287. Newbury Park, Calif.: Sage Publications.

fRoots. n.d. Press "Release 02–World Music". https://frootsmag.com/press-release-02-world-music. Accessed: 14 May 2019.

Giddens, Anthony. 1990. The consequences of modernity. Cambridge: Polity Press.

Gillett, Charlie. 2009. Playlist for week beginning 31 January 09. https://web.archive.org/web/20100325114855/. http://www.charliegillett.com/playlistnewformat.php?date=31January09. Accessed: 18 April 2019.

Glick Schiller, Nina. 2010. Transnationality. In A companion to the anthropology of politics, ed. David Nugent and Joan Vincent, 448–467. Malden, MA: Blackwell.

Glick Schiller, Nina, and Noel B. Salazar. 2013. Regimes of Mobility Across the Globe. Journal of Ethnic and Migration Studies 39(2): 183–200. https://doi.org/10.1080/1369183X.2013.723253.

Glick Schiller, Nina, and Ulricke H. Meinhof. 2011. Singing a New Song?: Transnational Migration, Methodological Nationalism and Cosmopolitan Perspectives. Music and Arts in Action 3(3): 21–39.

Goodwin, Andrew, and Joe Gore. 1990. World Beat and the Cultural Imperialism Debate. Special Issue "North and South Relation in the New World Order". Socialist Review 20(3): 63–80.

Guilbault, Jocelyn. 1996. Beyond the "World Music" Label: An Ethnography of Transnational Musical Practices. Beitrag zur Konferenz Grounding Music. PopScriptum. https://www2.hu-berlin.de/fpm/textpool/texte/guilbault_beyond-the-world-wusic-label.htm. Accessed: 8 November 2019.

Hall, Stuart. 1992. The West and the Rest: Discourse and Power. In Formations of modernity, ed. Stuart Hall and Bram Gieben, 275–332. Cambridge: Polity Press.

Harvey, David. 1989. The condition of postmodernity: An enquiry into the origins of cultural change. Cambridge, Mass.: Blackwell.

Le Coz, Sandrine. 2019. From desire for recognition to desire for independence: World music filtered in the market economy. In Music Practices Across Borders: (E)Valuating Space, Diversity and Exchange, ed. Glaucia Peres da Silva and Konstantin Hondros, 105–126. Bielefeld: transcript. https://doi.org/10.14361/9783839446676-005.

Lipphardt, Anna. 2015. Kein Zugang zum Backstage-Bereich? Methodologische Überlegungen zu biographischen Interviews mit hochmobilen Künstlerinnen und Künstlern. In Räumliche Mobilität und Lebenslauf, ed. Joachim Scheiner and Christian Holz-Rau, 57–76. Wiesbaden: Springer. https://doi.org/10.1007/978-3-658-07546-0_4.

Mendívil, Julio. 2017. Musikethnologie. In Global Pop: Das Buch zur Weltmusik, ed. Claus Leggewie and Erik Meyer, 35–42. Stuttgart: J.B. Metzler. https://doi.org/10.1007/978-3-476-05480-7_4.

Nettl, Bruno. 1985. The Western impact on world music. Change, adaption, and survival. New York: Schirmer Books.

Peres da Silva, Glaucia. 2016. Wie klingt die globale Ordnung. Die Entstehung eines Marktes für World Music. Wiesbaden: Springer. https://doi.org/10.1007/978-3-658-11813-6.

Peres da Silva, Glaucia. 2017. Weltmusik: Ein politisch umstrittener Begriff. In Global Pop: Das Buch zur Weltmusik, ed. Claus Leggewie and Erik Meyer, 9–16. Stuttgart: J.B. Metzler. https://doi.org/10.1007/978-3-476-05480-7_1.

Pries, Ludger. 1996. Transnationale soziale Räume: Theoretisch-empirische Skizze am Beispiel der Arbeitswanderungen Mexico - USA. Zeitschrift für Soziologie 25(6): 456–472. https://doi.org/10.1515/zfsoz-1996-0603.

Salmen, Walter. 1960. Der fahrende Musiker im europäischen Mittelalter. Kassel: Johann Philipp Ninnenthal.

United Nations. 2017. International Migration Report 2017. New York: United Nations.

Urry, John. 2012. Social networks, mobile lives and social inequalities. Journal of Transport Geography 21: 24–30. https://doi.org/10.1016/j.jtrangeo.2011.10.003.

Voswinkel, Stephan. 1985. Über die Vielfalt der Musik: Zu Berendts Artikel "Über Weltmusik". Jazz Podium, 24: 10–11.

Wicke, Peter. 2004. Über die dieskursive Formation musikalischer Praxis: Diskurs-Strategien auf dem Feld der populären Musik. In Festschrift Prof. Dr. Rienäcker zum 65. Geburtstag, ed. Stephan Aderhold, 163–174. http://festschrift65-rienaecker.de/Seite2.htm. Accessed: 8 November 2019.

Wimmer, Andreas, and Nina Glick Schiller. 2002. Methodological nationalism and beyond: Nation-state building, migration and the social sciences. Global Networks 2 (4): 301–334. https://doi.org/10.1111/1471-0374.00043.

WOMEX. 2019. http://www.womex.com/. Accessed: 8 November 2019.

List of authors and German abstracts

Lisa Gaupp, Alenka Barber-Kersovan and Volker Kirchberg

Biographies

Prof. Dr. Lisa Gaupp is Full Professor of Cultural Institutions Studies at the University of Music and Performing Arts Vienna (mdw). She was Interim Professor for Cultural Sociology at Leuphana University of Lüneburg, studied cultural studies (*Kulturwissenschaften*) at the Universities of Lüneburg and Barcelona and holds a PhD in ethnomusicology from Hanover University of Music, Drama and Media. Lisa's scholarly work has been supported by numerous grants, including from the German National Scholarship Foundation, and she received several awards for her publications, such as the Best Research Article Award 2020 (Journal of Cultural Management and Cultural Policy) for "Curatorial Practices of the 'Global'", the Best Early Career Research Paper Award in Sociology of Culture (ESA) for "Festival Curators as Gatekeepers for Sociocultural Diversity" and the SOPHIA Prize for University Graduates (Soroptomists International) for her dissertation. She is co-editor of among other "Diversity and Otherness. Transcultural Insights into Norms, Practices, Negotiations" (De Gruyter) and of a book series on "Urban Music Studies" (Intellect). Lisa has lived in the USA, Haiti, Guatemala and Spain and was executive manager of the 2009 Hannover International Violin Competition (Stiftung Niedersachsen).

Hon. Prof. Dr. Alenka Barber-Kersovan studied historical musicology, systematic musicology, psychology and aesthetics at the Universities of Ljubljana, Vienna and Hamburg, where she finished her studies with a thesis on Slovenian Punk and its contribution to the erosion of the socialistic value system. She worked as a music

L. Gaupp et al. (eds.), *Arts and Power*, Kunst und Gesellschaft,
https://doi.org/10.1007/978-3-658-37429-7

therapist at the Psychiatric Clinic in Ljubljana, program director of the Slovenian Musical Youth and scientific officer at the Institute for Sociology of Music in Vienna, Music Academy in Hamburg and the Institute for Music Education in Hamburg. Further she was a lecturer at the Institute for Musicology of the University of Hamburg and at the University of Music and Performing Arts in Vienna. Currently she is teaching sociology of music at the Leuphana University in Lüneburg.

Barber-Kersovan held for a long time the position of the managing director of the Gesellschaft für Popularmusikforschung/German Society for Popular Music Studies and was also active in music politics. Her main research interests are popular music, music and gender, music and politics and urban music studies. She is also a co-editor of the publication series Urban Music Studies of the Intellect Publishers.

Prof. Dr. Volker Kirchberg studied at the Institute of Sociology at the University of Hamburg. Following his diploma in 1986, he worked as a research assistant in urban sociology at this university. From 1988 to 1991, he was a graduate student at the Institute for Policy Studies at Johns Hopkins University and the University of Baltimore. In the following year, he received his doctorate from the Department of Social Sciences at the University of Hamburg, with a study on cultural behavior of Baltimore's urban population. As a German Marshall Fund post-doctoral fellow, he then conducted research on cultural sponsorship. As director of the Hamburg "Basica" social research institute in Hamburg he conducted several comprehensive audience studies for major museums in Germany. In 2003 he habilitated at the Institute of Sociology at the Free University of Berlin with an empirical thesis on "Social Functions of Museums". From 2000 to 2004, he held an Assistant Professor position in sociology at William Paterson University in New Jersey. In 2005, he became University Professor of Sociology of the Arts at Leuphana University in Lüneburg. He has been a board member of several international bodies in the sociology of the arts.

Power and Policies in and by the Arts—Introduction

Lisa Gaupp

Abstract

In dem Einleitungskapitel zu diesem Buch wird zunächst allgemein die soziologische Beschäftigung mit Konzepten der Herrschaft und der Macht in der Gesellschaft vorgestellt. In den folgenden Kapiteln werden dann (a) die Macht gesellschaftlicher Strukturen über die Künste und (b) die Macht der Künste über gesellschaftliche Strukturen abgehandelt. Die neueren Begriffe zum Verständnis dieser Wechselwirkungen stammen dabei aus der postkolonialen und intersektionalen Soziologie, da Ungleichheiten in den Künsten aus mehrfachen und verwobenen Diskriminierungsprozessen entstehen. Soziologische Analysen der Künste versuchen, diese machtbeladenen Prozesse aufzudecken, etablierte Strukturen zu dekonstruieren und künstlerische Praktiken auch über Konzepte der Dekolonisierung und eines queeren Bewusstseins als Machtausübung zu erkennen. Dabei ist es wichtig, zwischen Macht und Herrschaft zu unterscheiden. Entweder werden die Künste von kunstexternen Herrschern und ihren institutionalisierten Strukturen beherrscht, oder die Künste sind in der Lage, den Spieß umzudrehen und über ihren Legitimitätsvorsprung selber Herrschaft auszuüben, indem sie zum Beispiel Geschmack prägen. Die Politik der Kulturproduktion ist dagegen als Beziehung von Künsten und Macht zu untersuchen. Im nächsten Abschnitt dieses Kapitel wird dann in die Struktur des Buches eingeführt, mit Kapiteln in den drei Teilen „Etablierung und De-Etablierung von Macht in den Künsten", „Künste und die Macht sozialer Strukturen" und „Künste und die Dominanz politischer (Un-)Ordnung". Die wechselseitigen Perspektiven „Macht in den Künsten" und „Macht durch die Künste" werden durch ein zweidimensionales Modell dargelegt, dessen erste Dimension durch „Herrschaft versus Macht" und dessen zweite Dimension durch „Strukturzwang versus Handlungsfähigkeit" gekennzeichnet sind; alle Artikel dieses Buches können in diesem Koordinatensystem eingeordnet werden. Die gesellschaftlichen Funktionen der Kultur und der Künste können somit in einem Kräftefeld von Herrschaft (Struktur) und Macht (Agency) positioniert sind, unabhängig, ob sie für negative (hegemoniale, totalitäre, ungerechte) oder positive (gleichberechtigte, basisdemokratische, ausgleichende) gesellschaftliche Zwecke stehen.

Repurposing the Black Arts Movement on the South Side of Chicago

Julia Rothenberg

Abstract

In diesem Kapitel wird exemplarisch untersucht, wie sich für die städtische Wirtschaft zentrale kunst- und kulturgeleitete urbane Rebranding-Projekte in afro-amerikanischen Communities vollziehen. Unter der Leitung der University of Chicago und in Zusammenarbeit mit Theaster Gates, einem afro-amerikanischen Künstler, Pädagogen, Unternehmer und gebürtigen Chicagoer von der South Side, der sich eine starke Position als kreativer Schrittmacher und Kunstimpresario an der Universität erarbeitet hat, werden rund um die Universität und in den benachbarten verarmten Schwarzen Gemeinden eine Reihe solcher Projekte durchgeführt. Ich stelle diese Entwicklungen in den Kontext der Geschichte der Segregation in Chicago, der anhaltenden Rolle der Universität als wichtiger Partner der Wachstumskoalition und der Debatten über kreative Entwicklungsstrategien in einer neoliberalen Stadt. Das vermutlich wichtigste Unterfangen dieses Kapitels ist der Versuch zu verstehen, wie und zu welchem Zweck die Resignifikation der hochpolitischen Kunst des Black Arts Movement in der neuen Ära des Marktfundamentalismus, der seine ideologischen Grundlagen interessanterweise der Chicago School of Economics verdankt, funktioniert. Durch die Gegenüberstellung des städtischen Kontextes, in dem das Chicago Black Arts Movement entstand, und seiner heutigen Umsetzung hoffe ich, insbesondere die rassifizierten Dimensionen der kreativen Stadtentwicklungsprozesse zu beleuchten.

Biography

Julia Rothenberg is an Associate Professor of Sociology at Queensborough Community College, City University of New York, where she teaches courses in Introductory Sociology, Sociology of the Arts and Urban Sociology. Her research addresses questions in the sociology of art and culture and urban sociology. She has published articles on the arts, gentrification and social inequality in West Chelsea in New York and a book on the Sociology of the Arts. Her recent research includes an investigation of the role of the university in tandem with arts and arts entrepreneurialism in gentrifying historically Black neighborhoods on the South Side of Chicago a project on green development in Long Island City, Queens. She is also working with students in her Sociology of the Arts class on the impact of the Covid pandemic on community based arts institutions in New York City. Julia earned an MFA in Painting for Queens College and a PhD in Sociology from the Graduate Center.

Regulatory Power and the Power of Regulations. ArtReview's „Power 100" List's Understanding of Power in Contemporary Art

Paul Buckermann

Abstract

Warum gibt es ein Ranking der einflussreichsten Akteure der zeitgenössischen Kunst? Dieser Aufsatz fragt nach paradigmatischen Weltsichten und trägt so zu einem tieferen Verständnis von Epistemologien in Kunstwelten bei. Eine qualitative Rekonstruktion des Rankings *Power 100* zeigt, wie die Liste in sich Sinn ergibt, weil sie sich konsistent auf eigene Vorstellungen über die gesellschaftliche Rolle der Kunst, ihre internen Mechanismen und die Funktion von Macht in ihr bezieht. Anhand soziologischer Theorien von Macht (Weber, Bourdieu, Luhmann) wird gezeigt, dass die Liste eine interne Sicht von Kunstwelten darstellt. Die paradigmatische Weltsicht der Liste kombiniert einen tiefen Glauben an positive gesellschaftliche Potenziale der Kunst mit einem proto-soziologischen Verständnis von symbolischen (Macht-)Strukturen einer autonomen Kunst. In Verbindung mit Forschung zu Bewertung und Vergleichen haben die Resultate Implikationen für die Soziologie der Künste als auch der Rankingforschung.

Biography

Dr. Paul Buckermann is a postdoctoral researcher at Thematic Research Network "Validity of Knowledge" at Heidelberg University. His sociological research focuses on culture, knowledge, technology, and science. He conducted research at *NCCR-eikones* Basel/Luzern, at *Research Training Group Automatisms* at Paderborn University, and the sociological department at University of Lucerne. Recent publications include "Ranking Art: Paradigmatic Worldviews in the Quantification and Evaluation of Contemporary Art". In: *Theory, Culture & Society* (2021); *Die Vermessung der Kunstwelt. Quantifizierende Beobachtungen und plurale Ordnungen der Kunst*. Velbrück Wissenschaft (2020).

How Organisations Exercise Power through Peer Review: A Case Study from Contemporary Art Music

Annelies Fryberger

Abstract

Gatekeeper üben Macht aus, indem sie den Zugang zu Ressourcen in ihren jeweiligen Bereichen gewähren und diese Ressourcen normalerweise auf der Grundlage von Qualitätsbewertungen verteilen. Organisationen zur Finanzierung von Kunst fungieren als Gatekeeping-Organisationen und nutzen die Bewertung, um zu bestimmen, wer in einen bestimmten Bereich integriert und wer ausgeschlossen wird. Im vorliegenden Fall handelt es sich um eine amerikanische Organisation, deren Aufgabe es ist, zeitgenössische Kunstmusik zu fördern. Diese Art von Musik wird, wie auch im akademischen Bereich, in der Regel durch ein Peer-Review-Verfahren bewertet. Die Organisation muss die widersprüchlichen Prioritäten ihrer öffentlichen und privaten Geldgeber ausgleichen: Erstere fordern Offenheit und Vielfalt, während letztere kulturelle Hierarchien reproduzieren. Um diesen inhärenten Konflikt zu umgehen hat sich die Organisation dafür entschieden, nicht öffentlich zu definieren, welche Art von Musik sie unterstützt. Indem sie jedoch das Peer-Review-Verfahren auf eine bestimmte Art und Weise steuert, kann die Organisation dafür sorgen, dass das Ergebnis der Bewertung mit der Definition zeitgenössischer Musik übereinstimmt, die von ihren privaten Spender*innen vertreten wird. Anhand dieses Falles wird erkennbar, wie Kunstförderorganisationen die Macht ausüben können, Möglichkeiten und Grenzen von Kunstproduktionsbereichen zu definieren.

Biography

Annelies Fryberger is a sociologist of music. She obtained her PhD in 2016 at the EHESS (Paris, France) where she was a member of the Center for Research on Arts and Language (CRAL-EHESS) and the research team "Analysis of Musical Practices" at the IRCAM. Her research has been published in Poetics (2020), *Figures de l'art,* N. XXX (2015), *Circuit. Musiques contemporaines* (2016), *Proteus* (2018), among others. She was previously a DAAD postdoctoral researcher (University of Hannover), and conducted a project on sound art in the context of a postdoctoral research position with the Labex CAP (IRCAM/Pompidou Center). She is currently the Executive Director of the Society for the Advancement of Socio-Economics (SASE).

Art Perception and Power. A Plea for Relational Sociological Aesthetics

Nina Tessa Zahner

Abstract

Kunstwahrnehmen führt Fragen nach Partizipation, Gerechtigkeit und der Legitimität von Weltanschauungen mit sich und ist als solches immer Gegenstand von Machtkämpfen. Auch die Forschung zum Kunstwahrnehmen präferiert implizit oder explizit bestimmte Weltbilder und Kunstbegriffe gegenüber anderen und ist so tief in Konstellationen der Macht verstrickt. Der Beitrag diskutiert zunächst aktuelle Entwicklungen der Kunstwahrnehmungsforschung vor diesem Hintergrund und fragt, wie unterschiedliche Forschungstraditionen bestimmte Konzeption von Kunst und Kunstbetrachtung anderen vorziehen und diese in ihre Methodologien integrieren. Anschließend zeigt er auf, wie das feldanalytische Forschungsinstrumentarium Pierre Bourdieus helfen kann, um Prozesse der Machtreproduktion im Kontext der Kunstbetrachtung zu erfassen und sichtbar zu machen und ergänzt dieses abschließend mit Simmels Konzeption des Zusammenspiels von Form und Inhalt, um den Antisensualismus Bourdieus, der sich gerade für die Wahrnehmung zeitgenössischer Kunst als wenig hilfreich erweist, zu überwinden.

Biography

Zahner, Nina Tessa, Dr., is Professor for Sociology at The State Academy of the Arts in Düsseldorf Germany. Her focus of work lies in the Sociology of the Art, the Sociology of Perception and the Senses, the Sociology of Art Markets and in Relational Sociology. Her publications include: "*Wahrnehmen als Soziale Praxis. Künste und Sinne im Zusammenspiel*", ed. Christiane Schürkmann and Nina T. Zahner (Wiesbaden: Springer; 2021), "*Begegnung mit dem Materiellen in Architekturgeschichte und Soziologie*", ed. Cornelia Escher and Nina T. Zahner

(Bielfeld: trancript; 2021), "*The Economization of the Arts and Culture Sector in Germany After 1945*" in: Victoria D. Alexander et al. (eds): "*Art and the Challenge of Markets I*" (Cham: Springer, p. 95–124; 2018), "*Die neuen Regeln der Kunst. Andy Warhol und der Umbau des Kunstbetriebs im 20. Jahrhundert*" (Frankfurt am Main: Campus; 2006).

Art Organisations and Their Visitors. About Standardisation Methods of Non-members

Christian Steuerwald

Abstract

Die vorliegende Studie untersucht, wie es Kunstorganisationen gelingen kann, Einfluss auf die Verhaltensweisen ihrer Besucher zu nehmen. Wie die empirischen Untersuchungen nachweisen, haben Kunstorganisationen hierfür nur limitierte Möglichkeiten. Vielmehr sind es vor allem Nebenfolgen der Sozialbereiche Familie und Schule, die die Verhaltensweisen der Besucher in Kunstorganisationen prägen, sodass sich auffallend Ähnlichkeiten in den Verhaltensweisen der Besucherinnen und Besucher beobachten lassen.

Biography

Dr. Christian Steuerwald is currently employed at the Faculty of Sociology of Bielefeld University, editor of the series: Kunst und Gesellschaft as well as co-speaker of the working group: Soziologie der Künste. Studied sociology, political science and law in Mainz from 1995 to 2002. 2010 PhD in sociology with a thesis on the body and social inequality. Subsequently worked at the Universities of Mainz, Trier and Oldenburg.

Equality, History, Tradition: Gender-Political Issues in the Oxford Collegiate Choral Scene

Sigrún Lilja Einarsdóttir

Abstract

Der Fokus dieses Kapitels liegt auf dem Status und den Erfahrungen von Sängerinnen innerhalb der historisch männerdominierten akademischen Chortradition der Universität Oxford. Die akademische Chortradition steht in direktem Zusammenhang mit der liturgischen Chortradition innerhalb der anglikanischen Kirche, die bis in die frühen 1990er Jahre Mädchen und Frauen ausschloss. Bis 1850 waren die Universität Oxford und ihre Colleges in erster Linie klosterähnliche religiöse Einrichtungen, in denen die (ausschließlich männlichen) Stipendiaten nicht heiraten durften und ein zölibatäres Leben führen mussten. Seit den späten 1870er Jahren wurde Frauen erlaubt, in Oxford Kurse zu besuchen, jedoch schloss die Universität Frauen als Stipendiatinnen und Studentinnen bis 1920 aus. Bis Anfang der 1990er Jahre, als erstmals auch Mädchen als Chorsängerinnen zugelassen wurden, war das Musikleben an den Colleges und auf den Bühnen Oxfords historisch bedingt ausschließlich männlich geprägt: Männern sangen in den hinteren (Tenöre, Bässe und Countertenöre) und Sängerknaben in den vorderen Reihen, was die Tradition der Liturgie in der anglikanischen Kirche widerspiegelt. In den letzten Jahrzehnten haben die Oxforder Kollegiumschöre mit der Einrichtung gemischter Chöre einen drastischen Wandel erfahren. In diesem Beitrag werden die Ergebnisse einer musiksoziologischen Studie über die Chortradition der Oxforder Kollegienchöre vorgestellt, wobei der Schwerpunkt auf drei Hauptpfeilern liegt. Erstens die Frage der Ästhetik und des Erhalts rein männlicher Chöre als kulturelles Phänomen, zweitens die widersprüchliche Stellung der weiblichen Soprane und Altstimmen gegenüber den Knabenstimmen und Countertenören und drittens die institutionell bedingte geschlechtsspezifische Diskriminierung und der Mangel an Aufführungs-, Ausbildungs- und Karrieremöglichkeiten für Frauen innerhalb der Chor-Traditionen an den Colleges und Kathedralen.

Biography

Sigrún Lilja Einarsdóttir is associate professor at the Deparment of Social Science, Bifröst University, Iceland and part time lecturer at Reykjavik University. Sigrún holds a PhD in Sociology from University of Exeter (England) where her supervisor was Professor Tia DeNora. In 2016–2017, Sigrún was a Marie Sklodowska-Curie fellow at the Faculty of Music, University of Oxford. She has published various articles about her research in cultural management, gender studies and diversity management in international journals and presented her work in many international conferences.

"When she talks, I hear the revolution". Pop Feminism between Appropriation and Subalternity

Magdalena Protte

Abstract

Wenn wir über Popfeminismus sprechen, fallen vermutlich zunächst Worte wie: laut, phrasenhaft, Glitzer, Pink, entpolitisiert, sinnentleert, oberflächlich. Vor allem aus akademischer Perspektive wird Popfeminismus oft als nicht kritisch genug abgetan und der Vorwurf wird schnell laut, dass er feministischen Kämpfen sogar abträglich sei. Dieser Artikel nutzt Theorieansätze von Antonio Gramsci, um gegen diese Sichtweise zu argumentieren. Popfeminismus wird nach diesen Ansätzen als inhärenter Teil der Kräfteverhältnisse innerhalb eines Staates betrachtet. Er spielt daher eine nicht zu unterschätzende Rolle im Kampf um Hegemonie innerhalb der Zivilgesellschaft und kann den Bezug zum Alltagsverstand subalterner Gruppen herstellen. Dadurch kann Popfeminismus subversiven Charakter entwickeln – im Mindesten kann er Brücken zwischen Theorie und Praxis schlagen, ohne welche sich Feministische Theorie auf Dauer nicht wird durchsetzen können.

Biography

Magdalena Protte studied Political Science in Marburg. Her research focuses on the intersections of politics, civil society, and popular culture. She wrote her Bachelor Thesis about the Gramscian understanding of pop feminism and about the importance of the socialisation of critical truths, especially outside of academia. After reading and writing a lot about translating theory into praxis, she decided to focus on the latter and dropped out of her postgraduate classes in Political Theory in Frankfurt am Main. She is now pursuing a career in science communication and political education.

Female bodybuilding and patriarchal civilization. The intrusion of a practice in sport into artistic fields and visual culture

Isabel Fontbona Mola and Ulf Wuggenig

Abstract

Bodybuilding wurde von Loïc Wacquant (1995) als eine Praxis von „Agenten" charakterisiert, „deren ultimatives Ziel auf die Skulpturalisierung ihrer Körper für ritualisierte, kompetitive Darstellungen hinausläuft, und zwar für öffentliche Präsentationen, in denen sie ihre Muskulatur auf einer Bühne präsentieren". Als solche erlangte sie bereits in den 1970er Jahren gesteigerte Aufmerksamkeit auch Feld der visuellen Kunst.

Zu dieser Zeit wurde Arnold Schwarzenegger nicht nur von einem bereits zu dieser Zeit viel bewunderten Avantgarde-Künstler des Postmodernismus fotografisch porträtiert, nämlich Robert Mapplethorpe (1976). Die „österreichische Eiche" posierte im Februar 1976 zudem im New Yorker Whitney Museum of American Art vor rd. 2400 Besucher*innen aus dem Kunstfeld. Bei dieser Gelegenheit erklärte er auch, dass er seinen Körper als „Kunstmedium" verstehe.

In seiner Betrachtung von soziologischen Zugängen zu Praktiken der ostentativen muskelorientierten Körperpräsentation behandelte Pierre Bourdieu das Thema kurz in „La Distinction" (1979). Er interpretierte sie allerdings ausschließlich als Spielarten von Sport. Außerdem als eine typisch männliche Praxisform der unteren sozialen Klassen.

Bodybuilding war jedoch zu dieser Zeit in Kalifornien bereits als weibliche Praxis mit teilweise anderer soziale Konnotation und ersten überregional beachteten Wettbewerben aufgekommen. In den 1980er und 1990er Jahren fand Bodybuilding weiterhin Aufmerksamkeit im künstlerischen Feld – ein gutes Beispiel dafür gibt Andy Warhols Abbildung von Schwarzenegger auf dem Cover seines Magazins „Interview" (1985) ab.

Mit schrittweise stärkerer Inklusion von Frauen und der Ausdehnung dieser Art von Selbstpräsentation über das Sportfeld hinaus – wofür wiederum ein Film und ein Buch von Mapplethorpe (1984), welche ex-Kunststudentin Lisa Lyon Aufmerksamkeit verschafften, ein Beispiel abgeben – wurde die Praxis zunehmend auch als eine queere Überschreitung von konventionellen hetero-normativen Geschlechterstereotypen interpretiert. In jüngerer Zeit zählte er/sie/es

Heather Cassils dann zu jenen Transfiguren, die sich als Künstler*innen einen Namen über waghalsige Konstruktionen ihres Körpers machten.

Der Essay zu dem eine Person wie Isabel Fontbona Mola beiträgt, die auf eine erfolgreiche Karriere im Bereich des kompetitiven Bodybuilding blicken kann, zeigt ein spezielles Interesse an Prozessen der Ausdehnung und Aneignung des Muskelaufbaus – "pumping iron" – in Feldern jenseits des Sports. Er wendet sich deshalb auch den Feldern der visuellen und performativen Künste sowie des Designs und der Unterhaltung zu. Auf der anderen Seite behandelt er Importstrategien von Praktiken und Diskursen welche sich schließlich in umgekehrte Richtung bewegten: von künstlerischen Feldern in die von Sport und Unterhaltung.

Indem er sich auf Bodybuilding von Frauen und Transakteur*innen konzentriert wirft er schließlich auch die Frage auf, wie diese Angriffe auf die Reproduktion von Geschlechter-Binarität bzw. von „männliche Herrschaft“ (Bourdieu) aus einer politischen Perspektive zu beurteilen sind. In theoretischer Sicht versucht er zu explorieren, was eventuell für einen Diskurs zu gewinnen ist, zu dem bislang primär Autor*innen aus den Feldern von Cultural Studies, Geschlechterstudien, feministischer Theorie, Sportstudien und Medizin sowie poststrukturalistischer Theorie beigetragen haben, wenn eine feldtheoretische Perspektive, die an Pierre Bourdieu und Loïc Wacquant anknüpft, einbezogen wird.

Biographies

Isabel Fontbona Mola, MA is a Ph.D. candidate in the Department of Art History at the University of Girona (Spain). She has a BA in Philosophy and a BA in Art History, as well as a MA in Humanity Research. She is currently completing her dissertation on the artistic and gendered interventions enabled by female and trans bodybuilding practices. The foundation of her research lays out in gender studies, sociology, art theory, philosophy, queer theory, body modification, and sports. The dissertation involves theorizing body identity, representation, performance, and gender issues. In this academic frame, she also gives lectures and performative-conferences regarding her research in several university contexts around Europe and beyond.

She has been imparting some classes internationally: at L-Università ta' Malta, Malta (November 2019); at University of Alberta, Edmonton, Canada (2018/19), where she developed her doctoral research. And at Leuphana University, Lüneburg, Germany (2018) for the seminar Building the body, transgressing beauty standards, doing gender in the Faculty of Humanities and Social Sciences.

Fontbona is also a natural bodybuilder competitor—second place in the World Championship WNBB (2017), third place in the World Championship UIBBN (2015), second place in the European Championship UIBBN (2014)—and a performance artist as well. She explores the malleability of the body with her own skin, through bodybuilding, with the intention to question the binary gender categorization and body stereotypes, in some cases challenging herself to the limits of the endurance of her body.

Ulf Wuggenig, Professor of Sociology of Art, did his PhD at Vienna University after studies of sociology, philosophy, economics and political science in Vienna, and his habilitation in sociology at University of Erlangen-Nürnberg, Germany. He has been teaching in several universities and art schools in Austria (Vienna), Germany (Hanover, Nürnberg, Osnabrück, Hildesheim, Lüneburg) and Switzerland (Zurich). At Leuphana University, Germany, where he is also based, he is member of the Institutes of Philosophy and Sciences of Art as well as of Sociology and Cultural Organization. There has been Dean of the Faculty of Humanities and Social Sciences from 2014 to 2022. Among his publications are „Comments on Chan and Goldthorpe. Pitfalls in testing Bourdieu's homology assumptions using mainstream social science methodology" in: Poetics 35, 2007, 306–316; with Peter Mnich: "Explorations in Social Spaces. Gender, Age, Class Fractions and Photographical Choices of Objects," in: D. Robbins (ed.) (2000): Pierre Bourdieu. Vol. 4, London, 342–363; and with Gerald Raunig and Gene Ray (eds.) (2011): Critique of Creativity. Precarity, Subjectivity and Resistance in the Creative Industries. London: MayFlyBooks.

The Development of Blues in the US-American Field of Musical Production at the Turn of the 20th Century

Christoph Behnke

Abstract

In diesem Artikel wird versucht, bestimmte Aspekte der Geschichte von Blues Musik mit einem feldtheoretischen Ansatz zu analysieren. Es wird gezeigt, dass die Blues Musik - insbesondere in der Phase des „classic Blues" in den 1920er Jahren - in verschiedene Subfelder der US-amerikanischen kulturellen Produktion hineinwirkte, sei es in Form von „Race-Records" im Segment der populären

Musik oder im Rahmen der „Harlem Renaissance“, die im literarisch-künstlerischen Feld angesiedelt war. Grundlegend für alle Segmente der kulturellen Produktion in den USA der 1920er Jahre ist die strukturelle Wirkung der colour-line als Folge der rassistischen Segregation in der US-amerikanischen Gesellschaft. Zu den beschriebenen Feldeffekten gehört neben der musikalischen colour line mit ihren Ausschlussmechanismen und der damit zusammenhängenden kulturellen Appropriation jedoch auch die Generierung von "racial pride" und die Erfindung einer Musik, die im Verlauf des 20. Jhdt. von enormen Einfluss sein sollte. Es wird deshalb vorgeschlagen, für den von Bourdieu als heteronom bezeichneten Pol des Feldes der kulturellen Produktion eine erweiterte Perspektive einzunehmen.

Biography

Dr. Christoph Behnke worked at the Leuphana University in the fields of aesthetics and visual studies as well as cultural organization. His work is focused on the sociology of art and cultural economics, as well as cultural politics. He publishes in the fields of corporate collecting, the artistic field, cultural politics, and the role of artists in the creative industries, focusing on theoretical studies of funding the arts and the sociology of curation. Diverse empirical art studies include the audience in contemporary art, Expo 2000, Ruhr.2010 ('2–3 Straßen' by Jochen Gerz), and cultural sponsoring. He has been the project leader in the major EU project 'Innovation Incubator' on promoting regional art markets (KIM), concluded in 2015 with the publication „Art in the Periphery of the Center“ (Sternberg Press). Currently lecturer in the field of cultural organization at the Leuphana University Lüneburg.

"It's Almost Like Greeting"—Artistic Development Projects and Value Making of Traditional Art Practices in Ugandan Civil Society

Anna-Lisa Klages

Abstract

Dieser Aufsatz setzt sich mit den Verknüpfungen zwischen künstlerischen Praktiken und zivilgesellschaftlichen Dynamiken in Uganda auseinander. Er adressiert zunächst moderne Interpretationen, beschäftigt sich dann jedoch vor allem mit

„traditionellen“ Bildpraktiken. Vorläufige Ergebnisse deuten an, dass traditionelle Bildpraktiken häufig soziale und kulturelle Dimensionen verhandeln. Außerdem, und obwohl als unwichtig für den ökonomischen Wachstum des Landes wahrgenommen, werden sie von Projektleiter*innen und involvierten Künstler*innen als wichtig für ebendiese Entwicklung Ugandas verstanden. Auffallend ist, dass sich, obwohl weit verbreitet in Uganda, bisher wenige Studien mit den Verknüpfungen zwischen traditionellen Bildpraktiken und *Empowerment* und *Development* in Entwicklungsprojekten, oft aus dem Ausland finanziert, auseinandersetzten.

Biography

Anna-Lisa Klages, M.A.: Anna-Lisa is a Ph.D. candidate of BayWISS Verbundkolleg Sozialer Wandel at the Faculty of Art History at Ludwig-Maximilians-University Munich (LMU) and research associate at the University of Applied Sciences Würzburg-Schweinfurt. She holds a Master's degree in Social Work with Specialization in Crisis and Trauma Studies from Tel Aviv University (Israel), and a B.A. in Creative Arts Therapies from NHL Stenden University of Applied Sciences (former Stenden Hogeschool, Netherlands). Prior to returning to academia to pursue her Ph.D. she worked in international socio-cultural and humanitarian response organizations. Her research interests lay in the linkages between visual culture, civil society and (international) development from a post-colonial perspective, cultural hegemony and south-south migration.

Art Politics' Tendencies of the Documenta Exhibition Series

Lutz Hieber

Abstract

Besucher von Ausstellungen gehen davon aus, dass sie dort Kunst sehen. Tatsächlich aber sehen sie zugleich eine Repräsentation des Kunst-Begriffs der Ausstellungs-Organisatoren. Die Ausstellungsreihe Documenta, die 1955 begründet wurde, gilt für die deutsche Kunstwelt als eines der bedeutendsten Ereignisse internationaler Gegenwartskunst. Doch sie zeigt nur ein eingeschränktes Feld des Kunst-Spektrums. Der Vergleich von Documentas der ersten vier Jahrzehnte mit Ausstellungen der beiden großen New Yorker Museen, die Kunst aus eigenen Beständen desselben Zeitraumes zeigten, macht strukturelle Unterschiede deutlich. Die Documentas sind symptomatisch für die spezifische Ausprägung der deutschen Kunstwelt. Indem sie althergebrachten Paradigma der Beaux-Arts festhielten, schlossen sie Positionen der historischen Avantgarde aus. Das durchgängige Festhalten an dieser Grundeinstellung trug zur Nachhaltigkeit des kulturellen Aderlasses bei, den die nationalsozialistische Diktatur bewirkt hatte.

Biography

Lutz Hieber, born 1944, professor at the Institute of Sociology, Leibniz University Hanover, 1975-2010 (in Germany you have to be retired by law in the age of 66 years). Key focuses of his work: cultural sociology, media sociology, art sociology, political sociology, history of sociology, curating and lending material for exhibitions.

www.lutzhieber.de

The Aesthetic Subversion of Cultural Posters—Examples from the Former People's Republic of Poland

Karolina Kempa

Abstract

Wie in allen sozialistischen Ländern wurde auch in der Volksrepublik Polen die Kunst in den Staatsdienst genommen. Insbesondere in der stalinistischen Ära herrschte die Doktrin des Sozialistischen Realismus vor. Doch in Polen etablierte sich in einer Nische der Kunst eine ästhetische (Gegen-)Bewegung: die polnische Plakatschule, deren Künstler vor allem Kulturplakate entwarfen. Die Künstler erreichten es, mit ästhetischen Mitteln ein Gegenbild zu den vorherrschenden Semantiken des Sozialistischen Realismus zu entwerfen. Hierdurch lassen sie sich im Bereich ästhetischer Subversion verorten.

Weiterhin verbinden sich zahlreiche Plakate ästhetisch auf eigentümliche Weise mit den Widerstandsbewegungen gegen das Regime in Polen, vor allem dem KOR (Komitet Obrony Robotników, Komitee zur Verteidigung der Arbeiter) und der Solidarność-Bewegung ab Ende der 1970er Jahre. In den Plakaten verdichten sich symbolisch und ikonografisch Stimmungen und Themen der Zeit ihres Entstehungskontextes. Der Beitrag analysiert exemplarisch ausgewählte Kulturplakate von zwei der bekanntesten Plakatkünstler Polens: Jan Lenica und Franciszek Starowieyski.

Biography Dr. Karolina Kempa studied Philosophy and Sociology at Leibniz University Hannover. She pursued her PhD in Sociology of Culture and Arts by examining the social importance of Polish culture posters in the former People's Republic of Poland. Her research interests are sociology of culture, arts, film, gender and queer studies. Since 2009 she works at the Zentrale Einrichtung für Weiterbildung (ZEW) and the Kooperationsstelle Hochschulen und Gewerkschaften at the University Hannover.

"The Dead are Coming". A Case Study on the Communicative Power of the Political Art of Intervention

Aude Bertrand and Matthias Kettner

Abstract

Künstlerische Intervention in den öffentlichen Raum erscheinen zunehmend ohnmächtig. Interventionskunst wurde außerhalb des institutionalisierten Mainstreams des Kunstbetriebs geboren und war in den 90er Jahren die Kunstform der Wahl für diejenigen, die mit ihrer Kunst entschieden zu gesellschaftlichen Veränderungen beitragen wollten. Tatsächlich aber haben die meisten die politische Öffentlichkeit nur wenig beeinflusst, denn ihre Wirkungssphäre blieb lokal oder wurde bald vom um sich selbst kreisenden Kunstbetrieb vereinnahmt. Wir fragen, wieweit die starke Medienresonanz, die die Interventionskunstaktion „Die Toten kommen" des *Zentrums für Politische Schönheit* (ZPS) 2015 erzeugt hat, durch die rhetorischen Kräfte erklärt werden kann, mit denen diese Kunstaktion Kraftfelder der Argumente verändern kann. Über die große Medienresonanz konnte das ZPS viele Mitbürger*innen und Politiker*innen mobilisieren und polarisieren und hatte somit eine gewisse Wirkung auf das Regierungshandeln in Deutschlands „Flüchtlingskrise". Wie wir „diskursive Macht", rhetorische Kräfte und die „Kräfte guter Gründe" begreifen, erläutern wir im Rahmen einer Diskurstheorie: Es geht um Formen der Macht von Akteuren, durch kommunikative Mittel gezielt und gewollt Richtigkeitsüberzeugungen zu verändern. Richtigkeitsüberzeugungen sind diejenigen Meinungen von Personen, von deren Richtigkeit sie überzeugt sind, weil sie ihre Gründe für diese Meinungen für so gut halten, dass die betreffenden Personen sich persönlich und auch öffentlich mit diesen Gründen identifizieren.

Biographies

Aude Bertrand is a PhD Candidate at the Arts & Humanities Faculty of Witten/Herdecke University, Germany. Her dissertation deals with conflicting narratives in contemporary public arts. Over the last 12 years she has been working as a cultural manager for transdisciplinary projects and new public art institutions in Western Germany. She holds a MSc in International Management from French Business School ESCP.

Matthias Kettner is professor of philosophy at the Department for Philosophy, Politics and Economics at Witten/Herdecke University (www.uni-wh.de). Diploma in psychology and a PhD in philosophy (with Karl-Otto Apel and Jürgen Habermas) at Goethe-University, Frankfurt. Current research interests: Realistic discourse ethics; theory of socio-cultural pathologies as part of Critical Theory; digital culture, especially algorithm accountability and digital ethics; non-relativistic theory of cultural processes.

Musical Interface Agendas. Musical Appropriation via Technological Pre-configuration

Carsten Wernicke

Abstract

Der Einsatz und die Implementierung von Software und digital-materiellen Controllern in Verbindung mit für den Prozess unverzichtbaren Protokollen verändern die ästhetischen Erfahrungen und Fähigkeiten der musikalischen und kreativen Ausdrucksmöglichkeiten in musikalischen Praktiken. Was sich zunächst als konstruierte Natürlichkeit einzelner Musikpraktiken etabliert, wird durch die digitalitätsspezifische Vielfalt des musikalischen Interface-Komplexes (Hardware/Software/Protokoll) herausgefordert und gestört. Die Rolle der Leere-Welt-Hypothese als eskamotiertes Konstitutionselement dieser Geräte soll anhand einer filmischen Selbstdokumentation näher untersucht werden, welche die Verwendung des digital-materiellen Interfaces ROLI Seaboard durch einen Musiker aus einer arabischsprachigen Region zum Inhalt hat. Aufgrund seines dialektischen Rahmungscharakters muss das ROLI hier einerseits als ein kulturell expansiver Ermöglicher und andererseits als ein kulturkapitalistisches Artefakt beschrieben werden.

Biography

Carsten Wernicke is a research assistant in the research project 'Musical interface designs: Augmented creativity and connectivity' (MIDAKuK-TP2) funded by the German Federal Ministry of Education and Research at Leuphana University of Lüneburg, Germany.

Who Plays? An Analysis of Musicians' Global Mobility in the World Music Market

Glaucia Peres da Silva

Abstract

Dieser Artikel diskutiert die globale Musiker*innenmobilität am Beispiel vom World Music Markt. Um zu untersuchen, wie die globale Mobilität Teil der Musiker*innenkarriere innerhalb ungleicher weltumspannender Machtfelder wird, stützt sich diese Diskussion auf Debatten über Migration und Mobilität. Der Fokus liegt dabei auf den Ansatz der Mobilitätsregime von Glick Schiller und Salazar (2013). Eine empirische Analyse der Migrationsgeschichte der Musiker*innen, die zwischen 2008 und 2017 auf WOMEX–The World Music Expo aufgetreten sind, zeigt, dass ein großer Anteil der migrantischen Musiker*innen (18%) aktiv auf dem World Music Markt sind. Dennoch sind es die nicht-migrantischen Musiker*innen in ihren Tourneen, die anscheinend diesen Markt als einen transnationalen Raum jenseits der Grenzen der Nationalstaaten und ethnischen Gruppen aufrechterhalten. Die meisten von ihnen wohnen in Europa bzw. in Spanien, Frankreich, Brasilien, Vereinigten Königreich und den USA, was eine klare Machtungleichheit innerhalb des World Music Marktes impliziert.

Biography

Glaucia Peres da Silva is sociologist and currently coordinates the Global Awareness Education Program at the University of Tübingen. Her research focuses on economic sociology, sociology of the arts, and global and transnational processes. Her PhD thesis was published in a book entitled *Wie klingt die globale Ordnung. Die Entstehung eines Marktes für World Music* (VS Springer 2016). She is currently studying transnational labor markets for artists based on the example of the World Music market with a particular focus on the dynamics of trade fairs, festivals, and exhibitions.

The manufacturer's authorised representative in the EU is Springer Nature Customer Service Centre GmbH, Europaplatz 3, 69115 Heidelberg, Germany. If you have any concerns regarding our products, please contact ProductSafety@springernature.com

Printed and bound by CPI Group (UK) Ltd, Croydon, CR0 4YY
28/09/2026
02224087-0001